# BLOOD

# BLOOD

AN EPIC HISTORY OF
MEDICINE AND COMMERCE

## DOUGLAS STARR

ALFRED A. KNOPF   NEW YORK   1998

THIS IS A BORZOI BOOK
PUBLISHED BY ALFRED A. KNOPF, INC.

Copyright © 1998 by Douglas Starr

www.randomhouse.com

Appreciation is made to Bertram Bernheim Jr. and I. W.
Burnham II for use of quotations from *Adventure in Blood Transfusion*
by Dr. Bertram B. Bernheim, an American pioneer in blood transfusion.

Library of Congress Cataloging-in-Publication Data
Starr, Douglas A.
Blood : an epic history of medicine and commerce /
by Douglas Starr. — 1st ed.
p.   cm.
Includes bibliographical references and index.
ISBN 0-679-41875-X (alk. paper)
1. Blood banks—History. I. Title.
RM172.S728   1998
362.1′784′09—dc21      97-46815
CIP

Manufactured in the United States of America

First Edition

For Mitch

# CONTENTS

## PART THREE:  BLOOD MONEY

*Illustrations follow pages 78 and 174.*

# PREFACE

The drama ended, as do so many these days, in a courtroom. This particular chamber was long and low-ceilinged, with a wide dais at its front for the eight black-robed judges. Each of the four defendants sat flanked by tall policemen who gazed impassively from under the brims of their trademark pillbox hats. In keeping with the formality of French courts, the prosecuting and defense attorneys wore flowing black robes, which would dramatically sweep behind them as they rose to make a point. The only visible flaw in the decorum appeared among the audience members, some of whom wore T-shirts bearing inflammatory slogans. There were audible exceptions to decorum as well, as people would moan or shout "*Non!*" at a defendant's response, or when one man, the most vocal of the plaintiffs, would, as his doctor walked past, loudly hiss "*Assassin!*"

The plaintiffs in this trial were dying of AIDS. They charged that they had been infected through the negligence of the defendants—high officials in the French national transfusion service. In France, where the government until recently held a monopoly on blood and its derivatives, these men were supposed to ensure the safety of blood products. Instead, they allowed thousands of the nation's hemophiliacs to inject blood-derived clotting factors they knew to be contaminated. The defendants had done so because of a complicated mixture of paternalism, economics, and to some extent the limits of science, but the victims saw the incident more starkly. To them the affair was a matter of betrayal. The doctors on trial in the summer of 1992 were supposed to have embodied all that was noble in the French transfusion tradition—altruism, medicine, business, and technology. Instead, dur-

ing the years of the "contaminated-blood affair" they came to symbol-
ize the cynicism and expediency of a money-driven age.

The sense of betrayal surfaced in many places beyond the courtroom
in Paris. For more than a decade the theme has been sounded in one
locale after another throughout the world. In America, patients have
filed hundreds of civil suits against doctors, drug companies, and even
their own patient organizations, for abandoning their health to the
expediency of the marketplace. In England, AIDS-infected hemophilia
patients castigated their national transfusion service with reacting too
slowly to the threat of emerging viruses. In Japan, patients charged that
the government and drug companies criminally concealed the contamination of blood products; as a result, some of the nation's most
revered doctors have gone to jail. In Canada, the scandal of contamination spread so wide that the government held a series of hearings
across the country that convulsed the nation with anger and shame.

Why those scandals erupted is one of the underlying questions of
this book, a history of human blood as a resource and humanity's
attempts to understand and exploit it. Blood is one of the world's most
vital medical commodities: The liquid and its derivatives save millions
of lives every year. Yet blood is a complex resource not completely
understood, easily contaminated, and bearing more than its share of
cultural baggage. Indeed, the mythic and moral symbolism of blood,
which has been with us since ancient times, subtly endures. It clouded
professional judgments and public perceptions in the AIDS scandals of
France, Canada, and Japan, among others.

If one considers blood a natural resource, then it must certainly rank
among the world's most precious liquids. A barrel of crude oil, for
example, sells for about $13 at this writing. The same quantity of
whole blood, in its "crude" state, would sell for more than $20,000.
Crude oil, as we know, can be broken down into several derivatives,
including gasoline, distillates such as diesel, and petrochemicals. Blood
can be separated into derivatives as well. Spun in a centrifuge, it divides
into layers—red cells on the bottom, a thin intermediary layer of
platelets and white cells, and an upper tea-colored layer of plasma.
Each layer, in turn, can be used as various therapeutic products. Red
cells can be transfused directly. White cells and platelets can be used to
restore resistance or clotting ability to patients undergoing chemother-
apy. Plasma, a resource in its own right, yields albumin for restoring
circulation, clotting factors for patients with hemophilia, antibodies
for vaccine production, and several other reagents and pharmaceuticals. Taken as a whole, the value of the derivatives in a forty-two-gallon
barrel of crude oil would raise its price to $42. The price of the same

quantity of completely processed blood would increase its value to more than $67,000.

Of course, blood is not processed by the barrel or handled in quantities anywhere near those of oil. (Only about sixteen million gallons of blood and plasma are collected annually worldwide—the equivalent of thirty-two Olympic-size swimming pools.) Indeed, the world market for blood and its derivatives probably does not exceed $18.5 billion per year, versus $474.5 billion for petroleum. Yet one cannot avoid comparing the two resources. Just like the oil industry, the blood trade involves collecting a liquid resource, breaking it into components, and selling the products globally. Red cells, being perishable, tend to remain within national borders, but certain portions of blood—plasma in particular—are traded among multinational companies and on a worldwide spot market. Just as with oil, one region has become the premier harvesting ground, providing much of the resource for the rest of the world. The United States, with its liberal rules regarding collection, has become known as the OPEC of plasma.

No wars have been fought over blood as they have been for oil, but the movement of blood has played an important role in our wars. A major anxiety about D-Day, for example, was whether enough blood could be stored to supply all the wounded that military planners had projected. In preparation for the Persian Gulf War, the military shipped massive quantities of blood to the battle zone for what they thought would be thousands of casualties. (Good fortune proved them wrong.) Such collections have always been secret, since intelligence services know that the mobilization of blood is a sure sign of an impending attack.

If the analogy between blood and oil is provocative, it is where the comparison breaks down that the story of blood becomes especially compelling, and life-changing to those who have been caught in its sweep. For one thing, oil does not transmit disease, a critical consideration in the blood trade. A slip in quality control at a refinery may result in the loss of a few dollars, but a mistake in blood processing can infect thousands of people. Second, whereas oil companies pay handsomely for drilling rights, blood collectors pay nothing or very little for their raw material, since donating is thought of as an act of human kindness. Such an arrangement, however admirable, can distort people's judgments. Think, for example, how the leaders of the oil industry would react if Saudi Arabia provided crude oil for free: They would bend over backward (even more than they currently do) never to offend their benefactors. So it had been with the blood collectors: When faced with the necessity of refusing blood from certain people to minimize the spread of viral disease, they found themselves reluctant to offend their cherished donors. As a result, public safety was compromised.

The most telling difference between the two resources, however, is the one that reaches into our cultural past. Though oil serves as a critical resource, it carries no particular cultural baggage. Blood, in contrast, is laden with meaning. The descriptive cliché, "the elixir of life," barely touches on the liquid's mystical, religious, and patriotic significance. The Bible mentions blood more than four hundred times: "The life of the flesh is in the blood," says Leviticus, equating blood with life itself. Blood is considered so holy in the Old Testament that the law specifically forbids its consumption, which is why Jehovah's Witnesses, who interpret the Bible literally, refuse transfusions. The Egyptians saw blood as the carrier of the vital human spirit, and would bathe in the liquid as a restorative. It is because blood conveyed strength to the Romans that gladiators were said to have drunk the blood of fallen opponents. Doctors from the medieval to the Victorian era assumed blood to have fantastical powers, draining it to remove evil humors, transfusing it to pacify the deranged. Our own culture attaches great value to blood, with the blood of Christ as among the holiest sacraments, blood libel as the most insidious slander, the blood-drinking vampire as the most odious demon.

The symbolic power of blood does not confine itself to mythology, for it has affected the behavior of doctors in modern times. The Nazis, in their perversity, refused transfusions from non-Aryan blood donors—condemning their armies to chronic shortages—and composed intricate charts of the presumed blood-related traits of the various races. Even the democracies were tainted by blood prejudice: During World War II, as America fought a racist enemy, the military maintained separated blood stocks from black and white donors for fear of offending white soldiers' sensibilities. Most recently, the persistent belief that blood products collected among their countrymen had to be inherently pure contributed to bad decision-making in the tainted-blood scandals of France and Japan.

Thus, the story of blood cannot be limited to the twentieth century, when doctors began to use it for transfusions. The narrative reaches back into antiquity, as an undercurrent to the history of medicine and civilization. It spans the globe over the course of several centuries, periodically surfacing in dramatic ways, from the first blood experiments, during the Age of Enlightenment, to the genetic-engineering labs that one day may render transfusion obsolete.

The story of blood is one of metamorphosis, of a liquid that became symbolically transformed as society learned how to deconstruct and manage it. As such, the history divides itself into three eras, each reflecting the spirit of its age.

The first period, described in the section "Blood Magic," involves the transformation of blood from a magical substance to a component of human anatomy, capable of being isolated and studied. This section covers the period from antiquity to the early twentieth century, the time when the concept of blood moved from the magical to the biological; when blood became recognized as a therapeutic liquid transfusible from one creature to another. It is a measure of the symbolic power of blood that the first transfusions were used to treat not blood loss or anemia but insanity.

The second era, covered in the section called "Blood Wars," describes the transformation of blood from a scientific curiosity to a strategic materiel. During the first few decades of the twentieth century, medical scientists began to master the resource, learning the techniques of mass collections, storage, and the separation of plasma. These advances occurred just in time for World War II, the greatest spilling of blood that the world has ever known. That conflict decisively altered blood's cultural significance—from the mother liquid of all health and disease, to a strategic resource, devoid of mystical overtones yet essential to human enterprise. The change became irreversible when Dr. Edwin J. Cohn of Harvard, working under a military contract, found a way to fractionate plasma into its many constituents. This technology, analogous to the "cracking" of oil, along with the freeze-drying of plasma, gave the Allies an enormous advantage over the Axis powers, whose blood-related technology was primitive. It also set the stage for a postwar global blood industry.

The final section, "Blood Money," describes how the liquid that saved so many lives became the basis for a global industry. A small group of drug companies dominates the plasma business, analogous to the "Seven Sisters" of oil. In their quest to harvest the resource, those drug firms set up "plasma mills" in America's skid rows, buying from the residents, who often included drug addicts and indigents. Later, seeking new sources of raw material, they imported plasma from the Third World, notably Central America—a practice of dubious safety and morality. So politically explosive was the idea of harvesting the resource from the poorest of the poor that in one Central American country the populace rose up, destroying the facility and sparking a revolution. Meanwhile, the business of whole blood boomed as surgical advances such as open-heart surgery and organ transplants required ever-larger transfusions. (A single liver transplant may require fifty units of red cells.) Whole blood, collected on a nonprofit basis by the Red Cross and community blood banks, became the target of fierce competition as the "benevolent" collectors struggled for dominance.

If the global blood business has been tainted by an element of exploitation, it must also be seen as tremendously beneficial. Countless lives have been saved by transfusions, not to mention plasma-derived pharmaceuticals. People with hemophilia, who have been using clotting factors since the late 1960s, have seen their average life expectancy double. Yet the same therapeutics that brought life to so many have also transmitted disease: If blood and plasma products could be routinely distributed among millions, so too would any pathogens they harbored. During the blood-products boom of the 1970s, blood-related hepatitis rates soared, killing tens of thousands of hemophiliacs and transfusion recipients. By the end of the decade, doctors thought they had solved the hepatitis problem, only to be confronted by another virus that spread in an identical pattern—HIV. Though tainted blood products only caused a small portion of the AIDS epidemic (the disease was mainly spread through sexual contact and intravenous drug use), they took an enormous toll. More recently, another public health crisis has begun to unfold—the silent epidemic of blood-borne hepatitis C. It is ironic that, after all the transformations of blood wrought by modern medicine, HIV and other viruses revived the medieval image of blood as the bearer of evil humors and death.

Today we confront a resource simultaneously safer and more threatening than before. Many nations, having learned from the AIDS crisis, have instituted virus screening-and-removal procedures. This has made blood more expensive, an ominous development in an era of shrinking health budgets. Furthermore, we can no longer complacently assume safety, since new diseases threaten to emerge. Meanwhile, poor nations, with little access to modern equipment, face unprecedented risks of blood-borne diseases. In order to address the inherent risks of the resource, some companies are creating artificial blood substitutes, immune to the pathogens that afflict humans. Even if those products someday appear, they will likely be expensive, prolonging the disparity between nations that have modern blood products and those that do not. Thus blood distribution, like that of other critical resources, will continue to raise questions of equity and social justice.

This, then, is the story of blood—the chronicle of a resource, the researchers who have studied it, the businessmen who have traded it, the doctors who have prescribed it, and the lay people whose lives it has so dramatically affected. The book is also a challenge to those who distribute, regulate, and use the resource. Indeed, a lasting tension in its history is how we view this most human of commodities. Is it a gift of charity or simply a pharmaceutical? Can a single resource be both, and if so, what are the safest and most ethical ways to manage it? The answers to such questions will determine the future of this precious, mysterious, and hazardous material.

# PART ONE

# BLOOD MAGIC

# CHAPTER 1

# THE BLOOD OF A GENTLE CALF

In a village near Paris in the seventeenth century lived a madman named Antoine Mauroy. Little is known about this obscure and pathetic character—no physical description, virtually nothing about his station in life. We do know that he suffered "phrensies" during which he would batter his wife, strip off his clothes, and run through the streets, setting house fires along the way. His name would have been completely lost to history if he had not taken part in an experiment that forever changed the practice of medicine.

In the winter of 1667, a nobleman found Mauroy wandering naked through Paris. Taking pity on the man, he brought him to a friend who had been conducting some experiments—Jean-Baptiste Denis, a physician to Louis XIV, who had been looking at the effects of transfusing blood from animals into human beings. He sat Mauroy in a chair, surrounded by physicians, surgeons, and "many people of quality . . . too intelligent to suspect them of being capable of the least surprise." At precisely six in the evening on December 19, according to the doctor's report, an associate opened a vein in Mauroy's arm, inserted a silver tube, and drained off about ten ounces of blood. He then inserted the other end of the tube into the leg artery of a calf and allowed about a cupful of the calf's blood to flow into the man. The doctor hoped that the calf's blood "by its mildness and freshness might possibly allay the heat and ebullition of [the patient's] blood."

That the king's doctor would infuse a man with animal blood was not outrageous, given the state of medicine at the time. Seventeenth-century medicine was a haphazard mixture of folk cures, astrology, religious incantations, and lessons from the Greeks. Physicians would treat patients with a desperate assortment of remedies: roots, herbs, worms; powders made from precious stones, crabs' eyes, vipers' tongues, or "moss from the skull of a victim of violent death." Barbers operated as frequently as surgeons; both would destructively bleed patients at the first sign of disease, draining out the "bad humors" with the blood, often to the point of death.

Life was unhealthy, brutish, and short. The age of Louis XIV may summon images of foppishly dressed courtiers busying themselves with gambling, conspiracies, and sexual intrigue, but the masses, in truth, lived in less diverting circumstances. Fleeing rural poverty, they packed the poorer quarters of Europe's great cities, where streets ran with sewage and homes became nesting grounds for rats and other vermin. Contagions cyclically raced through the continent—malaria, yellow fever, the Black Death—emptying cities, killing economies, snuffing out tens of thousands of lives. During the latter half of the seventeenth century, plagues killed sixty-nine thousand in London, eighty-three thousand in Prague, and nearly half a million people in the Venetian republic. Only special physicians could enter these plague zones. The "pest doctors," as they were called, wore the era's equivalent of biological-containment suits: long leather robes and gauntlets; masks with glass-covered eye-slots and long, curved beaks saturated with fumigants. Wandering the neighborhoods like surreal birds of death, they would take pulses with a wand, pronouncing victims deceased and condemning their properties. Authorities hadn't a clue as to the source of the disease, and so would take whatever harsh measures they felt necessary, which generally meant charging victims with witchcraft, torturing them to death, and burning their homes.

Yet, as bleak as times were for the average individual, they held great promise for humanity as a whole. The period brought innovation in art, literature, philosophy, and science. It was the era of Rembrandt and Racine, Milton and Molière. Faith in human reasoning was challenging church dogma as never before; so much so that a philosopher like René Descartes could sum up the skepticism and confidence of the era with the resounding statement of human self-awareness: "I think, therefore I am." In science, superstition was yielding to detached observation. Nature, once mystical, was becoming quantifiable. Indeed, in the years preceding Denis's experiments, Isaac Newton had proposed his theory of gravity, Galileo had observed sunspots, and Robert Boyle had explained the behavior of gases. In France, Descartes had invented analytical geometry and, by applying strict mathematical concepts to the study of nature, created modern scientific thinking.

The times held promise for medicine as well. Now that the Church had relaxed its taboos on dissecting the human body, anatomists like Andreas Vesalius, William Harvey, and Marcello Malphigi were revealing the complexity of the human organism, with a surprisingly accurate knowledge of the organ systems' structure and function. They knew, for example, that the heart functioned as a pump, forcing blood outward though the arteries and allowing blood to flow back through the veins, and that the two kinds of vessels were connected by a system of capillaries. They knew that the pancreas, spleen, and digestive organs excreted corrosive juices, or enzymes; they also had a basic understanding of the workings of the eye.

Despite the doctors' sophistication, they (like everyone else in society) held fast to a core of ancient beliefs. They still believed that disease arose from an imbalance of invisible fluids or vapors in the body, called "humors." They also believed that blood somehow carried the essence of the creatures in which it flowed—a concept called "vitalism," which had survived unchanged for fifteen hundred years. According to this belief, a stag's blood might carry traits of courage and longevity; a calf's blood, tranquillity. Thus Denis's work, though misguided by modern standards, exhibited the mixture of science and superstition typical of his age.

Denis was a somber-looking man with large eyes, a prominent nose and forehead, and a hint of bourgeois jowliness surrounding the chin. Born to a modest family of artisans—his father had been a water-pump maker to the royal court—he studied theology in Paris and then medicine in Montpelier. In returning to Paris, he became a professor of philosophy and mathematics, as well as one of Louis XIV's physicians. An eager intellectual, he regularly attended the city's learned societies, or academies, to discuss the latest in physics, mathematics, medicine, and philosophy. He belonged to the academy sponsored by the count of Montmor, known for his progressive philosophies; it was the count, in fact, who brought the madman to Denis on that cold winter night.

Laboring over his patient, Denis watched for signs that the transfusion had taken effect. Minutes passed as the calf's blood flowed through the tube. He removed the apparatus when Mauroy complained of a great heat moving up his wrist; then he stitched up the wound and told Mauroy to go to sleep. Two hours later, the patient awoke. His pain gone, he ate a hearty supper and amused himself in whistling and in song.

Two days later, Denis gave him another transfusion, even larger than the first. As soon as the blood began to enter his veins, Mauroy complained about the same feeling of heat traveling up his arm. His pulse raced, then slowed, then raced again. "We observed a plentiful sweat all over his face," wrote Denis. "He complained of great pains in his Kidneys, and that he was not well in his stomack, and that he was ready to choak unless they gave him his liberty." Alarmed at Mauroy's erratic reactions, Denis and his assistant quickly removed the tube. "Whilst we were closing the wound, [Mauroy] vomited the store of Bacon and Fat he had eaten half an hour before," Denis wrote. The patient urinated goblets of fluid as black "as if it had been mixed with the soot of Chimneys." They put him to bed, only to find that, when he awoke the next morning, "he shewed a surprising calmness, and a great presence of mind . . . and a general lassitude in all his limbs."

Denis could not have known it, but his patient had just suffered a near-fatal episode of shock. Animal blood contains proteins completely foreign to human blood. When confronted by such substances, the human body reacts quickly and dramatically, mobilizing antibodies to destroy the invading cells. The reaction causes violent hemolysis (the physical destruction of the incoming red cells), inflammation, fever, and pain in the kidneys as they work to filter the toxic hemoglobin and cell fragments. Red blood cells die by the millions, and the oxidized hemoglobin turns the urine black.

Only by luck, then, did Mauroy survive. Staying with Denis for the next couple of days, he slept, prayed, bled from the nose, and continued to urinate coal-black fluid. No longer manic, he had little to say. On Friday, he was drained of two small porringers of blood. On Saturday, two and a half days after the procedure, Mauroy felt strong enough to go to confession. "That same day," wrote Denis, "his urine cleared up."

Meanwhile, Madame Perrine Mauroy, who had been searching for her husband from village to village, found him at last under Dr. Denis's care. She approached Antoine with trepidation, wary of his past brutality. To her surprise, her husband greeted her tenderly, relating "with great presence of mind all that had befallen him, running up and down streets; how the Watch [police] had seized him one night, and how Calfs-blood had been transfused into his veins." Denis barely could believe what he was seeing: The man who "used to do nothing but swear and beat [his wife]" had dramatically—almost magically—been cured.

Across the English Channel, Denis's competitors read his reports with concern and dismay. They did not question the truth of his experiments; what outraged them was the speed of his progress. As the English saw it, *they* had pioneered the technique of transfusion, having been first to transfuse blood among animals of the same species and from one species to another, and the first to propose human transfusions. How dare this newcomer transgress?

There was more to their concern than scientific jealousy. Great change was sweeping Europe: With Spain in decline and Germany preoccupied with destructive civil wars, England and France were rising as the two great European powers. Both nations had come through civil wars, had emerged with strong monarchs—Charles II of England and Louis XIV of France—and were becoming wealthy from colonies in the New World, Asia, and Africa. The nations competed on many fronts, including literature, music, the arts, and the sciences, and supported learned societies as well—the precursor of the Royal Society in England, and the French Academy in France. They competed ferociously for world domination in all areas, including medicine.

The English had reason to take pride in their research. They could trace a direct line from their experiments back to Harvey, who, forty years earlier, had first proved that blood circulated through arteries and veins.

From before the ancient Greeks, people had viewed the human body in a fundamentally different manner. Unlike now, they did not think in terms of systems—the digestive, nervous, and endocrine systems, for example—and knew nothing of hormones, genes, infections, or germs. Instead, they saw the body as a microcosm of nature. Since all natural phenomena were thought to result from the interplay of the four elements—air, fire, water, and earth—the Greeks assumed that four analogous factors must govern the body. These elements, or "humors," included phlegm, choler, bile, and blood. The linguistic remnants of this system can be seen in the words "bilious" and "choleric," for example. According to Greek medicine, good health depended on maintaining a balance of the humors, which later led to the practices of purging the digestive tract and the draining of blood.

This system remained unquestioned for centuries, as we shall see later, and was adopted almost intact by the Christians. Blood, as the Paramount Humor, was considered the bearer of life, carrying its vital spirit throughout the body, ebbing and flowing through arteries and veins and sloshing through imagined pores in the heart. As anatomists started performing dissections, however, they found the theory at odds with the evidence. The sixteenth-century anatomist Vesalius found it impossible to locate the pores. "We are driven to wonder at the handiwork of the Almighty," he wrote with understatement and humility, "by means of which blood sweats from the right into the left ventricle through passages which escape human vision." Harvey preferred to state it more directly. "But dammit! There are no pores!"

Harvey found other anomalies as well. In examining the veins of some eighty species of animals, from eels to lambs to man, he found them punctuated by numerous valves. He tried forcing water in backward through the vessels but, contrary to what humoral theory had suggested, could not get the liquid to slosh back and forth: The valves allowed flow in only one direction. After further anatomical studies, he found himself arriving at a startling conclusion: Rather than ebbing and flowing like a tide, blood flowed rather purposefully—out through the arteries, back through the veins—*circulating* through a closed, one-way system. As for the heart—the seat of the soul, the source of all life—it was a simple mechanical pump.

Harvey's conclusions, resisted at first, revolutionized the way people conceptualized the body, now seen as more mechanical than mystical. He also changed the practice of science with his quantitative methods: By actually *measuring* things like flow and volume, he gave birth to the field of experimental anatomy. (It should be noted that, although Harvey discovered the circulatory system, he never explicitly rejected humoral theory.)

Harvey worked in Oxford among a group of brilliant scientists who called themselves the Experimental Philosophy Club. His colleagues were so impressed by his methods that they undertook their own circulation work, even if they had been trained in completely different fields. Christopher Wren, the legendary architect, and Robert Boyle, the founder of modern chemistry, dabbled in circulation, using a hollow quill and bladder to inject opium and antimony into dogs. By injecting the drug and recording the symptoms—opium causes sleep; antimony, vomiting—they were able to show that the interventions had taken effect. That simple experiment yielded two striking results: the invention of the first intravenous syringe, and proof that the circulatory system, previously inviolate, could now be made open to interference from outside.

Anatomists began injecting all sorts of solutions into dogs in the laboratory, from urine to wine to milk to beer, often with fatal results. Finally, a talented young doctor named Richard Lower suggested injecting what he assumed would be the most compatible liquid of all.

Lower was born into a farming family in Cornwall, and had come to Oxford on a scholarship. A medical student when Wren and Boyle published their results, he had gotten to know the two men through the Experimental Philosophy Club and worked under their patronage and encouragement. In a series of experiments starting in 1665, Lower attempted to transfuse blood from one dog to another. Baring the jugular veins of two test animals, he would stitch them to opposite ends of a reed so blood could flow from one animal into the other. His experiments failed. Veins, unlike arteries, carry blood under low pressure as it makes its way back to the heart. A cut vein does not spurt like an artery, so, rather than rush from one animal to the other, the languid venous blood pooled and clotted inside the tube. Lower experimented for a year with various combinations until finally coming up with the successful procedure of connecting the artery of the donor animal with the vein of the recipient. The difference in pressure between the spurting artery and the passive vein forced the blood from donor to recipient. That simple advance would become critical to transfusionists in centuries to come.

Girded with this new information, Lower made ready for a "spectacular new experiment" in late February 1666. "I selected a medium-sized [dog] and drew off its blood from an exposed jugular vein," he wrote. He drained as much blood as he could without killing it. "The dog first set up a wailing but soon its strength was exhausted and convulsive twitchings began." Meanwhile, he had bound a large hound to a second bench, bared one of its neck arteries, and stitched it to a reed; then he attached the other end of the reed to the smaller dog's jugular vein. He opened a ligature and let the blood flow, until "neither blood nor life remained" in the hound.

Now he witnessed something that, given the beliefs and science of the time, must have been an astonishing sight. The smaller dog virtually came back from the dead, as though the hound's life force had filled it up again. Lower sewed up the jugular vein, loosened the shackles, and watched the little dog leap from the table. "Oblivious of its hurts, [it] fawned upon its master, and rolled on the grass to clean itself of blood; exactly as it would have done if it had merely been thrown into a stream, and with no more sign of discomfort or displeasure."

Lower's experiments excited his fellow philosophers as nothing had in years. Colleagues rushed to explore the implications. Robert Boyle wrote to Lower that he should consider a whole range of possibilities that could arise from transfusion. Would a fierce dog become tame "by being . . . stocked with the blood of a *cowardly* Dog . . . ?" Would a trained dog forget how to fetch if transfused from an animal that did not know how? Would the recipient's fur color somehow change to that of the donor?

Lower kept working, and later gave a progress report in a long letter to Boyle. He wrote that, after passing blood from one dog to another in quantities large enough to kill the donor, he transfused smaller amounts from several dogs into one, to preserve the donors' lives. He also mixed blood from different animal species, passing blood from a sheep into a dog. Aside from reinvigorating the recipients, transfusion did not seem to alter their dispositions. "The most probable use of this experiment," he concluded, is "that one Animal may live with the blood of another."

True to the best scientific traditions, other doctors tried to replicate Lower's experiments. At the Royal Society in London Dr. Edmund King conducted "a pretty experiment of the blood of one dog let out, till he died, into the body of another," according to Samuel Pepys, the noted diarist and president of the society. Afterward the men repaired to the Popeshead Tavern, where they philosophized into the night about the vast possibilities raised by transfusion. "This did give occasion to many pretty wishes, as of the blood of a Quaker to be let into an Archbishop, and such like."

The English work took place about a year before Denis transfused the calf's blood into Antoine Mauroy. In the interim, the French, dispensing with the quill or reed as a means of transfusing the blood, invented a new apparatus: a couple of silver cylinders connected in line with a small sack between them. With one tube inserted into a blood vessel of the first dog and another into the second dog, the sack could be squeezed in such a way as to direct blood forcefully from the donor to the recipient. Yet the French had none of Lower's luck: Time after time, for reasons they could not explain, the target dog would die or barely hobble away. One autopsy showed a dog's heart engorged with blood—the result, in retrospect, of either a heart attack or shock. Finally, on the seventh try, they succeeded in transferring two ounces of blood.

It was witnessing these experiments that inspired Denis to begin some work of his own. Over the next several months, he conducted nineteen transfusions among dogs. He seemed able to make the transfusion less traumatic by using the crural artery, in the leg, rather than the carotid artery of the neck. He expanded his repertoire, transfusing a calf's blood into a dog and the blood of four rams into a horse. For months, Denis enjoyed uninterrupted success: Blood, it seemed, was a universal nutrient, beneficial to all species.

He then presented his most daring idea in a carefully reasoned paper, in which he justified, step by step, what even now would seem an outrageous suggestion. He started with the philosophical assumption that nature must approve the principle of blood exchange—after all, fetuses share their mothers' blood through the placenta. Then he asserted the moral position that there was nothing wrong with taking nutrition from animals—did man not obtain milk and meat from the beasts? Finally, having shown by his experiments the benefits of transfusing blood from one animal to another, he now proposed that its blessings ought to be extended to man. "Pleurisies, the Small Pox, Leprosies, Cancers, Ulcers, St. Anthonies fire, Madness, Dotage, and other Maladies arising from the malignity of the blood"—all might be cured through the use of transfusion. Yet he could not sanction using men as a blood source: "It would be a very barbarous Operation, to prolong the life of some, by abridging that of others." Animals, on the other hand, did not seem to suffer unduly from giving blood, and farmers could provide a limitless supply. Beyond that, animal blood must surely be healthier than man's, which undoubtedly was debased by "debauchery and irregularities in eating and drinking." After all, "sadness, Envy, Anger, Melancholy, Disquiet . . . corrupt the whole substance of the blood," he wrote. His suggestion, after ten pages of methodical, sensitive reasoning, was to use the "mild and laudable blood" of animals to bring transfusion to man.

It did not take him long to put his notions to a test. In June 1667, he received a sixteen-year-old patient "tormented with a contumaceous and violent fever." Physicians had bled the young man twenty times, which only seemed to weaken him. "His wit seem'd wholly sunk, his memory perfectly lost, and his body so heavy and drowsie that he was not fit for any thing."

Denis decided that gentle lamb's blood could help. He bound the lamb, bared its neck, and allowed nine ounces of blood to flow into a vein in the boy's forearm. The patient felt a great heat rise up his arm, then took an hour's nap and awoke free of pain. "He executes nimbly whatever is appointed him, and he hath no longer the drowsiness nor heaviness of body," wrote Denis. Weeks later, he noted that the patient "grows fat visibly, and in brief, is a subject of amazement to all those who know him."

Next he transfused a burly, forty-five-year-old laborer. The man laughed and chatted throughout, mindless of heat in his arm. Afterward he refused to lie down and, eager to show his strength and former training as a butcher, seized the transfusion lamb and slaughtered it. He took a short nap, went out, gathered up his comrades, and took them to a tavern, "to drink part of the money given him for his day's business." He spent the rest of the day in heavy labor, wrote Denis.

The next day, the two met in the street. "I blam'd him of imprudence," Denis recalled. "But he told me in excuse of himself, that he could not be at rest when he was in health . . . that he had eaten, drank and slept very well, that he had more strength than ever before." Finally, the patient told him "that if we were minded to repeat the same experiment at any time, he desired we would choose no other person for it but him. . . ."

Denis published his report in the July 22, 1667, issue of *Philosophical Transactions*. Lower published a vituperative response. Transfusion "was first discovered by me," he protested, and accused Denis of stealing the idea. "As word of this newly devised blood transfusion was fluttering everywhere on man's lips, Dr. Dionys [*sic*] . . . attempted to deprive me of the credit of originating this famous experiment and appropriated it to himself."

Denis was not a confrontational man. As he pointed out later, he had given the English full credit for their discovery; he had merely advanced the technique. The English, however, did not want to share credit in the field.

Now the English hurried to regain the lead. On November 23, 1667, Drs. Lower and King paid 20 shillings to a thirty-two-year-old man named Arthur Coga to transfuse him with sheep's blood. Coga, a former minister, had by some unknown circumstance become "a little frantic . . . poor and debased," according to Pepys. The doctors had proposed to transfuse about twelve ounces of blood, which they estimated would pass in one minute by the watch. A week after the transfusion, they met for dinner in a public house, with Coga as their guest. "I was pleased to see the person who had his blood taken out," wrote Pepys. "He speaks well, and did this day give the Society a relation thereof in Latin." Pepys noted that Coga might benefit from yet another transfusion, for he still seemed "crack'd a little in the head. . . ."

The technique spread to Germany, Holland, and Italy, where doctors began transfusing among all sorts of creatures and presenting wildly exaggerated claims. Transfusion, some said, could cure scurvy, leprosy, and other "devouring eruptions." A German doctor saw it as a potential treatment for disagreeable personalities: A splenetic person might be calmed with blood from a mild one, or, better yet, marital problems could be solved by exchanging the blood of feuding husbands and wives.

By now Denis had completed two transfusions on Antoine Mauroy; he had also transfused a Swedish nobleman, who died, and a partially paralyzed woman, who survived. Yet the criticism from England faded beside the vilification he began to face at home. The French intelligentsia were highly political, corrupted by a desire to retain the pleasures of the court. In their eagerness to curry favor with the king, the elite had become expert at maligning each other, every achievement leaving a jealous throng in its wake. After months of listening to Denis's success stories, the mandarins at the French Academy—rivals to the Academy of Montmor—decided it was time to attack.

They began with a series of pamphlets maligning Denis, transfusion, and even the basic concept of blood circulation. "I could fill a book with every known malady, its nature and causes, and easily show why blood transfusion would be a useless way to cure it," wrote G. Lamy, a master of arts at the University of Paris. He then listed several diseases, like pleurisy and cancer, and explained, using the old humoral theory, why transfusion could never help. Then Lamy, a clever dialectician, took the opposite tack. Suppose transfusions *did* work—what then? All the world's sick people would demand them; there would not be enough blood in all the animals in the world. Some critics questioned why blood from a calf, if it conferred tranquillity, did not convey the animal's dumbness as well. Others dispensed with any logic at all. Pierre de la Martinière, another of the king's doctors and a French Academy member, labeled transfusion a "monster" methodology, a barbaric practice, reminiscent of cannibalism, issued directly from "Satan's boutique." Martinière, who had never accepted the theory of circulation, lobbied tirelessly to make transfusion illegal. In pamphlets and letters to ministers, magistrates, priests, noblemen, and ladies, he ranted that transfusion was antithetical to nature and threatened the very existence of the human species.

The argument, which pitted the conservatives of the French Academy against the more progressive scientists of the Academy of Montmor, spilled forth from the schools and academies, livening the king's court and spicing gossip in Paris. The English weighed in, supporting transfusion but deriding Denis. All Europe was listening to the debate. Denis, overwhelmed and shocked by the hubbub, refused to debase himself by participating. And then, just as the conflict approached an ugly crescendo, Denis answered a knock at his door.

There stood Antoine Mauroy and his wife, Perrine. They both looked tired and ragged, and she showed several bruises. Antoine was having his frenzies again. Ignoring the doctor's advice to conduct himself modestly, he had been visiting the tavern, smoking tobacco, and having sexual relations with his wife. Recently he had begun beating her as well. Perrine beseeched the doctor to consider transfusing her husband. Denis had doubts. The procedure was experimental, and Antoine was not looking strong enough to endure it. He seemed even more haggard than before, and trembled uncontrollably. Maybe the man should just rest for now. Perrine became desperate: The good doctor simply *must* perform the operation. Again Denis declined. When she recklessly threatened to petition the solicitor general to force him to perform the procedure, Denis refused and sent them away.

Sometime later, Denis received a conciliatory letter from Perrine: Would he "exercise the charity" to come to her home? When he arrived, he found his assistant, a set of tools, and a calf, all set up and ready for the transfusion. There sat the patient, twitching and shaking—clearly in no shape for the operation. Denis turned to leave, but Perrine fell to the ground and, "with tears in her eyes, and by unwearied clamor," begged the doctor and his assistant to stay. She had tried every means to help her husband, and now there was nothing left to do. Denis relented, and bound the calf and positioned the patient. Just as they inserted the tube, however, Antoine was seized by such violent fits of shaking that the cannula fell out. They ended the experiment without having transfused any of the calf's blood.

The next night, Antoine Mauroy died. Perrine refused Denis's request to examine the body. Suspicions aroused, Denis said he would return with several witnesses and, if necessary, conduct an autopsy by force. She buried her husband before they could return.

As soon as the news was "bruited abroad," according to an account written by Denis, his enemies closed in for the kill, publishing more defamatory books and pamphlets, decrying the doctor as a murderer and a fool. Soon after that Perrine visited Denis. She said that three physicians from the rival French Academy had offered her 50 louis d'or to charge him with murder resulting from the third attempted transfusion. If he financially supported her, she said, she would drop the issue; otherwise she would accept his enemies' proposition. He told her she and her doctor friends were mad, so crazy that they "stood more in need of transfusion than . . . her husband."

Denis had been keeping silent until then, but this latest threat unsettled him. Perrine was spreading slander and he wanted her to stop, so he filed the equivalent of a libel suit at the Criminal Court of Châtelet, outside Paris.

What followed must rank among the oddest reversals in judicial history. At first the case proceeded normally, with Perrine complaining about the inhuman treatment, the doctor defending himself with a parade of patients who attested to the effectiveness of his procedures. All agreed, according to trial findings, that the first two transfusions had managed to calm Mauroy but he then began having his frenzies again. But then a bizarre story emerged. One night, after he brutally boxed his wife's ears, Perrine began putting "certain powders" in his soup (quite an unpalatable broth, evidently, since the family cat drank some and died). By the time of the third attempted transfusion, Antoine was dying from arsenic poisoning.

In his findings of April 17, 1668, the lieutenant of the court confirmed the preliminary evidence against Perrine and ordered her and the three physicians to return for questioning. (The results of that inquiry were lost in a fire and never recovered.) He absolved Denis of any malpractice, and agreed that the third transfusion never took place. At the same time, he noted that transfusion was worrying the physicians of Paris. In deference to their concerns, he ruled that any doctor who wished to perform a transfusion must first seek permission from the Faculty of Medicine. That one small condition—virtually an afterthought to the judgment—was a devastating blow. With the faculty representing the most hidebound, hierarchical doctors in France, the more progressive physicians from Montpelier, Rheims, and other universities would sooner abandon the procedure than submit to faculty approval. And so, despite Denis's complete exoneration, the practice of transfusion faded away. Two years later, the French Parliament officially banned all transfusions involving human beings, with the English following suit. When two men died from transfusions in Rome, the pope banned the practice throughout most of Europe.

One and a half centuries would pass before doctors attempted another experimental human transfusion, and then only using the blood of other human beings. In time, doctors learned to work even more carefully, when it was discovered that human blood comes in several types, which must be matched precisely to avoid a fatal reaction.

It would be wrong to dismiss Denis and his contemporaries, naïve and haphazard though their work may have seemed. At a time when people saw blood as something magical, they understood it as a nutrient—a purely biological substance that could give life, as it were, from one creature to the next. They cracked the wall of humoral medicine, showing that the body was ruled not by vague humors but by chemicals, vessels, and pumps. They even cast doubt on the practice of bloodletting, in that blood could be a nutrient rather than an evil humor.

As for Mauroy, one might be tempted to dismiss Denis's reports of a temporary cure as wishful thinking: Mauroy may have been more exhausted than cured. Yet certain clues in the historical record suggest a tantalizing possibility. An Englishman who observed the experiments noted that Mauroy's madness was "Original of . . . Love." In other words, he probably had syphilis, which causes brain damage in later stages. Syphilis is caused by *Treponema pallidum,* a bacterium that cannot tolerate high temperatures. (Denis observed that his patient's mania abated after bouts of raging fever.) In the early twentieth century, before the development of antibiotics, doctors would treat syphilis by having the patient sit in a heated cabinet; sometimes they would administer a nonfatal strain of malaria to cause fever, driving up the body temperature and killing the bacteria. If these clues are accurate, then Denis might have triggered in Mauroy a strange but feasible chain of events: He gave the transfusion; the patient reacted; the fever that nearly killed him impeded the bacteria. And for a couple of months, the madman was sane.

# CHAPTER 2

# "THERE IS NO REMEDY AS MIRACULOUS AS BLEEDING"

The medical practice that Denis and his supporters attempted to defy was part of the longest-running tradition in medicine, one that would continue for centuries after his death. Phlebotomy, or bloodletting, originated in the ancient civilizations of Egypt and Greece, persisted through the medieval, Renaissance, and Enlightenment periods, and lasted through the second Industrial Revolution. It flourished in Arabic and Indian medicine. In terms of longevity, no other practice comes close. Germ theory, the basis for modern Western medicine, was formulated about 130 years ago. The modern practice of transfusion is about seventy-five years old. Bloodletting was faithfully and enthusiastically practiced for more than twenty-five hundred years.

Doctors bled patients for every ailment imaginable. They bled for pneumonia, fevers, and back pain; for diseases of the liver and spleen; for rheumatism; for a nonspecific ailment known as "going into a decline"; for headaches and melancholia, hypertension and apoplexy. They bled to heal bone fractures, to stop other wounds from bleeding, and simply to maintain bodily tone. Until the 1920s, country doctors in America would seasonally "breathe a vein" to keep patients in good health. And yet there was never any evidence that bloodletting did any good.

No one knows its origins—perhaps the ancients, seeing that menstrual bleeding seemed to alleviate discomfort in women, associated blood loss with the relief of symptoms. The Egyptians may have practiced bleeding as early as 2500 B.C., although it is not known why they bled or what they thought it would accomplish. (An illustration on a tomb near ancient Memphis depicts a patient being bled from the foot and neck.) Hippocrates, the father of Western medicine, who wrote in the fourth and fifth centuries B.C., gave an early account of bloodletting as it related to humoral medicine. He explained that, since all sickness resulted from an imbalance of humors, the cures involved restoring the equilibrium by causing vomiting, sweating, defecation, or bleeding. "Bleed in the acute affections," he wrote, "if the disease appears strong, and the patients be in the vigor of life. . . ."

Hippocrates' medical system passed to Aristotle, and through him to Alexander the Great, whose conquests spread it to the Persian and Hindu worlds. Centuries later, in Rome, Claudius Galenus, chief physician to the gladiators during the reign of Marcus Aurelius, adopted the beliefs. A talented and resourceful practitioner, Galenus (or Galen, as he later became called) wrote an estimated 120 books, most of which touted his cures in the absence of any true clinical histories. He recommended bloodletting for a wide array of maladies, and favored it as more precise than other methods of purging. After all, a physician could give potions to cause vomiting but could not predict how the body would react. In contrast, Galen could bleed to a precisely measured amount, or watch the patient blanch or swoon to know exactly when to stop.

In codifying humoral medicine for the Roman world, Galen added a doctrine of his own. Vitalism, as we have seen, asserted that blood was more than a nourishing liquid, embodying the spiritual essence of man: Flowing from the liver to the heart and brain, it acquired a trinity of spiritual characteristics from the combination of organs through which it passed. Galen was a pagan, but the spiritual implications of his theory inspired the Church to support it dogmatically for centuries.

When Rome fell, the Arabs, who adapted so much of Greek science, inherited the mantle of medicine as well. Giants such as Avicenna of Persia prescribed bleeding for practically every known malady. Unlike the Greeks, who bled from the same side of the body as the malady, the Arabs bled from the opposite side, or "revulsively." They would phlebotomize the right elbow, for example, to stop a patient's left nostril from bleeding. Bloodletting continued in Europe as well. Physicians at the medical school at Salerno, the greatest institution of its kind in the twelfth century, produced a rhyming Health Code that recommended bloodletting for a typically wide array of conditions:

> *Bleeding the body purges in disguise*
> *For it excites the nerves, improves the eyes*
> *And mind, and gives the bowels exercise*
> *Brings sleep, clear thoughts and sadness drives away*
> *And hearing, strength and voice augments each day.*

and so on, for eleven verses.

Bleeding was as trusted and as popular then as aspirin is today. Monks bled each other several times a year for general maintenance of health. (At an archeological dig of an ancient monastery in Scotland, scientists recently found a whole stratum of blood waste, accounting for an estimated three hundred thousand pints of blood.) The Talmudic authors laid out complex laws for bloodletting. Physicians obligingly drained the blood of feudal lords. Doctors devised elaborate charts indicating the most favorable astrological conditions for bleeding. The second medical text ever printed on Gutenberg's printing press was a *Bloodletting Calendar* in 1462. The first was a *Purgation Calendar* printed in 1457.

During the Middle Ages, a new category of practitioners emerged. The pope had forbidden the clergy to perform bloodletting (although they were welcome to undergo the procedure), and physicians were discouraged by the fact that feudal lords could kill them for malpractice. So bloodletting moved into the hands of barber-surgeons. More craftsmen than medics, they established their own guilds and competed for respectability with apothecaries and surgeons. They advertised with a symbol that endures to this day, a red-and-white-striped pole representing the stick patients would grab while being phlebotomized—the white stripes, the bandages; the red stripes, the blood.

Of all the doctors who supported bloodletting, few could surpass Guy Patin, a contemporary of Denis's, for arrogance, sophistry, and a cast-iron state of mind. As Dean of the Faculty of Medicine in Paris, Patin was an archconservative who wielded his influence over anyone who defied the medical orthodoxy. He left a rich and vivid correspondence that records the foibles of his day.

Patin effused about the benefits of bleeding. "There is no remedy in the world which works as many miracles as bleeding," he wrote to a colleague in 1645. "Our Parisians ordinarily take little exercise, drink and eat much and become very plethoric. In this condition they are hardly ever relieved of whatever sickness befalls them if bloodletting does not proceed powerfully and copiously. . . ."

In one letter he described the case of a colleague who "was attacked by a rude and violent rheumatism, for which he was bled sixty-four times in eight months, by order of his father . . . and in the end recovered." Another time Patin treated "a young gentleman of seven years, who fell into a pleurisy by over-heating himself when playing tennis. . . . He was bled thirteen times, and cured in fifteen days, as if by a miracle."

During Patin's time, some scholars began to inquire whether chemistry, not humors, influenced the body, which led them to question the effectiveness of bleeding. One was the Belgian chemist Jan Baptiste van Helmont, whom Patin cruelly castigated in 1645: "He was a wicked Flemish rascal, who died insane a few months ago. He never did anything of value. I have seen everything he wrote. This man only thought of one medicine, made up of chemical and empirical secrets. He wrote much against bloodletting, for lack of which, however, he died in a frenzy."

Patin was no hypocrite: He bled his wife twelve times consecutively for chest congestion, his son twenty times for a fever, and himself seven times for a head cold. He bled his father-in-law when the man was eighty years old. "We also bleed very fortunately children of two or three months, without any inconvenience," he wrote. "I can show more than two hundred of them bled at this early age. There is not a woman in Paris who does not think well of bleeding. . . ." In the face of such enthusiasm, it is worth repeating that, with the exception of rare medical conditions, bleeding does the body no good. The survival of any of Patin's patients had more to do with luck and the body's resiliency than any benefits they derived from his treatments. Like other physicians of the time, Patin kept no systematic records, basing his assertions entirely on his subjective impressions. He claimed credit for each success, and wrote off each failure as tragically inevitable. He wrote a typical apologia in 1659: "Our good M. Baralis has been bled eleven times in six days, which has prevented suffocation. . . . But he is in great danger of not being able to escape it. A continued fever, a bad lung besieged with inflammation, eighty-four years, are all signs which leave me with a gloomy suspicion."

Even though Patin was the most honored and powerful physician of his day, his position did not shield him from perceptive social critics. The novelist Alain-René Le Sage parodied the doctor in the person of Dr. Sangrado ("Dr. Bloody"), a "tall, withered executioner of sisters three," who bled his patients to the brink of death. Molière lampooned the crack certitude and foppish ways of physicians like Patin; his play *The Imaginary Invalid* is a virtual catalogue of their faults. Molière suffered a variety of ailments, and dealt with the medical profession with unwelcome frequency. One day Louis XIV asked how he was getting along with his latest physician. "Sire, we talk together," Molière replied. "He prescribes remedies for me; I do not take them; and I recover."

Still, bleeding remained a mainstay of medical care. Doctors developed elaborate systems for determining exactly where a vein should be opened, and under which astrological sign. Some said that the third, or "leech," finger housed a vein that led directly to the heart, which made it attractive for bleeding. Others described the "heart vein" or "meridian vein" as the optimal bleeding site. Most agreed that patients with serious illnesses should be bled rapidly in an upright position. Fainting was considered a positive sign.

Bleeders used an impressive array of hardware. Their mainstay was the lancet, a small, sharp, two-edged knife that usually occupied an exquisitely tooled case. The bloodletters would use a lancet to open a vein in the arm, leg, or neck. The procedure involved tying off the area with a tourniquet (excluding the neck, of course), having the patient grasp a stick, and, holding the lancet delicately between thumb and forefinger, striking diagonally or lengthwise into the vein. (A perpendicular cut might sever the blood vessel.) Bloodletters would collect the blood in measuring bowls, exquisitely wrought of fine Venetian glass. Many families kept bleeding bowls as heirlooms.

Wielding the lancet took quite a bit of skill: A false cut could slice a nerve or a tendon. To make the job easier, a Viennese inventor produced a spring-loaded lancet, called a *Schnapper* in German or, in English, a "phleam." It consisted of a case about two inches long with a spring-loaded blade emerging from the top. The bleeder would cock the blade, press the *Schnapper* against the skin, and push a release, causing the blade to snap down and across. The *Schnapper* had the safety feature of not cutting beyond a certain depth. German, Dutch, and American bleeders preferred it, but French practitioners used the simpler and more artful lancet.

Sometimes phlebotomists would use a "scarificator"—a spring-loaded box containing anywhere from twelve to eighteen blades—in conjunction with "cupping" to relieve local inflammation. The bleeder would place a glass cup against the skin with an edge slightly raised, and warm it with a torch. The heat would create a vacuum strong enough to suck the cup against the skin, raising a large blood-filled blister. The bleeder would pull off the cup, spring the scarificator, and then reapply the cup to draw out more blood. Leeches were also used. Derived from the Anglo-Saxon word *loece,* "to heal" (medieval doctors called themselves "leeches"), the leech removed blood from hard-to-reach places such as "the mouth of the womb, the gums, lips, nose, fingers," according to a medical text from 1634.

Some of the most poignant dramas in the history of bloodletting took place thousands of miles from its origins. In colonizing the New World, the Europeans took their practices with them. Furthermore, because no medical schools existed in the colonies, American medical students would study in Europe, even after the revolution. The old traditions remained intact.

The foremost American bleeder was the physician and patriot Benjamin Rush. Rush, known as the "Prince of Bleeders" by his contemporaries, was a scholar, humanist, social reformer, and signer of the Declaration of Independence. (His name appears just above that of his friend Benjamin Franklin.) A learned and ethical man, he spoke out against slavery, capital punishment, and cruelty to children. He wrote the first American textbook on the mentally ill. He founded the Philadelphia Dispensary for the treatment of the poor and, with Franklin, the Society for the Protection of Free Negroes. He served as surgeon general to the Continental Army, as chairman of the department of medicine at the College of Philadelphia, as supervisor of the United States Mint. So great was his influence on colonial medicine that some called him this nation's Hippocrates.

Rush studied at the University of Edinburgh, where he learned a modern variation on humoral medicine. In those days, patients did not suffer from diseases like "tuberculosis" or "strep," each caused by a specific pathogen; rather, they endured nonspecific ailments caused by humoral imbalances, such as "dropsy" (congestive heart failure) and "pleurisy" (chest pains), or "fits" and "convulsions," or "decay" and "bloody flux." They suffered an unending variety of fevers, including "intermittent," "continued," "remittent," or "putrid" fevers—all without a clue as to the cause. The doctors took pains to classify these symptoms and treat them in some logical way, which, lacking diagnosis, frequently was mistaken. Doctors believed that nature favored healing but that one must coax her vigorously along. They took extreme measures to adjust the body's humors, and to empty it of poisons. They would induce vomiting, diarrhea, and bleeding, and redouble their efforts if the patients became worse. They would change the treatments daily in response to the symptoms as they evolved.

The period became known as the Age of Heroic Medicine. Just how exaggerated those heroics could be is shown in the hospital records of a young man named Alexander Forbes, who was admitted to the Royal Infirmary in Edinburgh in 1795. Forbes, a twenty-two-year-old, was suffering from pains in the head and chest, which his doctors diagnosed as "excitability" of the blood vessels. For the first several days they gave him opiates, to kill pain; an emetic, to cause vomiting; and cathartics, to cause defecation. On the sixth day they blistered his chest. (Blistering meant applying a mustard plaster or some other substance caustic enough to create a second-degree burn, in order to draw out bodily poisons.) Then they gave him a mild stimulant called "tincture of guaiac" to strengthen his blood vessels and cause urination, after which they blistered him again.

By now he had been hospitalized for thirty-seven days. His doctors gave him Peruvian bark (a form of quinine) as a general tonic, and a week later they placed six leeches on his head overnight. Later they gave him more opiates and cathartics, and applied eight leeches to his head. Finally, after two months and five days, they released him, fever abated but still feeling pain.

Rush supported such vigorous methods, believing that all disease arose from excitation of the blood vessels, which copious bleeding would relieve. Rush taught that the body held about twenty-five pounds of blood, twenty of which could safely be drained. (Considering that the body actually holds less than half of Rush's estimate, woe to the patient who received his ministrations!) If the patient fainted, so much the better, for it meant that the harsh measures were taking effect.

Rush taught bloodletting to a generation of physicians. One former student, Dr. William Montgomery, wrote to Rush that he had treated a member of the South Carolina legislature by relieving him of 165 ounces of blood in five days (almost the man's entire complement of blood). "He died," wrote Montgomery. "Had we taken a still greater quantity the event might perhaps have been more fortunate."

Whatever the flaws in Rush's methods and theory, no one could criticize his courage or character. He demonstrated both during the great yellow-fever plague in Philadelphia in 1793—the worst in American history—during which he developed his vigorous techniques. Philadelphia, then the nation's capital, largest city, and commercial and intellectual center, sat on the lowlands at the convergence of the Schuylkill and Delaware rivers. The city was a hot, sultry, swampy place—an inviting breeding ground for *Aedes aegypti*, the mosquito whose bite spreads yellow fever. No one at the time understood the connection between the mosquito and the fever. All they knew was that, in August of a hot summer, after the wettest spring in memory, people began to die. They died in twos and threes, then twenties and thirties, and by the height of the contagion at a rate of more than a hundred per day. They all exhibited the same dramatic symptoms: skyrocketing fever, chills and pain, black vomit and jaundice, followed by death in just a few days. The mention of the disease was enough to cause panic. Philadelphia emptied and its government collapsed.

Rush stayed behind. He sent his wife and children to the countryside and—putting himself on a vegetarian diet as an intuitive protective measure—selflessly cared for the sick.

"Let anyone who desires to fully appreciate this great man . . . read and re-read Rush's account of this epidemic," wrote the provost of the University of Pennsylvania a century later. "For upwards of six weeks . . . he abandoned all precautions, and rested himself on the bedside of his patient, and drank milk and ate fruit in their sick rooms; he visited over a hundred fever patients daily, and his house was filled with the poor whose blood, from want of a sufficient number of bowls, was often allowed to flow upon the ground. . . ."

Rush strode through scenes medieval and surreal. The streets had practically emptied by then. Those people who did venture outside hurried by, not daring to look at each other as they chewed garlic, smoked cigars, and sprinkled themselves with vinegar to keep contagion at bay. They lit bonfires at street corners to "purify" the air. Church bells stopped ringing. The city's clock towers ran wildly off, for no one dared go out to reset them. Inside were scenes of unspeakable tragedy, as family members died in rapid succession.

At first Rush tried his usual cures—vomiting and catharsis followed by moderate bleeding—but his patients kept dying. Desperately poring over his references one night, he came upon an account of an earlier plague, in which the solution had been to purge in the most violent way imaginable, even if it drove patients to the brink of death. He tried the new regime on a man who had nearly expired, giving him a massive dose of mercury and jalap to produce diarrhea, and then bleeding him copiously. The man's condition seemed to improve.

Rush felt he had discovered a cure. Over the next few days he treated several people with the new regime, most of whom became "perfectly cured." Rush, like other physicians of his time, kept no records aside from financial ones, so had no statistics on his treatment. He worked purely on impression. Still, that impression was so strong that he posted his treatment for citizens on placards and urged it on the College of Physicians.

Many of Rush's peers disagreed. A French doctor who had opened a clinic specializing in gentle restorative methods reported an equally strong impression of success; others thought massive bleeding would fatally weaken the patient. These Rush dismissed. With the energy born of what he felt to be a celestial revelation, he raced throughout Philadelphia, administering his cure to all who required it. First he claimed to save four out of five lives, then eight out of twelve, then nineteen out of twenty. "Yesterday was a day of triumph to mercury, jalap and bleeding," he wrote to his wife, Julia, on September 13, 1793. "I am satisfied that they saved, in my hands only, nearly one hundred lives." To a friend he wrote: "At first I found the loss of ten or twelve ounces of blood sufficient . . . but I have been obliged, gradually, as the season advanced, to increase the quantity to sixty, seventy, and even eighty ounces, and in most cases with the happiest of effects. I have observed the most speedy convalescence where the bleeding has been most profuse, and as a proof that it has not been carried to excess, I have observed in no one instance the least inconveniences to succeed it."

We know now that yellow fever is caused by a virus: It runs its course and leaves, killing a rather small proportion of its victims. Doctors do not treat the virus specifically but, rather, offer nursing care and symptomatic relief. Dr. Rush would have served his patients better by making them comfortable and leaving them alone.

Still, no one could doubt his bravery or commitment. Working day and night, seven days a week, he managed to bleed more than a hundred patients a day. No one could forget how Rush, surrounded by death and terror, would stroll into a fetid room, examine a patient, and say reassuringly, "You have nothing but a yellow fever," as though it were no worse than a cold. When Rush's beloved sister died, he took a half-hour to mourn her and then hurried out to make his rounds. Rush himself caught the contagion and battled a violent infection for six days, all the while exhorting his assistants to keep bleeding and purging. As soon as he had regained a fraction of his strength, he rose and threw himself back into the fray.

"Thus you see that I have proved upon my own body," he wrote to his wife, "that yellow fever when treated in the new way, is no more than a common cold."

So many middle-class people had fled Philadelphia that Rush recruited free "Negroes" to help, training them and ordering them to fan out across the city. Historians recall one memorable scene in which Rush's driver was taking him through the neighborhood of Kensington. A huge crowd had surrounded the carriage, beseeching him for help, and Rush spoke to them sympathetically. Then he announced, "I treat my patients successfully by bloodletting, and copious purging with calomel and jalap—and I advise you, my good friends, to use the same remedies!"

"What?" someone cried out. "Bleed and purge everyone?"

"Yes!" he shouted. "Bleed and purge all of Kensington! Drive on, Ben!"

Eventually, the epidemic began to abate. The autumn frosts were exterminating the mosquitoes. Citizens began returning. Windows were opening, church bells were ringing, and life in the streets began to resume. Rush wrote to his wife and children, bone tired and yet deeply satisfied: "Never did I experience such sublime joy as I now felt in contemplating the success of my remedies. It repaid me for all the toils and studies of my life. The conquest of this formidable disease was not the effect of accident, nor of the application of a single remedy; it was the triumph of a principle of medicine."

Rush deserved to savor the moment. He had conducted himself with courage, dedication, and selflessness. Yet his science, as we now know, was tragically wrong. This brave, pious, genuinely saintly man undoubtedly killed more people than he cured.

Even though the epidemic was over, Rush could not rest. During the contagion, he had argued that the disease was local in origin—not imported, as some had said. (City fathers had wanted to blame a boatload of refugees from Santo Domingo. Rush said the fever arose from fetid conditions in the city itself—hardly the kind of news the city wanted advertised.) In November, Rush learned that the College of Physicians, which he had helped to establish, had written a report concluding that the disease was indeed imported. Disgusted, he resigned. Financial problems haunted him. During the epidemic, he had not sent out any bills, thinking it inappropriate to do so in a crisis; yet his landlord raised his rent. Then new attackers began to assail him about his treatment of the disease.

Rush's chief antagonist was a bilious young newspaperman named William Cobbett, publisher of the *Porcupine's Gazette,* the most widely read newspaper in the country for a time. Cobbett did not like Rush's politics or style—hardly surprising, since the two men were as opposite as people could be. Rush was pious, humorless, and high-minded, concerned with man's future and social reform. The newsman, in contrast, was a self-educated adventurer, a native of England who had traveled widely as a marine. Now settled in Philadelphia, he accused Rush of "puffing" his treatment and failing to report deaths. He decided to mount a campaign against the doctor, an ongoing series of personal attacks, often in the form of wicked rhyming couplets:

*The times are ominous indeed*
*When quack to quack cries purge and bleed.*

Cobbett sarcastically labeled Rush "the Pennsylvania Hippocrates." (He called Benjamin Franklin, another enemy, "Old Lightning Rod.") When Rush compared his treatment methods to the strength of Samson, Cobbett took the analogy and gave it a new twist: "I verily believe that they have slain more Americans . . . than ever Samson slew of the Philistines." He also published a satiric ad for the doctor: "Wanted by a physician, an entire new set of patients, his old ones having given him the slip. . . ."

Rush was beginning to tire of Philadelphia, and applied for a professorship at Columbia University, where one of the trustees had referred to the doctor as "a man born to be useful to society." When Cobbett got wind of this he gleefully rejoined, "And so is a mosquito, a horse-leech, a ferret, a polecat, a weasel: for these are all bleeders, and understand their business as full as well as Dr. Rush does his."

Cobbett's harping continued for months; the man seemed obsessed. Finally, against the advice of his friends, Rush sued Cobbett for libel. The trial, which encountered endless delays, did not move ahead until December 1799, when a jury found Cobbett guilty of slander and ordered him to pay Rush $5,000. On that very day, a monumental coincidence took place. As Cobbett recounted in his vituperative style: "On the 14th of December, on the same day, and in the very *same hour,* that a ruinous fine was imposed on me for endeavoring to put a stop to the practice of Rush, *General Washington was expiring under the operation of that very practice.*" It was true: Suffering from what appeared to be strep after a winter's ride on his farm in Virginia, Washington insisted that his doctors bleed him copiously. After two days of the treatment, the former president died.

Cobbett produced a series of pamphlets called *Rush-Light,* dedicated to the destruction of Benjamin Rush. Within a few months of Washington's death, *Rush-Light* ran a critique of his treatment, written by a doctor from Savannah, Georgia, named John Brickell. Brickell noted that, since "old people cannot bear bleeding so well as the young," the doctors should not have bled Washington so massively. (He suggested that they should have bled Washington gently instead, from under the tongue.) Meanwhile, it came out that Elisha Dick, the youngest of the

three doctors who had attended the president, had cautioned against
massive bleeding, but that the two others had overruled him. A few
weeks later, Dr. Gustavus Brown, another attending physician, wrote
that he was sorry they had not taken Dick's advice. Cobbett seized
upon the discussion with maniacal glee: What could be more perfect
than one of Rush's own students warning against his mentor's
excesses? "Don't you think it would be a good thing, Doctor, if the
names and places of abode of all Rush's pupils were published? If you
don't, I do."

The controversy did not continue for much longer, although scholars would debate Washington's treatment for years. Cobbett, looking
for more substantial targets to abuse, moved back to England. There,
he took on the entire class system, accusing it of injustice against small
landholders and tenant farmers. Rush, whose practice had declined
and who had suffered financial problems ever since the epidemic, nevertheless donated Cobbett's payment to charity. In recognition of past
service, President John Adams named him secretary of the Mint. Hearing of the appointment, William Loughton Smith, the American
ambassador to Portugal, wrote: "I have been much amused in reading
over some files of American papers by the last vessel. I see the old dispute revived with great violence [about] bleeding for fever and ague,
and that Dr. Rush is charged with bleeding many hundreds to
death. . . . I was not very much surprised at this charge, but I confess I
was surprized to see him appointed treasurer of the mint. I hope he
won't bleed that to death also."

Despite Dr. Rush's ill fortune and George Washington's death, bleeding remained popular for another half-century. With no real understanding of disease, doctors had few other remedies to which they
could turn. Besides, as any doctor could see by the swooning of his
patients, bleeding produced dramatic results. "Bleed *ad libitum*," rang
out in medical lecture rooms, "bleed in an upright posture to fainting."
France fell under the influence of Dr. François Joseph Victor Broussais,
a veteran of the Napoleonic campaigns with a rough field-surgeon's
disdain for half-measures. In 1833 alone, French doctors imported
41.5 million leeches. Broussais's disciple, Dr. Jacques Lisfranc, spilled
rivers of blood. Oliver Wendell Holmes, the American author and
physician (and father of the eminent jurist), studying in Paris at the
time, described Lisfranc as "a great drawer of blood and hewer of
members." He recalled seeing Lisfranc in a "phlebotomizing fit . . .
ordering a wholesale bleeding of his patients, right and left, whatever
might be the matter with them." In Italy, Giovanni Rasori, of the clinic

at Milan, routinely bled patients to the brink of death. British physicians held such respect for bleeding that they named their pre-eminent journal *The Lancet,* a title it bears to this day.

Like satirists before him, Charles Dickens took aim at the overzealous doctors. In the *Mudfog Papers,* a certain Dr. Kutankumagen presents a case to his associates. The patient is "stout and muscular," he reports, "his cheeks plump and red, his voice loud, his appetite good, his pulse full and round. He laughed constantly, and in so hearty a manner that it was terrible to hear him." The doctors treat him "by dint of powerful medicine, low diet and bleeding." The treatment continues for several weeks. Afterward everyone is relieved to see that the patient has to be "carried downstairs by two nurses . . . supported by soft pillows." He "ate little, slept little and was never heard to laugh by any accident whatever." The other physicians toast Dr. Kutankumagen on his "triumphant" cure.

Several developments eventually led physicians to abandon phlebotomy. One was a series of typhus epidemics that ravaged British cities in the early 1830s. Typhus, a rickettsial infection spread by fleas, is particularly unsuitable for bleeding. Some fevers, like malaria, cause excitability: The pulse races, the temperature soars, the patient sweats and goes into a delirium. In such cases bleeding, by weakening the patient, gives the illusion of symptomatic relief. But typhus is debilitating: The feverish patient goes into a decline. Even removing a few ounces of blood can trigger fainting. One would have to have been blind not to see that bloodletting made patients worse. "I altogether gave up blood-letting after the fever," wrote an Edinburgh physician.

The second development was a new field of medicine called medical statistics. Until then, as we have seen, doctors had worked mainly on impression—looking at the patient and, based on accumulated wisdom, deciding what to do. They did not keep records detailing the course of each treatment and the results. But in the 1830s, a doctor in Paris named Pierre-Charles-Alexandre Louis began gathering information more systematically. A familiar sight at La Pitié hospital, he was a tall, somber figure, with glasses on nose and notebook in hand—more like a bookkeeper than a doctor, really—moving from bed to bed taking copious notes. He would ask patients questions no one had asked before: When had they become sick? What had the symptoms been? Had they had a history of this kind of disease? Later he would produce tables containing hundreds of cases, documenting the course of diseases such as typhoid fever and tuberculosis, showing with statistics when a disease appeared, how it progressed, and how it responded to various treatments. False treatments cannot stand up to such scrutiny, and in one study Louis showed that bleeding might not be so effective as previously thought. His conclusions were equivocal, but their rigorous statistical basis gave them impact far beyond their intent.

Soon a new generation of clinicians arose, trained to take case histories in the emerging science of cell biology. Finally, the three giants of bacteriology—Louis Pasteur in France, Joseph Lister in Scotland, and Robert Koch in Germany—showed that microbes, not humors or other intangibles, cause disease. Germ theory became the basis of modern medicine. All these developments—closer observation, the use of statistics and detailed case histories, the development of cell biology, and the birth of germ theory—led to the decline of the practice of bleeding. In some places, to be sure, the old ways persisted. But for all practical purposes, phlebotomy was dead.

Twenty-five centuries is a long time for a useless practice to continue, but it is important to realize in retrospect how paltry were the tools that doctors had available. Disease had always been frightening, a mystery, and practitioners had no idea how to control it. Bleeding gave them a sense of control, a feeling that they could do something, anything; and the patient's accompanying lightheadedness—and the fact that most people actually recover from fevers—may have given the illusion of cure. Even in later years, when the evidence pointed to the futility of the practice, many doctors defended phlebotomy, perhaps to maintain their authority as much as anything. One gets that impression from reading the accounts of the doctors who, even as Washington weakened, continued to draw his blood. They wrote a defense that, given the subsequent history of blood, echoes eerily: "We were governed by the best light we had; we thought we were right, and so we are justified."

# CHAPTER 3

# A STRANGE AGGLUTINATION

The modern era of transfusion medicine began with a pounding on a door in the middle of the night. The year was 1908, the place was New York City, and the door led to the apartment of Dr. Alexis Carrel, a noted French researcher who had come to work at the Rockefeller Institute for Medical Research. Outside the door stood three ashen-faced men: Dr. Adrian V. S. Lambert of Columbia University and his two brothers, both surgeons. Lambert's wife had given birth to a girl who for the past several days had been oozing blood from her nose and mouth. The child seemed certain to die unless someone could introduce fresh blood into her system. Lambert had been following Carrel's work, in which he had grafted arteries and veins in experimental animals, and was convinced that Dr. Carrel could somehow employ the technique to save the life of his newborn daughter.

"*Alors,* how can I do that?" Carrel replied. "I have no license to practice surgery in the state of New York. My patients are only my dogs and cats."

Lambert promised that he would take full legal responsibility, if only Carrel would please come—at once. After stopping at the institute to pick up his instruments, Carrel rushed through the cold night to Lambert's apartment.

The drama that was about to unfold took place against the backdrop of an adventurous and hopeful period in history, a time of scientific discovery and dynamic social change. In the first few years of the twenti-

31

eth century, Einstein published his theory of relativity, Max Planck his theory of quantum physics, and Marie Curie her descriptions of radioactivity and the elements that produce it. The Wright brothers flew their heavier-than-air ship, and Guglielmo Marconi transmitted radio messages across the Atlantic. The mysteries of behavior began to reveal themselves as both Freud and Pavlov published their studies. Discoveries were cascading in medicine as well. Yellow fever, the disease fought so vainly by Benjamin Rush and others, now became less of a dread mystery as the American doctor Walter Reed and his colleagues, working in the recently acquired Panama Canal Zone, demonstrated that the disease was transmitted by mosquitoes, thereby paving the way for its control. In Germany, another scourge fell before medicine, as Paul Ehrlich discovered a cure for syphilis. Hospitals began to employ X-rays to create "shadow pictures" of human anatomy, and pharmaceutical companies released a new drug called aspirin.

Remnants of the old era certainly remained. Most hospitals were still not keeping case histories of their patients, and as many quacks treated people as did trained doctors and surgeons. Even qualified doctors found themselves helpless against a variety of diseases whose modes of transmission they did not understand, including typhus, malaria, and bubonic plague. They knew nothing about vitamins, or about insulin, the treatment for diabetes. Yet doctors no longer saw disease as an insurmountable adversary to be treated, however ineffectually, by the remedies of old. In medicine, just as in science, industry, and technology, the dominant themes of the era were progress and change.

Excitement prevailed as the new century got under way, and nowhere was it more palpable than in the United States, a nation that seemed to exemplify the age. Plunging into the new era of industry and production, America was generating more than a third of the world's steel, pig iron, and silver and half the world's cotton, corn, copper, and oil. More than a million miles of telephone wire linked the nation's people, introducing an age of instant communication, and more than two hundred thousand miles of railroad track moved people and goods from coast to distant coast. The U.S. Navy, having grasped the significance of Marconi's wireless communication, set free their flocks of carrier pigeons. An oil boom had erupted in Beaumont, Texas, ushering in the fuel of the twentieth century. Henry Ford was forming the world's largest automobile company, where he would soon produce the most popular car of all time—the "Tin Lizzie," or Model T Ford. This vast continent of people and resources seemed an enormous market, with fortunes just waiting to be made. Industrial magnates like John D. Rockefeller, J. P. Morgan, and Andrew Carnegie built empires based on oil, banking, and steel respectively, creating trusts that crushed all competition. Yet it was a nation that cheered the underdog as well. At the same time the "robber barons" reached the peak of their power, the public enthusiastically supported their president, Theodore Roosevelt, in his trust-busting efforts to rein them in. Later the magnates, in an effort to rehabilitate their image, established philanthropies for the advancement of research and learning, such as the Ford Foundation, the Carnegie Foundation, and the Rockefeller Institute, which became powerful forces for the advancement of the human condition, especially in medicine.

If America exemplified the new era's belief in growth and progress, then nowhere were these qualities found in greater abundance than in the city of New York. A raw and rapidly growing metropolis, New York—with a population of four million, surpassing those of Paris, Berlin, and Tokyo—rushed into the century with excitement, optimism, and dizzying speed. By 1908, the city had already been strung with streetlights, electric trolleys were replacing the old cable cars, and the first tunnel had been dug from Manhattan to New Jersey. As the gateway to the nation, the city was attracting tens of thousands of immigrants each year—Irish, Jews, Italians, Russians, Poles. Amid this teeming and eclectic mix of peoples, New York became home to some of the world's most energetic and innovative medical institutions, including Mount Sinai Hospital, the Lenox Hill Hospital (then called the German Hospital), and Bellevue Hospital, each vying for the best technology in New York and the world. In such a competitive atmosphere, any new advance was viewed as something to be embraced, even seized upon. This may explain why, during the first fifteen years of the century, transfusion was revolutionized in New York.

Alexis Carrel, like many others, had come to the city seeking freedom and opportunity—in his case, freedom from the rigidity and closed-mindedness of French medicine. Born to a prosperous family near Lyon, Carrel was a short, intelligent, opinionated man with one blue eye and the other brown, both obscured behind thick, semispherical glasses. Educated by Jesuits, he earned his medical degree at the University of Lyon. Soon afterward, in 1894, the president of the French Republic, Sadi Carnot, was visiting Lyon when an Italian anarchist stabbed him in the abdomen. The blade cut the portal vein, the large blood vessel leading from the intestines to the liver, an injury that surgeons assumed to be fatal. Shortly after Carnot bled to death, Carrel loudly asked why the best surgeons in France had not even attempted to stitch the vein closed. The question did not endear him to the medical establishment, but it inspired a lifelong interest in vascular surgery.

Carrel embarked on a self-imposed regimen to make himself the finest surgeon in France. Developing new techniques at his laboratory in Lyon, he produced ever more delicate needles and sutures, and sharpened his stitching skills under one of the premier embroiderers in Lyon, Madame Leroudier. (Any colleagues who might tease him about his needlework stopped when they witnessed his surgical dexterity.) Along with these abilities, Carrel cultivated his spirit, having always felt people existed on a spiritual as well as a material level. In 1903, he accompanied a group of pilgrims to the cathedral in Lourdes, where miraculous healings were said to occur. With them was a young girl suffering from tubercular peritonitis, a bacterial inflammation of the walls of the abdomen. Examining the girl, Carrel felt certain she would die, and advised the congregants to hospitalize her immediately. Instead, they sprinkled her with holy water and prayed. Hours went by, during which the girl hovered near death. And then the symptoms suddenly abated. Carrel examined her again. Something—perhaps, he thought, the power of suggestion—had brought the child back to life.

Carrel, a diligent and meticulous observer, described what he saw in a report to the Lyon medical community, and candidly answered questions from the press. The ensuing publicity created a furor. How could a surgeon as prominent as Carrel ascribe to the medieval notion of faith healing? Rebuke poured in from the medical establishment for considering that the cure had a mystical source, and the clergy for attributing the miracle to the power of suggestion rather than to God. As the weeks passed and the uproar continued, and it became clear to Carrel that it would permanently impede him, he yielded to a feeling that had been nagging him for years. He concluded that French medicine had become hopelessly ossified. If he wanted to make any real contribution, he would have to go to the New World.

Carrel worked in Canada and then accepted a position at the University of Illinois in Chicago, where he further developed his surgical techniques. There, experimenting on laboratory animals, he developed a technique called "anastomosis," in which he stitched arteries and veins end to end. Few had been able to accomplish the procedure. Blood vessels are so small that it is almost impossible to sew one wall of a vessel without poking through the other. Carrel developed a triangular method in which he made small stitches on three points around the vessel. He would then pull the vessel taut from one point to another, creating a straight line that he could stitch without poking through the vessel's rear wall. This technique was so effective, he found, that he could use it to reattach limbs of animals that had been experimentally severed. Standing for hours over his experimental animals, he wore special magnifying spectacles to guide him through the minute opera-tion—thus presaging the development of microsurgery.

As Carrel performed surgery on the tiniest of scales, one of the world's wealthiest men was attempting to change medicine with a big and grand design. The oil baron John D. Rockefeller had purchased a farm on a bluff overlooking the East River in New York City, and was clearing the grounds to produce a world-class center for medical research. There, in the words of one of its administrators, doctors "could give themselves to . . . uninterrupted study and investigation, on ample salary, entirely independent of practice." Simon Flexner, a pathologist at the University of Pennsylvania, was hired to direct the Rockefeller Institute for Medical Research and recruit the most innov-ative medical researchers he could find. Shortly after the construction crews broke ground, Carrel came to visit. Flexner hired him in 1906.

At the Rockefeller Institute, Carrel continued his work in blood-vessel anastomosis, at one point replacing a segment of the blood vessel of a cat with that of a dog that had been removed and preserved several days before. He was in the midst of this research with the grafting of blood vessels when, sometime before dawn on a Sunday in March 1908, Dr. Adrian Lambert and his brothers came to his door. They knew he had only employed his techniques with laboratory animals, yet they also knew that the moment was desperate.

Carrel sped to the Lamberts' comfortable apartment on West 36th Street. There he beheld a sight that was "pitiful indeed": Adrian's wife, a young mother, was lying in bed, distraught and almost too weak to move; her baby lay next to her, unconscious and sheet-white. "I feared that it would die before I could prepare for the operation," Carrel recalled. He bound the baby girl to an ironing board and placed it on the dining-room table. Then he asked who wished to give blood. He told the Lamberts it would be a dangerous operation; he had never done this on humans before, and the donor could well lose the use of his hand. At that point, "a real quarrel ensued. A disagreement arose from the father, the mother and the two uncles each insisting that he was the proper one to give blood."

Carrel decided to use the baby's father. Dr. Lambert lay down next to the child, his wrist bound firmly to her leg. No anesthetic was employed. There was only one vein in the baby large enough to use—the popliteal vein, in the crook behind the knee. Carrel sliced through the skin with a scalpel, snipping through a layer of glistening pink con-nective tissue and exposing the tiny vein, matchstick thin and with a reddish-blue tinge. He bared an artery in the father's left wrist (the father was right-handed) and, using his famed three-point technique, attempted to suture the father's artery to the tiny vessel in the baby's leg. He failed several times, ripping through the walls of the tissuelike vein, before finally connecting them. The vessels sprang to life as the father's blood raced into the child, yet the baby remained still and deathly white. Soon one of the uncles noticed a change—"a little pink tinge at the top of one of the ears. . . . Then the lips, perfectly blue, began to change to red." Suddenly "a pink glow broke out all over its body," almost as if she had been placed in a hot bath. The baby began to wail.

"You'd better turn it off or the baby will bust," her uncle Samuel remarked. Carrel clamped off the blood vessels and tied them. The baby survived, and circulation in the father's hand was also restored. Later the mother sent a letter to John D. Rockefeller in which, writing in the imagined voice of her baby, she thanked him personally for making it possible, through the institute, for Dr. Carrel to have attended her. "I shall always feel that you saved my life by making it possible for Dr. Carrel to successfully operate when all the other doctors had given up hope of keeping me alive." She also sent a photograph of the baby to Dr. Carrel, who thereafter regarded it as a *"souvenir precieux."* Rockefeller's publicists released the story to the press, which soon made Carrel a celebrated man, praised for performing "one of the most remarkable surgical successes ever achieved in this country," in the words of one of his fellow physicians. In 1912, he won the Nobel Prize. Years later, even the French acknowledged his brilliance, and made him a Knight of the Legion of Honor. Yet Carrel never let go of his resentment. One night, at a celebratory dinner party in Paris, the conversation turned to the new field of vaccination. "What does Dr. Carrel think about this?" asked one of the guests. "They are too ignorant in France to plan for the future," Carrel replied to a table gone uncomfortably silent; shortly thereafter, he returned to the United States.

For a long time after it had been banned in the seventeenth century, blood transfusion had remained the object of a strictly theoretical curiosity. Charles Darwin's grandfather Erasmus Darwin suggested in 1794 that blood transfusions might alleviate fevers and malnutrition, although there is no evidence that he actually tried it. The first transfusion of human blood did not come until 1818—nearly 150 years after the *de facto* ban—and even then the patient died. James Blundell, a physician, physiologist, and obstetrician at Guy's and St. Thomas hospitals in London, was distressed at the high mortality rate from hemorrhages in birthing mothers, and began to think about replacing the lost blood. After systematically experimenting on animals, Blundell made two pivotal decisions: (1) only human blood should be employed, and (2) transfusions should not be used to cure madness or change character, but only to replace blood. Armed with these ideas, and with blood-injecting mechanisms of his own design, he prepared to attempt several human transfusions.

On December 22, 1818, Blundell injected twelve to fourteen ounces of blood he had taken from several assistants into a patient who was dying of internal bleeding. The man rallied for two and a half days before expiring. Next Blundell tried to revive a young woman who moments before had died of placental hemorrhage while giving birth. The fresh blood had no resuscitative effect. He gave two more transfusions without apparent benefit, making four failures in a row. Finally, Blundell and a colleague injected six ounces of blood into a woman who was bleeding to death after a uterine hemorrhage. "I feel strong as a bull," she told them, and recovered. Altogether, Blundell transfused ten patients over an eleven-year period, five of whom survived.

Blundell's results, although far from completely successful, exhibited a rare degree of scientific rigor, and were not too discouraging by the standards of the time. As such, they led to a revival of interest in transfusion, in which doctors throughout Europe felt free to experiment. Like the bloodletters before them, the modern transfusionists invented a variety of instruments to withdraw and inject blood, such as the Scannel Apparatus, the Rotanda, and two of Blundell's called the Gravitator and the Impellor. These crude, unsterilized injectors required the user to cut open the patient's wrist and insert the nozzles into a vein. Dr. J. H. Aveling, senior medical officer at the Sheffield Hospital for Women in England, invented an apparatus consisting of two silver tubes connected to a length of rubber tubing with a squeeze bulb in the middle "to act as an auxiliary heart." For eight years he carried the device in his pocket, waiting for the right occasion to employ it. Finally, he was called to save a woman suffering a postpartum hemorrhage. He sliced open her wrist, teased out a vein, and transfused her with some blood from her coachman. (The coachman felt well enough during the procedure to offer helpful suggestions all the way through.) After a few moments she gained just enough consciousness to tell the doctor she was dying. "The mental improvement was not as marked and rapid as I anticipated," wrote Aveling, "but this was perhaps due to the quantity of brandy she had taken." A few hours later, she gained consciousness again, "spoke, took nourishment and began her fresh lease of life."

Dr. Alfred Higginson, consulting surgeon to the Liverpool Dispensaries, transfused seven patients from 1847 to 1856. Five of them died, yet Higginson interpreted the results as showing that "transfusion may fairly be said to be of use." One survivor was a thirty-three-year-old woman who had exhausted herself by "oversuckling" her twins. Higginson injected her with twelve ounces of blood from a healthy female servant. Suddenly she was "seized with a rather severe rigor," during which, in a profound state of "reaction and excitement," the patient "sang a hymn in a loud voice." Later she recovered, completely unaware of her "vocal exertions."

By the second half of the nineteenth century, transfusion was becoming popular again, with hundreds reported throughout Europe. American doctors recorded two transfusions during the Civil War, both for leg amputations; one of the patients survived. During Canada's great cholera plague, some doctors gave milk transfusions in the belief that the "white corpuscles of milk were capable of being transformed into red blood corpuscles," according to their reports. Most doctors followed Blundell's advice and restricted themselves to using human blood, but their efforts generally failed. They knew nothing about sterile procedure, blood groups, or how to prevent blood from clotting in their needles and tubes. It was no wonder that, when the Polish doctor F. Gesellius compiled statistics about transfusions in 1873, he found that 56 percent had ended in death. Eminent physicians such as Friedrich Wilhelm Scanzoni of Würzburg and Theodor Billroth of Vienna denounced transfusion as a showpiece that brought attention to the clinic at the expense of the patient. By the end of the century, the brief renaissance in transfusion was causing so much human suffering and death that the procedure threatened to slide into another long eclipse.

In 1900, a young researcher at the Institute of Pathological Anatomy in Vienna was mixing some blood samples in a test tube when he noticed an unusual phenomenon: Under certain conditions the blood cells would clump. Others had seen this reaction before while mixing the blood of animals and people or of sick patients with healthy ones; they assumed it was triggered by fundamental differences in the bloods of various species or between healthy and sick blood. But the researcher, Karl Landsteiner, noticed that the agglutination occurred among blood samples from healthy individuals.

When Landsteiner decided to explore the phenomenon, he did so in a series of experiments that exemplified some of the classic work of science in their precision, simplicity, and clarity. The style came naturally to Landsteiner, who felt that research must be conducted in the most rigorous yet uncluttered way. With few outside interests, he performed his experiments during the day and wrote them up at night. Tall, slim, and handsome, with a sweeping dark mustache and sensitive brown eyes, Landsteiner was a painfully shy man who completely submerged his ego for science. He claimed to have inherited his austerity from his father, a prominent Viennese journalist known for his severe style. The father died when Karl was eight, leaving him to be raised by a kindly but reclusive mother. The two remained devoted for life. Originally Jewish, they converted to Catholicism, the state religion of Austria, when Karl was twenty-two. Landsteiner cared so single-mindedly for his mother that he delayed his engagement until after her death, and kept her death mask on the wall of his bedroom.

As a student, Landsteiner showed a talent for chemistry, which he specialized in while studying at the University of Vienna Medical School and later at universities in Munich and Zurich. He found work in Vienna at the Institute of Pathological Anatomy as a coroner—or "prosecutor," as they called it. It could have been a less-than-demanding job, but not to someone with Landsteiner's zeal. In ten years at the institute, he performed 3,639 autopsies, accounting for one-fifth of the institute's total workload, and wrote at least seventy-five scientific papers, all in his exacting prose. He also taught classes at the university and gave instruction in pathological anatomy to the foreign doctors who flocked to Vienna. This was an exciting period for medical researchers, during which new bacteriological causes for diseases seemed to reveal themselves every few weeks.

In 1900, Landsteiner published a scientific paper on certain chemical properties of blood, plasma, and lymph fluids in which he mentioned, as a footnote, the tendency of one man's blood to cause another's to agglutinate. He added that it remained to be seen whether the clumping was caused by bacterial contamination or individual differences in human blood. Then he proceeded in a typically straightforward and rigorous way. He withdrew blood samples from himself and his associates, let the blood settle until the red cells separated from the plasma, and, in a series of test tubes, mixed the plasma from each person with the red cells of every other. Sometimes the red cells would clump, or even burst; other times nothing unusual would happen. He recorded the results in simple tables, with the names of plasma donors listed vertically on the left, the names of red-cell donors written across the bottom. Results were marked with a "+" or "−" depending on whether agglutination had occurred.

Looking at the columns of pluses and minuses, Landsteiner saw "a peculiar regularity in the reaction of the 22 tested blood specimens," in which the blood seemed to fall into three distinct groups. The plasma from one group, which he called "A," caused the red cells of another group, "B," to agglutinate. Similarly, the plasma from the "B" group clumped the cells of group "A." Neither caused the red cells of a third group to agglutinate. He called that third group "C," although he later changed the designation to "O." Landsteiner, whose red cells reacted with no one else's plasma, belonged to the blood group that came to be known as O—the universal donor. A couple of years later, two of his colleagues, running a larger cross-match at Landsteiner's behest, discovered a fourth group, which reacted to both kinds of plasma; they labeled this "AB."

The results reminded him of a typical antibody-antigen reaction, which scientists had revealed just a few years before. In such reactions, protective substances in the body attack invading organisms. But here, instead of reacting against a bacteria or virus, the antibodies in plasma were attacking the foreign red cells, as though the corpuscles themselves had become a harmful antigen. Landsteiner deduced that normal human blood must exist in several varieties, each of which may be seen as "foreign" by the others. That would explain why many transfusions ended disastrously. With doctors knowing nothing about blood types or agglutination reactions, their patients' survival rested entirely on the odds of receiving compatible blood. Landsteiner concluded that his findings "may assist in the explanation of the various consequences of therapeutical blood transfusions." He also found that he could detect the grouping of a blood sample even after it had been left to dry on a cloth for two weeks, thus laying the groundwork for the use of blood types in forensic science.

Landsteiner would win the Nobel Prize for his experiments, but for years no one paid attention to his work. The world was a bigger place than today—news traveled slowly, and sometimes not at all—and a great gulf existed between laboratory experiments and their practical application. It may also be true that the idea of blood typing was so revolutionary that it had to "wait" for the rest of science to catch up, just as Gregor Mendel's laws of heredity remained virtually undiscovered for more than thirty years. At the same time, transfusionists faced other, more critical problems. They were still habituating to the practice of sterilizing their instruments, which probably saved more lives during transfusions than any other advances at the time. They also faced the most vexing technical problem of all—developing equipment that would enable them to inject the blood before it had time to clot in their syringes.

When Alexis Carrel operated on the baby Mary Lambert, he used a direct approach to avoid clotting: By suturing her father's artery to her vein, he created, in effect, a continuous vessel that prevented the blood from making contact with the air and never allowed the clotting process to begin. (He, like his colleagues, knew nothing about blood types; her survival meant that her father luckily had a compatible blood type.) Carrel won immediate acclaim for his technique and inspired widespread imitation, although few surgeons possessed his skill. Dr. George Washington Crile, a prominent Cleveland surgeon, made the procedure more accessible by designing a metal ring, or "cannula," through which the recipient's vein could be drawn and cuffed, making it easier to attach the donor's artery. He performed sixty-one transfusions by this method, which he called the Carrel technique. By the end of the century's first decade, surgeons were performing some twenty transfusions a year at Mount Sinai Hospital in New York alone—with Crile's and other, related techniques—and charging a handsome $500 fee.

For several years, "direct transfusion"—in which the surgeon would directly connect the circulatory systems of donor and recipient to avoid clotting—remained the procedure of choice. In these ordeals, the doctors would position the patient and donor side by side on adjacent tables so the radial artery of the donor would be as close as possible to the vein of the recipient. The doctors would make small talk with them both as they administered anesthetic and sliced into their arms to dissect out the blood vessels. (Crile wrote of the need to minimize the "psychic factor" during the operation, in which, in addition to administering morphine and cocaine anesthetics, "a nurse places over their eyes a wet towel with the diverting explanation that the eyes must be protected from the bright light to prevent headache.") Slipping the cannula over the recipient's vein, the doctors would cup back the vessel over the metal ring, providing a rigid surface over which they could ligate the donor's artery. Then they would release the clamps on the blood vessels and, watching as the blood flowed into the recipient, judge by dead reckoning when the patient had enough. Then they would stitch up donor and patient again, hoping that the donor would not lose the use of his arm. They conducted no preliminary tests, either for blood type or disease—their primary focus was to maintain the blood flow. "In looking backwards I marvel at our recklessness," wrote Dr. Bertram M. Bernheim, a distinguished surgeon at the Johns Hopkins Hospital in Baltimore, Maryland, and a leading transfusionist of his day. "All [we] were interested in was the actual running of blood from one person to another and keeping it running. How it was to act when it got there, what it was to do . . . whether there would be reactions . . . and possibly death . . . concerned [us] little. . . ."

Bernheim wrote an unusually frank memoir about the early years of transfusion, the trial-and-error period, in which doctors were first learning, often by unfortunate experience, how to care for their patients. He portrays the operation as a scene of great drama, fully justifying the word "theater" in reference to surgery—a balancing of a potential miracle against its often fatal results.

He recalled one case of a late-middle-aged woman who suddenly became ill and needed a transfusion. The doctors brought in another patient, who had volunteered to donate. They had no way to measure the blood they transfused, nor had they considered that an overdose might be fatal. They cut open the skin on the donor's left wrist, separated an artery, and connected it with a special tube of Bernheim's design to a vein in the woman's wrist.

> Slowly, deliberately, the clamp was removed from the sitting donor's full-beating artery. Instantly the blood leaped through the tube. It swelled the patient's veins, almost to bursting. It raced on its way to the heart, throbbing, jumping, full-fledged and high of pressure. And it literally knocked that woman's heart out the same as if it had been hit with a sledge hammer!
>
> Within one minute—hardly much more—from the time the blood started to flow the patient had begun to exhibit signs of distress in breathing and within another minute she had ceased to breathe. And all the time, quite happy, contented and much pleased with myself, I sat there watching the beautiful functioning of my little apparatus. Fool, fool that I was! The anaesthetist first gave the alarm. I still sat there. He said the patient was acting queerly and more queerly then sharply called out that she had stopped breathing. They started giving artificial respiration and then it dawned on me what had happened—and I shut off the blood flow.
>
> I had run fluid under unusually high positive pressure into a chamber (the heart) so depleted that there was little or no pressure left in it. And since the walls of this chamber were made of living, somewhat debilitated and rather thin heart muscle, they had stretched and stretched under the strain and had been unable to pump out the onrushing blood fast enough. Perfectly simple, easy to understand. There was no doubt about it because the surgeon hurriedly opened the woman's chest, found the huge dilated heart and massaged it to relieve the strain. But it was no good, it was just no good. The woman was dead. I had killed her.

Such incidents were not rare. Dr. Reuben Ottenberg of New York, an eminent transfusionist of the period, wrote, "One never knew how much blood one had transfused at any moment, or when to stop (unless the donor collapsed). I remember one such collapse in which the donor almost died—and the surgeon needed to be revived." Eventually they learned to mediate the transfusion by keeping a finger on the blood vessel, near the connection.

Not only the patient suffered during transfusions: The procedure caused so much trauma and pain that doctors found it difficult to locate donors. Doctors first turned to the immediate family, although not all would submit to the procedure. Bernheim recalled one case involving a "highly educated, cultured woman—a member of a prominent family" whose children had gathered at the hospital to give blood. While waiting for the doctor, they fell into a discussion about various ways she had wronged them in the past; by the time he arrived, they all had deserted her. The doctors had no choice but to buy blood from a professional donor. When the woman awoke, she was furious to learn that she had received blood from a boxer. "Just why she should have objected to a prizefighter's blood no one of us could understand," Bernheim remarked, "but possibly her children could." Even the best-intentioned relatives could not provide sufficient quantities, given the amounts required by the crude operations of the time. And so doctors increasingly purchased blood from strangers. They tried advertising in the newspaper, but the ads drew unruly crowds—not to offer blood, but to watch the spectacle. Bernheim preferred to solicit blood from two local establishments, called the Levering House and the Wayfarer's Rest. There, after generously tipping the desk clerks, he found a stable of down-and-outers who cheerfully gave blood for a $50 fee, which, more often then not, they squandered on liquor. "It wasn't the most admirable method, perhaps, but it was the only one available at the time."

Nobody liked transfusion as it existed—not the patient or the donor, not even the doctors, who spent more time performing the transfusion than the operation they were using it to support. In 1913, Dr. Edward Lindeman of Bellevue Hospital in New York eliminated the need to cut open the patient's arm with a "multiple-syringe" method of transfusion. He slipped a sharp, hollow needle into the arms of patient and donor, puncturing the skin and entering the veins. The needles remained in place while he shuttled back and forth from donor to recipient, withdrawing and reinjecting blood with a syringe. The doctor could measure the blood, because he withdrew it in graduated syringes. Disadvantages remained: The surgeon had to act quickly before clotting began (although immersing the syringes in ice water delayed the process), and the operation required a trained team of personnel and up to a dozen expensive glass syringes. Still, his method was so successful that Lindeman became the first full-time specialist in transfusion.

Soon afterward Dr. Lester J. Unger of Mount Sinai Hospital designed a stopcock that eliminated the need for multiple syringes. Like Lindeman, he inserted a needle in the arm of the patient and of the donor, but instead of shuttling back and forth with a syringe, he connected the needles with rubber tubing and a four-way valve. By changing the position of the stopcock, he could draw the blood into a graduated cylinder, measure it, and then direct it to the recipient. He delayed clotting by spraying the apparatus with ether, which kept it chilled.

A lot of money and prestige went along with transfusions, and Unger and Lindeman found themselves competing over whose method would become the procedure of choice. They described their respective devices at the 1916 conference of the American Medical Association in the seaside resort of Atlantic City. After their presentations, Dr. Edward Libman, an internist of international reputation, pronounced Unger's device superior. Lindeman was incensed. The meeting took place on a hot night in June, and a colleague helpfully suggested he cool off with a dip. Lindeman swam out into the ocean, suffered a cramp, and drowned.

The new technologies made transfusion easier and less traumatic, but they did nothing to address the problem of blood types, which, as the operation proliferated, came into clearer view. More and more doctors were describing a mysterious post-transfusion reaction of fevers, chills, vomiting, kidney pain, black or bloody urine, and, sometimes, death. Crile, who by now had done hundreds of transfusions, found that 35 percent of his patients were experiencing post-transfusion reactions. Not having read Landsteiner's studies, he and his peers glossed over the dangers of mixing incompatible blood types. Preparing for a transfusion, they would sometimes mix the bloods of donor and recipient and watch for hemolysis—the breaking of blood cells that occurs when one of the samples carries a disease. But they did not look for telltale agglutination, which could be equally harmful. Fortunately, at least one man, Dr. Reuben Ottenberg, had read Landsteiner's work. A dozen years after Landsteiner's study, Ottenberg became the first person to use cross-matching to determine the blood types of donors and recipients. Employing the technique in 125 cases at Mount Sinai Hospital, Ottenberg reduced the accident rate to zero, enabling him to conclude that accidents "can be absolutely excluded by careful preliminary blood-tests." Yet he could not seem to get the attention of surgeons who, jealous of their turf in the operating theater, "felt that . . . they ought to be free from influence by the laboratory men. I offered to do compatibility tests but many of the surgeons did not accept the offer." A gentle, scholarly man, Ottenberg traveled to Vienna to present his findings to the great Landsteiner himself. But Landsteiner was so immersed in his current research—the transmission of the polio virus—that the subject of transfusion never came up. It would be well into the 1920s, after years of "campaigning, experimenting and a few accidents," before blood typing would become a standard procedure.

Half a world away and in circumstances unimaginable to the Americans, a father watched his newborn child with a growing sense of fear and despair. His wife had given birth to a beautiful baby, a plump little boy with curly blond hair and blue eyes. But something was wrong. "A hemorrhage began this morning without the slightest cause from the navel of our small Alexis," the father wrote in his diary. "It lasted with but a few interruptions until evening. We had to call . . . the surgeon Federov who at seven o'clock applied a bandage. The child was remarkably quiet and even merry but it was a dreadful thing to live through with such anxiety." The next day the baby bled again, and for weeks afterward he developed hideous bruises whenever he bumped himself in his crib. Soon it became clear that the newborn was suffering from hemophilia, a dreaded condition in which the blood fails to clot.

It is a historical coincidence that, even as transfusionists struggled against the blood clot, which clogged their instruments and disrupted the procedure, the absence of that phenomenon condemned thousands, like Alexis, to short, pain-filled lives. (People with hemophilia can enjoy relative longevity today, but when Alexis was born the condition was untreatable.) Had this been any other infant, the condition would have been desperate for him, his parents, and everyone else who loved him; but the circumstances surrounding this child imbued his affliction with a historic significance. The baby was His Imperial Highness Alexis Nicolaievich, sovereign heir tsarevich, grand duke of Russia, son of Tsar Nicholas II. And through a strange and tragic sequence of events, this baby's hemophilia sped the downfall of an empire.

Known since ancient times, hemophilia is a gender-linked and hereditary condition, appearing in about one of every ten thousand males, and genetically transmitted from mother to son. People with hemophilia lack a protein critical to the clotting of blood. They suffer prolonged bleeding from small cuts, internal hemorrhaging from minor accidents, and episodic bleeding in the knee, elbow, and hip. That final affliction proves the most agonizing: The patient suffers crippling pain in the joints as enzymes in the hemorrhaging blood corrode cartilage, bone, and nerve. History's most famous hemophilia carrier was Victoria, queen of England and grandmother to most of the royalty in Europe. One of her four sons, Leopold, had hemophilia—he died at the age of thirty-one from internal bleeding after a fall—and two of her four daughters carried the gene for the disease, although they did not know it until they gave birth. A grandson named Frederick died at the age of two; another grandson died after surgery; and four great-grandsons, of the English, Spanish, German, and Russian royal families, died prematurely. "Our whole family seems persecuted by this awful disease, the worst I know," Victoria wrote.

One of Victoria's granddaughters, Alexandra (Alix of Germany), married the Russian Tsar Nicholas II and bore four healthy daughters and the afflicted son, Alexis. A highly emotional and religious woman, Alexandra tormented herself over her inability to relieve her son's suffering and pain. (Indeed, during one spell in which her son bled internally for eleven days, her blond hair was seen to begin turning gray.) In this desperate state she turned to Gregory Rasputin, a monk with a reputation as a mystic and healer. "God has seen your tears and heard your prayers," he cabled her during one of her son's episodes. "Do not grieve. The Little One will not die." Two days later, the hemorrhaging stopped.

From then on, Rasputin became a presence in the court. Gruff and grasping, filthy and unethical, offensive to most of the tsar's men, he nonetheless possessed a hypnotic quality that he exercised not only among the ladies of St. Petersburg, but to the apparent benefit of little Alexis. The monk, intense blue eyes set deep in his bearded face, would tell stories to the boy, hypnotically calming him and shortening his bleeding spells. As far as Alexandra was concerned, God had sent her family a savior.

Rasputin arrived at a precarious time, when the Russian people, long suffering from gross inequalities and a failure of their country to modernize, increasingly agitated for reform. Nicholas made some preliminary concessions but, goaded by his wife and Rasputin, increasingly stiffened. "Be more autocratic," the empress wrote to her husband during World War I, when he was away. "Be the master and lord, you are the autocrat." Rasputin cared nothing for politics; his sole motivation lay in satisfying his needs for wealth and power. Under his influence, Alexandra pressured Nicholas to make certain ministerial appointments, based solely on the applicants' feelings toward the monk. "He likes our Friend. . . . He venerates our Friend. . . . He calls our Friend Father Gregory. . . . Is he not our Friend's enemy?" Another time she cabled: "Gregory earnestly begs you to name Protopopov. . . . Please take Protopopov as Minister of the Interior." The tsar demurred— "Our Friend's opinions of people are sometimes very strange"—but gave in. The minister proved a disastrous appointment, and presided over the nation's economic collapse. Indeed, Rasputin exerted such a corrosive effect that people assumed that he was a paid German spy and that the empress was his lover, a German sympathizer, or both. Finally, in the waning days of 1916, seeing their government crumble around them, a group of royal conspirators assassinated Rasputin. Within three months, the tsar was forced to abdicate his throne. A year and three months after his resignation, the Bolsheviks murdered the tsar and his family.

It may seem improbable that the illness of a single child contributed so much to an empire's destruction. Surely one must not overlook "the backwardness and restlessness of Russian society, the clamor for reform, the strain and battering of a world war," as Robert Massie wrote in *Nicholas and Alexandra,* an intimate and careful study of that period. The tsar might have saved himself and his empire if he had continued his early, liberal path, allowing Russia to come into step with the modern world. But he abandoned that early instinct when confronted by circumstances that ultimately would doom his family and his empire: "Fate introduced hemophilia and Rasputin."

Mount Sinai Hospital sprawls over several square blocks adjacent to Central Park on the Upper East Side of Manhattan. Built in the Italian Renaissance style—five stories of brick rising over a fortresslike base of rusticated stone—it conveys the reassuring qualities of strength, grace, and permanence. Founded by German Jews in 1852, the hospital was planned in the European tradition, as a center for research as well as patient care, fitted with the most modern equipment of its day. A physician visiting from Italy in 1909 described it as a model "of luxury, cleanliness and order" and marveled over its operating rooms—"veritable amphitheaters of marble and crystal, stocked with every scientific necessity." Mount Sinai could never be a haven to kindly, old-fashioned physicians, personable men who reveled in their roles and could work in contentment. No—this was a hospital strictly for strivers. Populated by brilliant but often irascible people, it was a competitive locale where "there was only one rule—you'd better be right," in the words of a longtime staff member. It was not without reason that so many advances took place within its walls.

One of the typically assertive physicians at Mount Sinai early in the century was Dr. Richard Lewisohn. Born to a prominent and wealthy New York family and educated in Germany, Lewisohn retained a characteristic American impatience with the *status quo*. In this case, the *status* involved certain objectionable properties of blood. Researchers at the time had overcome two of the three major hurdles to transfusion. They had simplified the procedure, through the apparatus of people like Lindeman and Unger, making it less traumatic and more practical to the donors and recipients. Then they reduced the chance of dangerous reactions, by cross-matching blood types, following Landsteiner, Ottenberg, and others. But they still faced a third and formidable hurdle—the inevitable blood clot, which after three to five minutes halted the transfusion by blocking up their needles and tubes. They altered their procedures to make transfusions run faster, but the immutable clotting time prevented them from transferring enough blood to their patients. Their desperate attempts both irritated and inspired Lewisohn, who decided that the time had come to take a different approach. He said so in a series of challenging questions: "Must we accept this coagulation time as an unchangeable law? Any transfusion in which the normal coagulation time of the blood is considered an unalterable factor, is apt to be difficult and apt to require a good deal of personal experience and skill. Might it not be possible to inhibit the danger of the clotting of blood during its transfer without diminishing the clinical value of blood to the recipient?" In short, rather than streamline transfusion to conform to the clotting time, Lewisohn would attack the coagulation itself.

He had not been the first to try. In the 1860s, the English obstetrician John Braxton Hicks had experimented with an anticoagulant, and lost three patients as a result. Other doctors tried hirudin, a chemical found in leeches, that also in quantity proved toxic to humans. Others used oxalates, bicarbonates, and phosphates, all of which delayed the blood clot but were poisonous.

Lewisohn began working with sodium citrate, a commercially available anticoagulant for laboratory use. (The chemical delays clotting by absorbing calcium, a free-floating element in the blood whose presence is critical to coagulation.) The standard laboratory concentration of sodium citrate was a 1 percent solution—one part sodium citrate to ninety-nine parts donor blood—a concentration known to be dangerous. But Lewisohn wondered if citrate would be toxic in all concentrations. "Nobody had ever followed the simple thought of carrying out experiments to ascertain whether a much smaller dose might not be sufficient. . . ."

For four years, Lewisohn ran experiments, first in dogs, later in humans, trying to find a concentration high enough to prevent clotting yet low enough to remain nontoxic. Varying the dose by fractions of a percent, he found that one-fifth the dose formerly thought effective—a .2 percent concentration—would keep the donor's blood liquid while posing no danger to the recipient. Doctors could run the donor's blood into a glass jar, stir in some sodium-citrate solution, and then reinject it into the recipient—all at the careful pace transfusion deserved. "Technic very simple," he wrote in his notes. "Requires no special apparatus. . . . Any country physician could use it." He must have felt as if he had found the Holy Grail:

> How beautifully this drug responds to human needs!
>
> I remember very well the surprise of my colleagues when this method was published in January 1915. They could not believe that the technique of blood transfusion which had been so very complicated up to then, was suddenly made as simple as an ordinary saline infusion. . . . Naturally I thought that the safe and simple sodium citrate method would be adopted universally. . . .

Doctors waited years to accept his solution. Transfusion had developed as a complicated operation, requiring surgeons of the highest price and skill; not gladly would they surrender the practice to every "country physician," as Lewisohn quaintly put it. Furthermore, as the number of citrate transfusions increased, disturbing side effects began to appear, including fever and chills. Jealous colleagues attributed the symptoms to "citrate toxicity"; some undertook a whispering campaign to impugn Lewisohn's results and discourage the chemical's use. Lewisohn knew the citrate did not remain in the body long enough to trigger toxic effects. Within a few minutes of its entry with transfused blood, the liver metabolized the chemical into harmless by-products. Over the course of the next eighteen years, he proved that the side effects resulted not from the citrate, but from trace contamination in the water and glassware, which hospitals could eliminate by careful sterilization.

In time the advantages of citrate had become so clear that its use became standard procedure, as it has remained to this day. With Lewisohn's discovery, doctors had overcome the three major hurdles impeding successful transfusion. They had the equipment to make transfusion bearable for the patient, the blood-typing tests to make the procedure safe, and the anticoagulant to make it all practical. "A small amount of sodium citrate in solution . . . and presto! the blood remained in a fluid state," wrote Dr. Bertram M. Bernheim of the Johns Hopkins Hospital. "It was almost as if the sun had been made to stand still." The science had come a long way since those first grim and bitter transfusions—indeed, it was a science now, no longer a craft— and Bernheim exulted over the changes he saw in the first couple of decades of the twentieth century.

It took me seven hours to do my first transfusion; I wore out utterly one donor and all but killed the second. . . . Yet here comes a man who makes it all look silly with his few drops of clear, watery looking solution that takes all the fight out of blood. . . .

And how does he finally use it? Simply by sticking a thin little hollow needle through the skin into the patient's vein—arm, leg, neck, head—attaching it by rubber tubing to a bottle and letting the fluid blood run in. Any amount, little or big, as much as needed, slow or fast, at ordinary temperature—not even warmed. Most amazing.

And what did he do? Nothing but make transfusion available to every sick person in the world, high and low, rich or poor, black or white, man, woman or child. What else did it do? Nothing but knock into a cocked hat great transfusion specialists like me and others who could be mentioned. . . . Young doctors, old doctors, interns, nurses, technicians, all do . . . transfusions without the slightest difficulty.

PART TWO

# BLOOD WARS

# CHAPTER 4

# BLOOD ON THE HOOF

Sodium citrate gave surgeons the time they needed to bleed a donor into a flask and slowly infuse blood into the patient. But transfusion remained a laborious affair. No one was storing blood in a blood bank, with bottles available for immediate use; rather, doctors viewed citrate as a short-term anticoagulant to keep blood liquid long enough to complete the transfusion. When a patient needed blood, the doctor still had to recruit a compatible donor, bring him to the hospital, and use the blood right away. True, transfusion had become an "indirect" procedure—blood no longer had to flow directly from donor to recipient; it could be collected in a flask—yet it was still collected from donors "on the hoof."

That was the situation that Percy Lane Oliver confronted in London in 1921. Oliver, a balding, bespectacled, bookish-looking man, was secretary of the Camberwell Division of the British Red Cross in Southeast London when a call came in urgently requesting blood. With no other resources from which to draw, Oliver and three co-workers rushed to the hospital; one had a compatible blood type, and the patient survived. That dramatic rescue gave Oliver an idea. Doctors were depending on an unreliable donor supply, recruiting from the patient's family or their own personal network, including indigents. Often they could find no compatible donors, or did so only after dangerous delays. But what if a citywide bureau existed to provide prescreened and pretested volunteers? Such a service could save hundreds of lives by providing donors of any blood type at a moment's notice.

Oliver took his first tentative steps by recruiting twenty volunteers among professional acquaintances. Doctors called him thirteen times during his first year of operation, 1922. Word quickly spread that Oliver had gathered a safe and reliable donor pool. Hospitals called him 428 times in 1925, and nearly double that number of times the following year. Clearly Oliver had tapped into a need. To widen his donor base, he called the YMCA, the Rover Scouts, and other service groups, and received such an overwhelming response that he felt confident in establishing a new organization: the Greater London Red Cross Blood Transfusion Service, the world's first municipal donor panel.

Each volunteer would receive a physical exam, undergo tests for blood type and syphilis, and be entered into a phone log, ready to rush to a hospital at any hour if a call for his blood type came in. Oliver ran the organization from his home, where, aided only by his wife and secretary, he worked seven days a week recruiting, keeping paperwork, and phoning volunteers. "[The] work is constant, and . . . it is never possible to leave the office for a minute," Oliver wrote in 1926. "Considerable use is made of the telephone, [with] over 2,000 outgoing and 3,500 incoming calls being made during the year. . . . Apart from the actual supplying of the donor, which may entail as many as eight calls, expenses of donors have to be paid, the receipts booked and filed, medical reports written . . . copied and sent to the doctors. . . . [His daily obligations included] donations sent, visitors interviewed, volunteers canvassed, grouped and enrolled, lectures prepared and given, and scores of letters explaining the Service, or smoothing over little frictions. . . ."

Oliver's accomplishments grow more impressive when one considers the spotty communications of the day. Not everyone in London owned a telephone, or could be counted on to be near a phone at all times. Sometimes Oliver would call a neighbor to track down the donor—"Clergymen are very courteous and obliging in this way"—or, as a last resort, the local police, although "most people have a natural objection to a police constable hammering at their door in the dead of night." The Service was free—no charge was made to the patient or the hospital. To finance the operation, Oliver raised money from charitable donations and the collection of waste tinfoil, which was redeemed for cash.

The growing demand for blood in London in the 1920s could be laid at the feet of Dr. Geoffrey Keynes, brother of the economist John Maynard Keynes and one of the most prominent British surgeons of his day. Dr. Keynes became a proponent of citrate transfusions during World War I, having learned the techniques from American physicians stationed in Europe. Their work was extremely limited—of the millions of casualties in the war, doctors transfused perhaps a few hundred—yet the results amazed Keynes and the other British surgeons.

> Transfusion naturally provided an incomparable extension of the possibilities of life-saving surgery. . . . A preliminary transfusion followed by a spinal analgesic enabled me to do a major operation single-handed. A second transfusion then established the patient so firmly on the road to recovery that he could be dismissed from the ward without further anxiety. The possibility of blood transfusion now raised hopes where formerly there had not been any, and I made it my business during any lull in the work to steal into the moribund ward [where patients considered hopeless were kept], choose a patient who was still breathing and had a perceptible pulse, transfuse him, and carry out the necessary operation. Most of them were suffering primarily from shock and blood loss, and in this way I had the satisfaction of pulling many men back from the jaws of death.

On returning to civilian life, Keynes was astonished to find doctors in London ignoring transfusion. The surgeons he worked with considered it an encumbrance, even during operations involving great blood loss. "There were plain reasons for believing that patients would be much helped by having the blood replaced as soon as possible after it had been lost, but my superiors were afraid my activities would 'get in the way' of the operators." Struck by the need to educate doctors as well as the public, Keynes became an advocate for transfusion, promoting it in journal articles, radio addresses, and his 1922 book, *Blood Transfusion,* the first modern British text on the subject. As a surgeon at St. Bartholomew's Hospital, he organized a small panel of medical students as donors. Later, when Percy Oliver asked him to serve as one of the Service's medical advisers, he happily agreed.

Now that Oliver had formed an organization, he wanted to do more than merely amass donors; he desired to create a sense of community, of shared humanity in the new enterprise. Writing in his quarterly newsletter, he emphasized the personal nature of the donation, stressing the link between donor and recipient by reporting the results. Sometimes he would print letters from patients to thank and inspire his volunteers. "It was supremely noble of that gentleman to give his blood to save a fellow-creature and an utter stranger," read one letter, from the brother of a recipient. ". . . He will have the satisfaction of knowing that the sacrifice he made was not in vain." Seeking to enlighten and entertain his readers, Oliver supplemented his reports with new information about the emerging practice:

Q: How much blood is there in the human body?
A: . . . approximately 8 to 10 pints.
Q: What is the record number of blood transfusions given by a human donor?
A: The world's record is believed to be held by a Frenchman, Raymond Briez, who . . . has given blood 459 times . . . and a total of 125 litres (109 pints) in all.

He established himself as an advocate for donors, who rarely received the credit they deserved. Many surgeons assumed that the donors had been paid for their services (as many were, elsewhere in England) and treated them with a measure of contempt, keeping them waiting for hours, or sending them home without clear explanations. Some employed the technique of "cutting down"—surgically opening a vein to insert the needle—or baring the vessel and "levering it up." Oliver excoriated the doctors for their carelessness, and forbade any extraction method other than a needle. He chided the press for their overly heroic depiction of transfusion, which, although well intentioned, created the impression that donation was best reserved for the brave. Sometimes, after a particularly florid press account, Oliver would be inundated with telegrams from members, their relatives, or even their doctors, imploring him to remove their names from the list. "An occasional obstacle is the old family physician," wrote Oliver. "His style of argument is: 'I have no objection to blood transfusion, but you, my dear Madame, with your peculiar constitution, are the Last Person in the World to undergo it. . . .' " One volunteer wrote: "I mentioned to my wife just after dinner on Sunday that I had joined. She promptly fainted. When she came to, she telephoned for our Doctor, and the two of them argued with me for the rest of the evening. When they finished I had come to the conclusion that . . . I shall have to withdraw my enrollment. . . ."

Despite the lingering fear and superstition, the size of the donor panel continued to grow, serving 160 hospitals in Greater London by the 1930s and answering more than three thousand calls per year. Oliver was invited to Buckingham Palace, and received a letter from the duke of York recognizing his organization as an "inspiring service to humanity." He pointed with pride to the fact that every one of his nearly twenty-five hundred donors acted strictly as a nonremunerated volunteer.

Inspired by Oliver's example and compelled by pressing need, donor panels were established throughout the world during the 1920s and '30s. Doctors organized them in Germany, Austria, Belgium, Australia, Russia, even Siam. In Japan, Dr. H. Ijima established the Nippon Blood Educational Society, with about two hundred donors "ready to be sent to any hospital in Tokyo or vicinity of Tokyo within a short notice."

In France, Dr. Arnault Tzanck established the Emergency Blood Transfusion Society (L'Oeuvre Transfusion Sanguine d'Urgence), with its own examining rooms and laboratory at the Hôpital St.-Antoine in Paris. He and his staff kept card files of panel members who were triple-tested for blood type and examined four times a year to make sure they were in good health. Donors received small allowances to compensate for the inconvenience, most of which was paid by public assistance. Tzanck, a small round man with glasses and a pipe, became admired in his country for his commitment and energy, as he traveled the provinces helping local doctors establish transfusion societies—bringing his own equipment, offering his own blood, and enthusiastically recruiting hospital staff, medical students, and police. (His fame spread internationally as he helped found the International Society of Blood Transfusion.) Tzanck seemed indefatigable: A prototypical humanist-physician, he trained a generation of admiring transfusionists, performed important hematological research, designed apparatus, and even wrote philosophical treatises exploring the moral basis of his work. "Tzanck was an incomparable motivator," wrote one of his disciples, Dr. Jean Pierre Soulier, who later headed the French transfusion service for thirty years. Tzanck touched on the wellsprings of his energy in his book *The Creative Conscience* (*La Conscience créatrice*), a series of reflections on what he had learned from his career. As a fundamental principle, he stated, "The man is truly poor who does not know how to give."

If Tzanck represented the height of humanity and Oliver the epitome of unselfishness, then the Americans, with their practical and market-driven methods, embodied cool professionalism in their mobilization of donors. They saw nothing wrong with trading blood for money, as long as they took adequate precautions to safeguard the donors' and recipients' health—although, like many enterprises in the *laissez-faire* world of American capitalism, the blood-selling business got quickly out of hand. "Our 1,001st Way to Make a Living" is how

*The New York Times* described the practice in 1923, and told of a $35–50 standard fee. "The professional donors follow other lines of permanent employment and answer calls in order to increase their income," the paper explained. "The present list of available donors includes men of many occupations, usually those who do manual labor, for the donor must keep his blood in the pink of condition." The donors were more varied than the newspaper imagined, from college students (in 1925, more than 150 University of Michigan undergraduates sold their blood to pay tuition) to down-and-outers selling blood to buy liquor. The practice became more common during the Depression, when jobs became scarce; by the end of the decade, the New York City Health Department reported that professional donor bureaus were fanning out across the city's parks, recruiting indigents who likely carried syphilis and other diseases. Some men dangerously depleted themselves, one selling fifty-six pints in a single year.

Horrified at the situation, a group of doctors in New York established an organization to raise the standards of professional donation. The Blood Transfusion Betterment Association, as they called it, was managed by the most eminent doctors in the field, including Reuben Ottenberg, Lester J. Unger, the famed immunologist Arthur F. Coca, and even Karl Landsteiner, who had emigrated to New York a few years before.

Simon Flexner had achieved quite a coup in luring Landsteiner to the Rockefeller Institute, although the offer was more akin to a rescue. In the decades since his discovery of blood types, Landsteiner had become world-renowned, not only for his blood work, but for his research in cancer and polio. Yet his life in Europe had become difficult. Vienna, once a world center for research, had fallen into economic chaos and moral despair, along with the rest of Austria, in the wake of its catastrophic defeat in World War I. This once bustling and glamorous city had become gray, ramshackle, dispirited, and destitute. Industry shut down as the nation ran out of coal, streetlights blinked sporadically, and trolleys sat idle for days at a time. Research became impossible, as refrigerators and equipment failed. Everyone suffered from hunger and the cold; so little canned milk was available that Landsteiner bought a goat to nourish his infant son. Then, one day, when he saw his neighbors dismantling his front fence for firewood, he knew it was time to leave.

In 1919, at the age of fifty-one, Landsteiner moved his family to Holland, where he took a job at a small Catholic hospital in The Hague. The position was menial—performing routine autopsies and laboratory tests in a small room assisted by a manservant and a nun who also had the job of making coffee for the staff—but at least it established him outside of Austria. Yet, as grateful as he was, he could not conceal his impatience. "His position would be a fine one for a young man without any ambition . . . but for a man like Landsteiner it is perfectly horrible," wrote one of his associates in a letter to Flexner. "He has to do all the routine work, postmortems, examinations of tissue, bloode [*sic*], urines, Wasserman tests and so forth, and he has to do that with the assistance only of one roman catholic nurse who is at the same time the organist of the hospital church. . . . There is only one room for Landsteiner, where he works with that nurse and one man-servant; that room is used for several other purposes, every doctor who wants to examine a urine or who wants to have a cup of coffee or who wants to talk to Landsteiner comes into that room. The trouble is that all these people are good and decent people; they all like Landsteiner very much and drive him to despair. . . . That Landsteiner can under such circumstances do any scientific work at all is simply admirable."

Flexner knew this was his chance to save a world-class scientific mind. He met with his board of directors and in 1922 was authorized to offer Landsteiner a position. This would be a dramatic change for the shy scientist—new colleagues, a new language, a new life. And so, like many immigrants before them, the Landsteiner family packed a few cherished possessions in a trunk, along with a barrel of coal tar that Landsteiner used to induce tumors in the skin of laboratory animals, and steamed across the Atlantic. Arriving in Manhattan, they were greeted by Peyton Rous, another renowned Rockefeller researcher, who asked what kind of quarters the family preferred. Landsteiner replied wistfully, "I would like a little cottage by the sea with a rose garden, such as I had in Scheveningen [Holland]." The best they could offer was an apartment above a butcher shop on Madison Avenue—the first of many shocks Landsteiner would receive in his new country.

Nevertheless, he quickly settled into life at the institute, where he pursued his research in immunology and became a beloved if somewhat rigid and aloof figure. Known for his unwavering seriousness, he abhorred all forms of attention or recognition. In 1930, when he received the Nobel Prize (one year after becoming an American citizen), he fended off reporters, blurting, "Please, I have not done anything. I am just working, and that is all." Indeed, he felt so timid about public speaking that, when his turn came to address a celebratory dinner in Stockholm, he asked Sinclair Lewis, the Nobel laureate for literature, to speak on his behalf. Lewis was eloquent: "You may call me the master of words, but what is he?" he said, gesturing to Landsteiner. "He has been in a thousand cases the master of death." But Landsteiner truly cared nothing for recognition—only for stability, regularity, and an environment in which he could work unencumbered. In return, he gave himself wholly to his research, serving as a mentor to younger scientists and generously donating his expertise. When his colleagues organized the Blood Transfusion Betterment Association, he agreed to serve as a special medical consultant.

Under Landsteiner, Ottenberg, Coca, and the others, the association became a professional donors' panel in the most positive sense, with rigorous standards of discipline and hygiene. All donors were required to register at the city health department with proof of a recent physical examination and syphilis test, which had to be repeated at least four times a year. They excluded any donor with a history of half a dozen communicable diseases, or with alcohol or drug problems. Donors without telephones could not participate—no chasing down the donor with clergymen for this group. Each donor received a green registration book in which he kept an up-to-date record of all donations, payments, tests, and examinations. A donor could not sell his blood without passbook in hand.

Becoming a professional donor in the city required a continuing commitment to good health and discipline. The association guidelines specified that every donor "keep himself in good physical condition. [He] should be particularly careful about the cleanliness of his body . . . have plenty of sleep in a well ventilated room [and] daily exercise. . . ." No donor could sell his blood more than once in five weeks. Any sign of weight loss or anemia put a halt to the transaction. Donors were expected to be prompt and reliable, and, lest anyone doubt the association's resolve, they could look at the case of donor William Davidson. Davidson appeared at the hospital one day and, after one needle puncture failed to produce blood, declined to submit to a second, even though he was offered an anesthetic. Working through the authority of the health department, the association dispatched a policeman to his home to confiscate his passbook.

It may seem that, when compared to the London Service, the Blood Transfusion Betterment Association operated too commercially—certainly the British thought so and made that quite clear—but the enterprise produced commendable results. Under this commercial system, transfusion, once dangerous and exploitive, became safe and professional, leading hospitals to abandon the fly-by-night donor bureaus. Transfusions soared—by 1937, the association was answering more than nine thousand calls a year, well above the average in London. Furthermore, by charging money for blood (donors received $35, and the bureau got a $6 commission), the association could fund a more professional approach, hiring a medical director, developing its own high-quality blood-typing sera, and distributing grants for serological research. One might make pronouncements about the moral dimensions of trading blood for money, but in terms of providing what society needed—clean blood from healthy individuals who were not being harmed or exploited in any way—the New York professionals and the London volunteers were virtually the same.

By the late 1920s, blood was poised to enter a new era, not only in medicine but in the legal world as well. Landsteiner had predicted that serology would become useful in forensics and paternity cases, and his successors ensured that his vision came true. In Italy, Dr. Leone Lattes developed reagents to determine a grouping months after a bloodstain had dried, and began to use them to dramatic effect. In one famous case, he secured the release of a man who the police believed had murdered his wife, pointing to a suspicious bloodstain on his coat. Lattes showed that the blood type did not match the man's wife's but was his own, confirming the man's story that it had come from a nosebleed. In Russia, two men were acquitted of murder after it was found that blood on the dagger belonging to one of them did not match that of the victim.

The heritability of blood types likewise gained legal standing. Landsteiner and his colleagues found that blood types moved from one generation to the next by simple Mendelian inheritance, in which some genes dominate over others. In blood types the genes for groups A and B dominate over O, producing certain predictable patterns. Thus, if a mother charged a man with fathering her child, and the blood types of the mother and child were known, one could construct a family tree to determine to which blood groups the father might belong.

The ability to rule out certain combinations became important in paternity cases, first in Europe and then in the United States. Germany accepted serological tests for paternity exclusion as early as 1924; within five years, German courts had accepted blood types as evidence in more than five thousand cases of disputed paternity. In one case, a court sentenced a woman to six months in jail for perjury after blood testing proved that she had falsely accused a man of being the father of her child. Thousands of trials took place in Denmark and Austria in which blood types were accepted as the overriding proof.

The procedure rose to prominence in America during a notorious baby-switching incident in the summer of 1930. On June 30, Mrs.

William Watkins and Mrs. Charles Bamberger, who shared a room in Englewood Hospital in Chicago, both gave birth to boys. The women joyfully coddled their babies, nursed them, bathed them, and became friendly over the course of their own convalescence. Everything was fine until a couple of weeks later, when Mr. Watkins, watching his wife give the baby a bath, noticed a piece of adhesive tape stuck to the child's back. Peering closer, he saw the name "Bamberger" printed in red. He telephoned the Bambergers, who, retrieving the label they had removed from their baby's back, found the name "Watkins." Somehow the babies must have been switched—or had they? Had the mothers gone home with the wrong babies, or had they taken the right babies but with the wrong labels on their backs? The parents called the hospital, whose staff members denied any wrongdoing or responsibility, only increasing the parents' anxiety. The dispute was threatening to flare into a scandal when the health commissioner of Chicago, Dr. Arnold H. Kegel, decided to intervene.

Kegel promised to use the most modern techniques of empirical investigation, such as fingerprinting, reflexology, and phrenology (the study of head shapes). The case had no need for lawyers or judges—he would let *scientists* decide. And so he assembled a body of thirteen sages from various disciplines to determine which baby rightfully belonged to which parents.

From the University of Illinois came Dr. Gerhardt von Bonin, a medical anthropologist, who compared the babies' head shapes to their parents'. Exactly what he hoped to learn from this perplexed some observers, who noted that the babies presented identical profiles, with rounded foreheads, button noses, and receding chins—rather like all babies, come to think of it. Nonetheless, von Bonin decided, based on his careful study of their profiles, that the babies must have been switched. He hastened to add that the babies might have inherited their grandparents' profiles, in which case he would have to reconsider. After all, "the laws of heredity are very imperfectly known in the case of man."

Next came Dr. Cleveland White, a dermatologist at the Northwestern University Medical School. He had intended to use skin tone to differentiate the babies, but found them too similar, since both were a typically Caucasian baby-pink. He attempted to study their "racial" characteristics but found them too similar as well, since the parents' backgrounds were virtually identical: The Bambergers were Austrian and Irish-English, the Watkinses German and Welsh. He did find a birthmark on the Bamberger baby's left heel, just as on the left heel of his supposed father. Unfortunately, not even that could do him much good. "Besides," he added, the father's birthmark is "round and hard. It might even be a callus. . . . The baby's birthmark is one of blood vessels and is oblong, rather the shape of a bean."

The babies were examined by a reflexologist, Dr. Harold S. Hulbert, who with a rubber hammer gently tapped their knees. He discovered an interesting similarity: Mrs. Bamberger and her baby displayed "lively" reflexes, whereas the Watkins mother and child showed a more "normal" response. The mothers and babies must have been properly matched.

Then came the fingerprint expert, Ferdinand Watzek, late of Vienna and director of the scientific crime laboratory at Northwestern University. He had a bit of trouble at first, these being the youngest people whose prints he had ever had to take. He dusted the babies' fingertips with charcoal, but could not get the infants to open their hands. "Ooch, such work this is!" he said as he struggled with the tiny fists. When he and his assistants finally obtained the twenty baby fingerprints and forty adult ones, they removed themselves to their lab to classify them according to twenty-four hundred possible convolutions. After two days, they emerged with their findings: The babies had been placed with the right families after all. A skeptical Chicago police captain watching the procedure said, "Why, I wouldn't call these fingerprints at all. They're more like a collection of smears."

Finally came the blood expert, Dr. Hamilton Fishback of Northwestern University, whose deductions, as it turned out, represented the only clear and irrefutable proof. Fishback took blood samples from the four parents and both babies, determined their blood types, and presented his findings. Mr. and Mrs. Watkins both had group O blood, which, under the laws of genetics, meant that they could only have a baby with the same blood type. Yet the baby in their custody was type A. The Bambergers' blood types were AB and O, which meant that they could give birth to a child with a blood type of A or B. Yet the baby in their custody was Type O. There was no denying or getting around it: The infants had been switched. Persuaded by the evidence that Fishback presented, Commissioner Kegel concluded that the Watkinses and Bambergers should swap babies, and said so in a meeting at City Hall.

It was one thing for the commissioner to decide, quite another for the traumatized parents. Mr. Bamberger stormed out of the meeting shouting, "I'm sick of this science business!" Later, from his doorstep, his wife inside in a state of collapse, he declared, "I know my own baby and nobody is going to take him away from me. My wife is convinced the baby is ours. I guess a mother's instinct is as good as the opinions of a lot of medical experts." Then he shouted "Humbug!" and slammed the door.

Across town, Mr. Watkins told reporters that he would go and take his baby by force. "I'll sue Bamberger for a writ of habeas corpus. And I'll sue the hospital. . . . Talk about Bamberger's wife being hysterical. Here's Bamberger chasing around in the hot sun with *my* baby, and my wife sitting at home nursing *his* baby."

The Bambergers had retreated to a relative's house in the country. Later they endeavored to have the Almighty ordain what all the city's scientists could not and, taking the baby to Our Lady of Solace Church in Chicago, had him baptized with the name of George Edward Bamberger. "I don't know nothing about those blood tests," remarked Mr. Bamberger. Meanwhile, reporters at the Watkins home found Mrs. Watkins in a "hysterical condition."

The drama unfolded in newspapers across the nation, and across the pages of Chicago's broadsheets and tabloids. The news aroused the city's citizens, who felt this case involved more than just two star-crossed families, but threatened the foundations of motherhood itself. In Chicago, several hundred mothers formed a committee to intervene, which they called the Associated Mothers of Illinois. "This can be solved by mother's instinct where science has failed," the chairwoman, Mrs. Grace E. Dibrell, said at the group's organizational meeting. "And we not only want to settle the question for these poor parents, but we want to make it safe for our future children, the future children of Chicago. We want to pass a state law protecting mothers against this kind of thing."

Weeks passed. The Watkinses vacillated, unable to decide which baby was theirs, and the Bambergers refused to entertain any doubts. The Watkinses slapped the hospital with a $100,000 lawsuit for their son's "lost identity." Then, about six weeks after the babies were born, the Bambergers changed their minds. Their lawyer called Commissioner Kegel to say that his clients had reconsidered the evidence and that the blood tests convinced them the commissioner was right. The dispute resolved itself so quickly and peacefully that the end came as something of a quiet shock. On August 19, the Bambergers drove to the Watkinses' home. There, with Commissioner Kegel smiling in the background, the two mothers undressed their babies, exchanged them, and tearfully embraced.

The Watkins-Bamberger baby exchange represented a landmark in medicolegal practice and was reviewed in medical journals for years.

Yet the blood-typing precedent did not rapidly spread throughout the nation, especially in cases where juries were involved. Although quick to adopt new gadgets and technology, people were slow to accept new *theories* of science, especially those that involved the mysterious subjects of biology, heredity, and evolution. Only a few years earlier, for example, a jury trial had found a schoolteacher guilty of teaching evolution in the famous Scopes trial in Tennessee. (The conviction was later reversed on a technicality, but the law remained in place until the mid-1960s.) And so it took years for blood types to be accepted as proof of nonpaternity, much to the despair of men who may have been falsely accused.

The most famous such case involved Charlie Chaplin, who was sued by a tormented young woman named Joan Barry, with whom he had had a long love affair. Long after the affair had ended, she gave birth to a baby girl and sued Chaplin for child support. The judge convened a triumvirate of doctors who concluded that, given the blood types of the parties in question, Chaplin could not possibly have sired the child. Faced with irrefutable scientific evidence, Barry's attorney, Joseph Scott, took an emotional approach, drawing out the story of how Chaplin had seduced the innocent young lady, almost like a villain of one of his own films. He called Chaplin a "reptile," a "cad . . . a little runt of a Svengali," and a "lecherous hound." (At one point Chaplin, stung by the abuse, cried out, "Your Honor, I've committed no crime. I'm only human. But this man is trying to make a monster out of me.") As a final tactic, Scott had the jury stare at Chaplin and the baby for a full forty-five seconds in order to recognize a family resemblance. "Wives and mothers all over the country are watching to see you stop him dead in his tracks," Scott intoned. The lawyer's histrionics deadlocked the jury. The jury at the retrial ignored the medical evidence and declared Chaplin the father.

Fortunately, such outcomes were highly unusual. In fact, countries throughout the world legislated blood groups as the determinant factor in paternity cases, as did many states in the union. California did not do so until 1953.

In the historic heart of Moscow, near Kolkhoz Square, stands a long, massive structure known as the Sklifosovsky Institute. Built in the neoclassical style, with a majestic central cupola framed by sweeping colonnades (a bulky modern edifice has since been added), it was named for Nikolai Vasilyevech Sklifosovsky, the Russian pioneer of emergency and battlefield surgery. Muscovites know the place as the

"Sklif." It serves as a central trauma ward for the city, with thousands of beds and dozens of operating theaters specializing in emergency care. Day and night, ambulances roar through the city to the Sklif, delivering their burdens of the gravely ill and wounded.

It was here that Serge Yudin conducted one of the strangest and most far-reaching experiments that the field of transfusion had ever seen. On a night in March 1930, an ambulance crew brought in a young man who had attempted to commit suicide by slitting his wrists. The patient had slashed deeply, losing so much blood that he barely had a pulse. To survive, he would need a massive transfusion, yet there was no time to call for a donor-on-the-hoof; no time to do anything, really, other than exercise the option that Yudin had been anticipating and yet dreading for more than a year. Yudin did his best to stabilize the patient, then rushed to a nearby room in the hospital, where a freshly dead cadaver awaited him.

Yudin belonged to a long and particularly Russian tradition of boldness, innovation, and sacrifice in medicine. Russian medicine is rich with a folklore of doctors who suffered along with their patients in the frontlines of epidemics or war. "Physicians forward" is how they described it—*vrachebnaya etika,* or the physician's ethic—and Russians still use the phrase to represent the self-sacrificing tradition in their medicine. They use the phrase in a conceptual sense as well, meaning that they should not be too timid to experiment on themselves or even their patients. It is all part of their service to the state, of their commitment to the greater good.

Yudin could see such commitment in his predecessor, Dr. Alexander Bogdanov, who died for the advancement of Soviet transfusion. Bogdanov was a Bolshevik Renaissance man. Born humbly to provincial schoolteachers, he became a doctor, revolutionary, Marxist theorist, and writer, and for a time one of Lenin's closest comrades. In 1908, he wrote a prescient science-fiction novel about a utopian society on Mars where, in addition to providing free education, health care, and meaningful work, the government extends the lives of its citizens through the mutual exchange of rejuvenating blood. "How is it that our medicine on Earth does not employ this method?" says the incredulous visitor from Earth. "I don't know," the Martian replies. "Perhaps . . . it is merely due to your predominantly individualistic psychology, which isolates people from each other so completely that the thought of fusing them is almost inconceivable."

Bogdanov became fascinated with the idea of sharing blood to facilitate universal good health. Having visited London and seen Oliver's organization, he determined that the Marxist state could do the same on a larger scale. He initiated a program of transfusion centers throughout the republics—where workers were paid handsomely for their blood—linked to scores of state-supported institutes where scientists performed research and training. In 1926, he established the Central Institute of Hematology in Moscow, the world's first center for transfusion research.

Even as Bogdanov moved forward in organizing a national transfusion infrastructure, he remained fascinated with the act of the blood exchange itself—most notably, in its possible life-extending properties and how they might apply to communist philosophy. Practicing what he called "physiological collectivism," he gathered a circle of his students with whom he exchanged blood through mutual transfusions. Eleven times he exchanged blood with them, each time experiencing a revitalizing effect. What he did not know—what no one knew, at the time—was that each additional transfusion of foreign blood cells sensitized his immune system, gearing up its production of antibodies. (This state of high immunological readiness, with a high concentration of antibodies, is known as "high titre," and in some people occurs naturally.) To describe it in basic military terms, his system was on red alert and waiting for an attack. In April 1928, he underwent a twelfth transfusion, injecting himself with blood from a student who, although the same major blood type as Bogdanov, had a factor in his blood that Bogdanov's body recognized as foreign. A massive reaction ensued, in which Bogdanov's immune system destroyed the foreign red cells, littering his bloodstream with their broken fragments. For fifteen days he lingered with uremia as pieces of the cells disrupted his kidneys. He finally died in the best Soviet tradition, narrating his symptoms to the doctors who surrounded him.

Despite Bogdanov's death, transfusion became popular in the Soviet Union, where, just as in England and America, doctors had been relying on donors-on-the-hoof. But the Soviet doctors soon became impatient. They found it frustrating to have to hope that the donor with the right blood type would make himself available at the time of greatest need; furthermore, in a country as big and, in many areas, sparsely populated as the Soviet Union, one could not always hope to find the right donor in the right locale. Those difficulties led one Soviet doctor, Dr. V. N. Shamov of the Institute of Blood Transfusion in Kharkov, in the Ukraine, to consider another alternative. Shamov knew that certain tissues, such as the muscles and glands, continued to function hours after an animal's death. He wondered whether blood would remain viable as well. If so, he might be able to remove blood from a dead body to sustain a live one. The procedure could be dangerous, since the tissues of a dead creature could become toxic as bacteria proliferated. Yet, if it could be made safe, millions might benefit.

Beginning in 1927, Shamov conducted a series of experiments to determine if cadaver blood was safe. He took hundreds of blood samples from freshly killed dogs, searching for bacteria throughout the body at intervals from fifteen minutes to twelve days after death. He found that the degree of bacterial contamination depended on the time and location of the withdrawal. Bacteria spread outward from the gut—blood from the abdominal region became infected more quickly than blood from the muscles, joints, marrow, or brain—and blood from those distant locations remained "sterile" more than a week after death. The next step was to determine if the blood retained its oxygen-carrying ability. In a series of experiments with laboratory dogs, he drained 90 percent of the blood from their bodies—a level "absolutely incompatible with life." When he replaced it with cadaver blood, the dogs recovered. "These data . . . confirm our fundamental thesis," he wrote, "that the blood in a dead body preserves its vitality for about ten hours after death, and within this period may be used for transfusion into a live organism."

Yudin was sitting in the audience when Shamov presented his findings to the Ukrainian Surgical Congress in September 1928. Yudin knew that, with as many as ten thousand patients passing through the Sklif every year, he could not provide an adequate blood supply. Cadaver blood might present a solution. Still, he hesitated to embrace it. It was not just the cultural abhorrence of dead bodies—as a doctor he had learned to overcome such feelings, and he assumed his patients could as well—but he worried about unforeseen medical consequences, such as "cadaveric intoxication" (the chance the blood might be toxic to humans) or syphilis, which can be transmitted by blood but would not have shown up in Shamov's experiments because it does not infect dogs. He resolved eventually to try the procedure, but not until a patient came in who had no other chance of surviving; only then would his conscience permit him to act. He waited a year and a half for the opportunity. And then, on the night of March 23, 1930, the ambulance arrived with the dying young man.

> I was called out to the receiving room and shown a young engineer who had severed [his] blood-vessels . . . in an attempt to commit suicide. I saw I had the conditions that I desired. He was dying of acute anemia [blood loss], but otherwise he was a strong and robust man. At the same time in the receiving room lay the corpse of a 60-year-old man who had died six hours previously . . . due to being knocked

down by an omnibus. The blood of the old man was of the same group as that of the young engineer. Lastly, my moral responsibility in the event of failure would have been minimal, in that the patient himself had courted death. I ordered the corpse of the old man to be transferred to a laboratory. . . .

Painting the corpse's torso with iodine, Yudin sliced into the body and found the inferior vena cava, the large vein that carries blood from the body back to the heart. He plunged in a needle and began withdrawing blood, but the vein quickly emptied and went flat. Yudin's assistants raised the corpse's hands and feet and, massaging the limbs while pumping vigorously on the chest, drove more blood toward the heart. Yudin withdrew several more syringefuls. Then the emergency-room surgeon burst in to tell him that the patient was expiring. Yudin rushed over to check the young man's vital signs. "The radial pulse was impalpable, and he was deathly pale and unconscious. . . . The pupils reacted feebly to light." No time to lose: Yudin thrust the syringe into the radial vein in the crook of the patient's elbow, pushed in the plunger, and watched.

> After the injection of the first 250 c.cm. [cubic centimeters] of blood the radial pulse began to be clearly felt, and after the next 150 c.cm. of blood the patient began to breathe evenly and deeply. Soon consciousness returned, and by the end of the transfusion the patient had a good pulse and his colour had improved. The transfusion itself passed off without giving rise to any general reactions and there were no other complications.

There was one more question Yudin felt he must address. Returning to the corpse, he conducted an autopsy to search for signs of syphilis and malaria. Fortunately, there were none. Two days later, Yudin discharged the young patient "fully recovered and with the wound sutured." Two days after that, he took blood from another corpse and gave it to three patients—two with cancer preparing for surgery and one with an amputated leg. "All these transfusions had excellent effects and passed off without complication," he reported. Soon afterward he performed three more.

Yudin's results persuaded the state attorney's office to grant him a special permit to collect blood from cadavers. Soon ambulances were delivering bodies from all over Moscow—victims of heart attacks, accidents, murders, and strokes. Tram accidents proved particularly productive. Only cases of sudden death were accepted, because that blood, after clotting, soon liquefied again. (Yudin never understood why the blood of sudden-death victims unclotted whereas the blood of those who suffered lingering deaths did not. Hematologists subsequently learned that only in cases of sudden death does the body release enzymes that dissolve blood clots.) Yudin would tie the body to an operating table, insert glass drainage tubes into the jugular vein, and tip the table steeply back. The blood would flow down toward the head, out the tubes, and into a bottle. He learned that he could force more blood out of the body by injecting salt solution into the femoral artery, in the thigh. That done, he tested the blood for blood group and syphilis, and performed an autopsy on the donor to look for signs of disease. He developed a small "bank" of cadaver blood, which experience proved to be relatively safe. By 1938, he had transfused twenty-five hundred people with cadaver blood, of whom seven died and 125 experienced nonfatal reactions such as fever and chills.

Yudin's transfusion results astonished his overseas colleagues, who were both impressed and repelled by what he reported. "The danger of impurities in the donor's blood is practically negligible," wrote Dr. Fritz Schiff of Berlin, although, assuming that patients would find the method objectionable, Schiff added that it "is not too applicable on the German scene." In London, Percy Oliver wrote, "British temperament has a strong aversion to making use of a corpse, and were it even suggested there would be a huge protest. . . ." Yet others could not suppress their curiosity, and in the ensuing decades scientists in Canada, India, and the United States quietly experimented with cadaver blood.

Beginning in 1935, two doctors at an unnamed private hospital in the Chicago area performed about thirty-five cadaver transfusions over a period of two years (the exact number is unknown, because the doctors did not keep precise records). Fearful of condemnation if their experiments became public, they kept the work secret, not even telling the patients or their families. Twenty-five years later, in a retrospective article, one of the doctors, Donald F. Farmer, wrote that cadaver blood had been every bit as "adequate and safe" as the blood they collected from donors-on-the-hoof. In 1961 and 1964, doctors at Pontiac General Hospital in Michigan performed two series of cadaver transfusions. The lead investigator was Dr. Jack Kevorkian, who would become known as "Dr. Death" for helping terminally ill patients take their own lives. Kevorkian reported that of seven patients who received cadaver blood five recovered and only two died. In no case did he see evidence of any reactions or ill effects; the two patients who died probably expired from the conditions for which they were hospitalized, having been near death at the time of the trials. "It would be presumptuous to recommend widespread institution of this novel procedure," he concluded. "Nevertheless we are now convinced of its merit and practicality, and are indebted to the Russian investigators who blazed the trail."

Cadaver blood was not the only kind of blood that Yudin and his colleagues harvested. Inspired by the possibilities of stockpiling blood, they began storing blood that they had drained from placentas and from walk-in donors as well. Indeed, the notion of stored blood became so attractive that, using the citrate techniques Lewisohn had developed, they initiated blood storage on a national scale, shifting away from donors-on-the-hoof. They did not proceed delicately in their quest, but infused blood that had been stored for weeks, despite post-transfusion reaction rates that often exceeded 50 percent. By the mid-1930s, the Soviets had set up a system of at least sixty large blood centers and more than five hundred subsidiary ones, all storing "canned" blood and shipping it to all corners of their union—by far the largest such service in the world. Soviet doctors transfused more than ten thousand quarts of blood in 1937 alone.

News of the Soviet experience traveled to America, where doctors, still relying on donors-on-the-hoof, were transfusing patients by the dozen rather than the thousand. Even though American physicians knew about Lewisohn's citrate technique and used it to mediate walk-in transfusions, it had not occurred to them to use it for blood *storage*. An extremely conservative group, they thought that using stored blood would increase the rate of post-transfusion reactions. But the Soviet experience, imperfect though it might be, lowered a conceptual barrier. In 1937, after reading about the Soviets, Bernard Fantus, a doctor at the Cook County Hospital in Chicago, established a facility at his hospital where he bled the donors into a flask containing a small amount of sodium citrate, tested it, sealed it, and stored it in a refrigerator. Recipients of this blood sometimes developed fever and chills, but at a much lower rate than in Russia, probably because of the cleanliness of American equipment. Fantus called the facility the Blood Preservation Laboratory but, given the system of deposits and withdrawals, soon came up with a snappier name that immediately became part of the popular vocabulary. He called it a "blood bank." The development could not have arrived at a better time, for the world was about to enter a period of global war and unprecedented bloodshed.

# CHAPTER 5

# PRELUDE TO A BLOOD BATH

In 1935, in the village of Niederlungwitz in eastern Germany, a physician performed an act of quiet heroism. A patient needed blood. With no available donors-on-the-hoof, the doctor, Dr. Hans Serelman, sliced into his own arm, opened an artery, and donated his own blood, which happened to be of a compatible type. It was the kind of small drama that had occasionally played out in other countries, bringing praise from the community and a news article or two. But Serelman's sacrifice brought a different response. The patient was an Aryan and Dr. Serelman a Jew; the doctor was sent to a concentration camp for six months, charged with defiling the blood of the German race.

At a time when many countries were advancing in the exploration of blood, the Germans, who had practiced the highest levels of medical research, fell backward into myth and superstition. Under the Nazi regime, blood became not only a resource, as in the other advanced countries, but a liquid rich in allegorical meaning, a symbol of racial purity. That belief—alarmingly literal at times—seriously hampered German medicine and cost the nation thousands of lives.

The Nazi ideal of racial blood-purity comprised a toxic mixture of anti-Semitism, a belief in social Darwinism, and a desperate longing for power. Aside from its murderous ramifications, Nazi philosophy was especially reprehensible in the extent to which the intelligentsia distorted science in its behalf. German anthropologists spent years performing research on how to distinguish between a Jew and an Aryan by means of such features as the shape of the forehead, lips, and eyelids or the "typical Jewish posture." Mixing propaganda and scholarship, many of the nation's leading anthropologists and doctors sought to convince the public of Aryan superiority. In 1926, for example, they sponsored a national competition for "best Nordic head," with cash prizes for the man and woman whose photographs most exemplified the Aryan ideal.

The scientists subverted blood research as well. Led by Otto Reche, one of the premier racial theorists of his time and founder of the German Society for the Study of Blood Groups (Deutsche Gesellschaft für Blutgruppenforschung), they attempted to use serological research as proof of Aryan superiority and territorial claims. Reche had carved out a notable career in the field of racial purity, having served as professor of ethnography at the University of Vienna, where he founded the Viennese Eugenics Society; then at the University of Leipzig in Germany, where he directed the Institute for Race and Folk Anthropology and the State Institute for Folk Culture, and edited two racist journals—*Folk and Race* (*Volk und Rasse*) and the *Journal of Race Anthropology*. Reche and his colleagues produced a body of work to justify racism on "biological" grounds by showing the relationship between blood types and assumed racial characteristics. They based their work on a distorted version of the work of Ludwig Hirszfeld, a Polish serologist who was second only to Landsteiner in the regard of Europeans. Hirszfeld had spent many years working in Germany—it was there that, following on Landsteiner's work, he demonstrated the heritability of blood groups. He felt he owed his allegiance to Serbia in World War I and served as a medical officer on the Macedonian front with his wife, Hanna, also a doctor. For two years the Hirszfelds were trapped in what became known as the "cage," an unbreakable German cordon that had bottled up tens of thousands of Allied troops in Salonika. Confined with men from more than a dozen nationalities—African and Asian as well as European, for soldiers from the colonies had been mustered—the Hirszfelds reacted with characteristic resourcefulness and, finding themselves in what amounted to a museum of human diversity, began collecting thousands of blood samples.

The Hirszfelds conducted their research at a time when anthropologists were seeking new tools to study human evolution. Skull shape, skin color, hair texture—all had been considered guides to the human races, but scientists had begun to question their value as true racial indicators. Certain traits, such as skin tone or hair texture, could change over time in a given individual, making them less-than-permanent markers. Other traits, such as skull shapes, had been found to vary broadly from one generation to the next. It was becoming increasingly difficult to define the human races at all, much less speculate how they evolved. The Hirszfelds thought blood-type research might prove more reliable than the more obvious physical clues, based as it was in biochemistry and statistics. After all, an individual's blood type remains constant for life. Admittedly, the blood types changed from one generation to the next; but taking a broad-based statistical approach, and looking at the relative frequency of blood types among thousands, might reveal certain group characteristics.

Marooned in Salonika, the Hirszfelds collected about eight thousand samples from nineteen ethnic groups in all. As they correlated the blood types with the geographical origins of the groups, they found that certain patterns emerged. The English, for example, had a relatively high percentage of A blood: 43.2 percent of type A versus 7.2 percent of type B. The Indians, on the contrary, had relatively more B blood: 41.2 percent, versus 19 percent A. The French and Serbs, among others, had intermediate proportions. Plotting the blood types on a map, the Hirszfelds found larger patterns as well. As one moved generally west to east, the proportion of the B group increased, growing higher in France and Italy, still higher among the Turks, and peaking in India. Conversely the relative frequency of A blood increased as one traveled west, peaking in North Central Europe.

Those patterns of blood types suggested to the Hirszfelds that the human species may have originated in two different locations. One prototype—"biochemical race B"—arose in India, where the gene for that blood type still appeared with greatest frequency. Another group—the A race—must have originated somewhere in North Central Europe. From the day the two progenitor races arose, the theory went, they spread across the globe, intermingling their gene pools and producing the mixtures of blood types the Hirszfelds revealed. Thus, the distribution of blood types could serve as an evolutionary road map of sorts, revealing the origins and journeys of the various nationalities.

The Hirszfelds made fundamental mistakes. The most obvious was the omission of type O blood from their analysis, which in many places occurs more frequently than A or B. They justified the exclusion by reasoning that, whereas the presence of A and B types was determined by a blood-type gene, type O appeared in the *absence* of a gene; we now know this to be wrong. To their credit, however, they avoided the racist fallacies that held sway at the time. To the Hirszfelds, blood types said nothing about any stereotyped characteristic, be it a crouched walk or noble bearing; they only conveyed *geographical* information, free of all other values, simply revealing where the races had been. As they wrote in their study in 1919: "The serological formula for a particular race is in no way affected by the anthropological characteristics. . . . The Russians and the Jews, who differ so much from each other in anatomical characteristics, mode of life, occupation, and temperament, have exactly the same proportions of A and B. On the other hand, it is clear that the distribution of A and B corresponds with surprising accuracy to geographical situations."

Seven years after the Hirszfelds released their findings, Otto Reche and his colleagues established the German Society for the Study of Blood Groups, as the very embodiment of the fallacy the Hirszfelds had wished to avoid. Staffed by about four dozen Austrian and German doctors and anthropologists, the society embarked on a campaign to correlate blood groups with assumed racial characteristics. They distributed thousands of questionnaires, asking doctors to fill out their patients' blood groups and races, then correlated their findings in their journal *Volk und Rasse* and in their 1932 publication, *Handbook for Blood Groups* (*Handbuch der Blutgruppenkunde*).

The results seem almost comic in retrospect. Because blood type B appeared with slightly greater frequency among Eastern Europeans and Jews (although still not in the majority), Nazi doctors identified it as a "Slavic" or "Jewish" marker. To them, B became the blood type of the dark, Asiatic races and of the *Minderwertig* in Germany—the undesirable elements. Researchers correlated B blood with a host of negative racial traits, such as dark hair and a broad Slavic face. They linked it to "bearers of Polish names," rather than "bearers of German names"; to urban as opposed to rural dwellers; to violent instead of nonviolent prison inmates; and to uncoordinated people versus graceful athletes. Whereas the A type, slightly more common among Germans, became linked with positive traits such as intelligence and industry, B was the mark of the retrograde population—imbeciles, alcoholics, those who were more prone to infection, and those suffering nervous disease, most of whom, incidentally, had their origins in the East. One researcher correlated B blood to the length of time spent on the toilet. According to his study, A types took just a moment to defecate, B types forty minutes or more. A French researcher sympathetic to the Nazis wrote that B blood makes a person "better to retail trade than bearing arms."

The research was absurd. As anyone who was not a Nazi knew even then, blood types bear no more relation to temperament or habits than eye color. Furthermore, even if one did believe that type B represented certain characteristics, it was impossible to avoid the evidence that blood types were related to geography, not ethnicity. The relative frequency of A and B bloods differed very little among the inhabitants of a given locale. Even in Berlin, the very capital of Nazi Germany, Aryans exhibited more B blood than Jews. Yet the Nazis moved forward, corrupting the Hirszfelds' thesis to such a degree that in 1938 Ludwig Hirszfeld himself felt compelled to denounce them: "I wish to separate myself from those who attach the blood groups to the mystique of race. . . . The actual distribution of blood groups on the earth reflects the crossing of races and constitutes further proof that humanity presents a mosaic of races."

But it was too late. The racists had gotten hold of his theories and used them to justify their goals. Reche's colleague Paul Steffan had distorted the research to prove the existence of two "agglutination poles"—one near the Harz Mountains of Germany, the other in China—where two great warring blood-races evolved. As they bred and intersected, they produced the mongrel races of Europe (most notably the Jews and Slavs), who concentrated along Germany's eastern frontier. Later the German Society for the Study of Blood Groups produced vividly colored wall maps depicting this situation, with color-coded islands of orange-red German blood awash in a yellow Slavic sea. The import was unmistakable: The Aryans must conquer and resettle the Eastern lands, and liberate the islands of pure German blood.

At the same time the Nazis were dehumanizing their enemies with bizarre anthropological theories, they took measures to physically remove all signs of "alien" ethnicity and blood. In 1935 they passed the infamous Nuremberg Blood Protection Laws denying citizenship to anyone who could not prove himself "of German or related blood." Anyone who consorted with a Jew—this included not just sexual union, but virtually any social contact—faced prison. Jews who violated the laws were charged with committing "an attack on German blood"; Germans faced sanctions for "treason against their own blood."

Early in their racist campaigns, the Nazis focused on removing the "Jewish influence" from medicine. Jews had achieved prominence in German health care, since it was one of the few fields that traditionally had been open to them. Starting in the mid-1930s, Germany passed a succession of laws limiting the rights of Jewish doctors to complete courses, to obtain medical degrees, and finally to treat any but Jewish patients. Prominent German medical journals launched vehement anti-Semitic attacks and advertised for Aryan replacements. Thousands of the country's most able practitioners wrote to their overseas colleagues for positions, attempting to salvage their lives and careers. Dozens contacted Karl Landsteiner, who, although living in America, remained known and respected throughout the German world. In April 1933, a Danish doctor wrote to Landsteiner on behalf of their mutual friend Dr. Fritz Schiff, one of the world's three greatest serologists, along with Landsteiner and Hirszfeld. The letter, written in broken English, offers an intimate view of the times:

> A few days ago I was in Germany and had there a talk *with Dr. Fritz Schiff.* He is a good friend to you, I know, and me too. Probably you know of the conditions in Germany at present, particularly as to the jews. . . . The conditions are worse than we think about. . . .
>
> Dr. Schiff has lost all his medico-legal examinations and researches. . . . He is now almost without money for his living. He is going to give up his flat in Hohenzollerndamm and look out for some cheap rooms. . . .
>
> Our friend, who has seen his whole existence threatened, beg me to ask you, if you would do something for him, would try to find a position for him in U.S.A. or anywhere, a position, whre he could act according to his qualifications as a serologist and bacteriologist. The conditions in Germany are so incertain now, that he can be forced to leave his motherland. . . .
>
> Please do not mention this letter, if you write directly to Dr. Schiff. Almost all letters from abroad are opened. . . .

Eventually, Schiff found work at New York's Beth Israel Hospital, where he continued to perform serological research. Most Jewish doctors were less fortunate. A letter from a friend of a Dr. Rothinger in Vienna, for example, informed Landsteiner that the man "has been imprisoned in Austria. . . ." Landsteiner, a sensitive and humane man, did all that he could to help his foreign colleagues, judging from the correspondence he left behind, making inquiries throughout North America and on at least one occasion sending money when he found he could do nothing else. Meanwhile other American scientists set up an *ad hoc* committee in New York to find places of refuge for German intellectuals.

The Nazis accomplished their "cleansing" of German medicine with great dispatch. Within five years of Hitler's election, the number of Jewish doctors in Germany plummeted from nine thousand to less than seven hundred. About half had escaped, an estimated 450 committed suicide, and thousands disappeared in the concentration camps. By 1942, Dr. Rudolf Ramm, who oversaw Nazi medical education, could triumphantly declare, "No man of German blood is treated by a Jewish doctor." The achievement proved self-defeating, however: Ger-

man medicine, once the envy of the world, fell into a precipitous
decline. With the nation running short of doctors, medical schools hurriedly trained Aryan students, mostly as paramedics. At the same time,
the medical establishment developed a tolerance for quackery and folk
remedies. Wounded soldiers returning from the front found themselves
treated by too few doctors, using outmoded equipment and techniques. The Germans recognized their tactical error and ordered the
Jewish doctors remobilized, but the regime's efforts had been too
effective. The Nazis' self-inflicted wound, as we shall see later, would
prove grievous in the areas of war surgery and transfusion.

The world was rapidly dividing now between fascists and nonfascists,
Aryans and other races (although racial distinctions conveniently disappeared whenever they hindered the Nazis' strategic goals). In 1936,
Hitler allied himself with Francisco Franco, a Spanish general and fascist who was leading a rebellion against a weak leftist government in
Madrid. Franco had gained Mussolini's support, in the form of tens of
thousands of troops; now Hitler added tanks and the Luftwaffe. Seeking to halt the fascist incursions, the Soviets sent men and machines to
the leftist, Republican side. As the war's ideological dimensions
became clear, volunteers from more than sixty countries poured into
Spain for what would become a bloody rehearsal of ideas and machinery for the great war to come.
     One of the volunteers who traveled to Spain was Norman Bethune,
a Canadian surgeon, Communist, and self-described man of action. A
strapping, handsome, and egotistical man, Bethune did not want to
serve in a hospital anonymously like other foreign medical volunteers;
he intended to make a direct and dramatic impact. In Madrid, he visited the wound wards, where he witnessed young soldiers dying for
lack of blood. Such folly, he thought: Why bring the bleeding men *back*
to the hospital when the blood should travel *forward* to them?
     Bethune was suggesting something that had not been attempted
before: collecting blood from civilians, storing it in bottles, and shipping it to the front. The doctors who performed transfusions during
World War I had relied mainly on donors-on-the-hoof—the blood
came forward, in effect, in a person. Most doctors considered stored
blood too novel and dangerous for war. Yet Bethune, a thoracic surgeon, knew little about the complications associated with stored blood,
such as its limited shelf life or the chance of rupturing red cells in transport; he simply had an idea and decided to try it. Funded by a Canadian organization called the Committee to Aid Spanish Democracy,
he traveled to London to purchase a van and fitted it with a roof

*Man receiving blood from a lamb. This fanciful illustration, from a 1692 German medical textbook, depicts the experiments of those who attempted to transfuse lamb's blood for the treatment of insanity.*

*Three illustrations from a 1679 treatise "concerning the Origin and Decline of Blood Transfusion." The author of the work maintained that animal-to-man transfusions, depicted in the top panel, had been "shown to be wrong," while those from man to man, shown in the bottom two panels, should be "left to the test of experience."*

Georg Abraham Mercklinus
De
Ortu et Occasu
Transfusionis Sanguinis.

Transfusion techniques and equipment, as pictured by German surgeon J. S. Elsholtz in 1667. Elsholtz believed that mutual transfusions between a husband and his wife could transfer personality traits, thereby easing marital discord.

Even as doctors began to understand the benefits of infusing blood, the ancient practice of bloodletting persisted. In this illustration, The Bloodletting, by the French artist Abraham Bossé (1602–1676), a barber-surgeon treats his wealthy patient.

*A collection of bloodletting instruments. Bloodletting as therapy endured for millennia, although there was no evidence to show that it worked.*

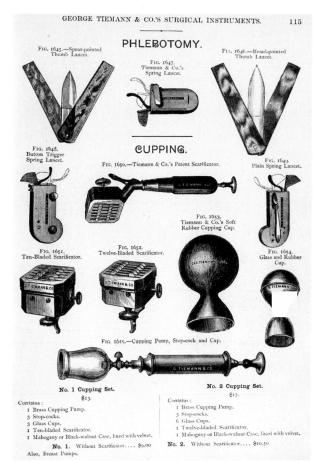

## PHLEBOTOMY.

FIG. 1645.—Spear-pointed Thumb Lancet.

FIG. 1647. Tiemann & Co.'s Spring Lancet.

FIG. 1646.—Broad-pointed Thumb Lancet.

FIG. 1648. Button Trigger Spring Lancet.

FIG. 1649. Plain Spring Lancet.

## CUPPING.

FIG. 1650.—Tiemann & Co.'s Patent Scarificator.

FIG. 1653. Tiemann & Co.'s Soft Rubber Cupping Cup.

FIG. 1651. Ten-Bladed Scarificator.

FIG. 1652. Twelve-Bladed Scarificator.

FIG. 1654. Glass and Rubber Cup.

FIG. 1655.—Cupping Pump, Stop-cock and Cup.

No. 1 Cupping Set.
$13.
Contains:
1 Brass Cupping Pump.
3 Stop-cocks.
3 Glass Cups.
1 Ten-bladed Scarificator.
1 Mahogany or Black-walnut Case, lined with velvet.
        No. 1.   Without Scarificator.... $9.00
Also, Breast Pumps.

No. 2 Cupping Set.
$15.
Contains:
1 Brass Cupping Pump.
3 Stop-cocks.
6 Glass Cups.
1 Twelve-bladed Scarificator.
1 Mahogany or Black-walnut Case, lined with velvet.
No. 2.   Without Scarificator.... $10.50

*A transfusion at La Pitié hospital in Paris, 1874. This illustration is at least partially imaginary. It is doubtful that the blood would have fountained so neatly into the cup, or that it would have reached the woman's vein before clotting.*

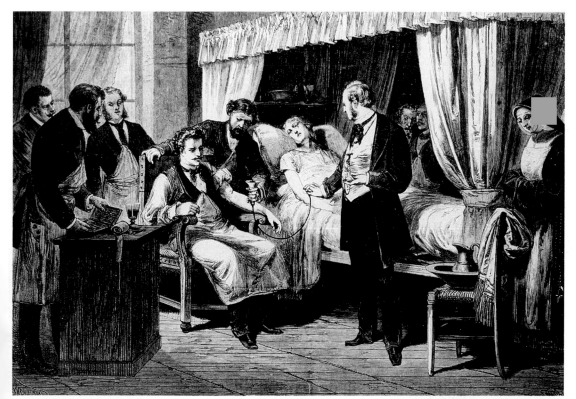

*Dr. Karl Landsteiner's discovery of blood groups made transfusion safe and predictable.*

*In 1908 Dr. Alexis Carrel, a brilliant surgeon who left France for the United States, performed the first modern transfusion by suturing the vein of a baby's leg to an artery in her father's arm. Carrel became one of the most popularly known scientists of his day, only to die isolated and depressed, wrongly accused of Nazi collaboration.*

*Dr. Richard Lewisohn, of Mount Sinai Hospital in New York. His development of anti-coagulants made the storage of blood, and blood banking, possible.*

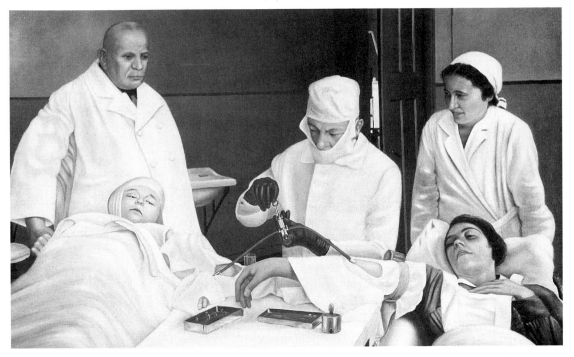

*By the 1930s arm-to-arm transfusions were becoming common. Above, a depiction of Dutch doctors transfusing blood directly from a nurse into a patient.*

*During the Spanish Civil War, Dr. Norman Bethune, a Canadian surgeon and revolutionary, took blood transfusion to the Republicans in Spain. He ran a mobile transfusion service, speeding along the front lines delivering blood to the wounded. He is shown here assisting refugees during the 1937 evacuation of Málaga.*

During World War II the Allies mobilized blood and its constituent parts on an industrial scale. Here, an American medic administers plasma to a wounded soldier in Sicily.

Dr. Charles Drew pioneered the industrial processing of plasma during the early years of World War II. He was lionized by the government as an exemplary "Negro" American, at a time when blood donations were segregated by race.

*Dr. Edwin J. Cohn's fractionation of blood plasma was one of the most important medical advances during the war. Here Cohn lectures to colleagues while separating the blood components of two volunteers.*

# PLASMA FRACTIONATION (Cohn Method)

**PLASMA**
ethyl alcohol
pH, salt, & temp. adjust

*(Centrifuge)*

**SUPERNATANT LIQUID**
ethyl alcohol
pH, salt, & temp. adjust

**FRACTION I POWDER**
fibrinogen (useful in clotting)

*(Centrifuge)*

**SUPERNATANT LIQUID**
ethyl alcohol
pH, salt, & temp. adjust

**FRACTION II & III POWDER**
rich in globulins (antibodies)

*(Centrifuge)*

**SUPERNATANT LIQUID**

**FRACTION IV POWDER**
immune agents, cholesterol

*(Filter, centrifuge)*

**SUPERNATANT LIQUID**
*(Discarded)*

**FRACTION V POWDER**
albumin

*(Treatment, processing)*

**PURIFIED ALBUMIN**

*Cohn fractionation: By this series of steps—akin to the cracking of petroleum to yield petrochemicals—Cohn was able to break down plasma into its constituents. Particularly valuable at the time were albumin and gamma globulin. His process would yield more than a dozen pharmaceuticals and give birth to a global postwar industry.*

БУДЬ ДОНОРОМ!

КРОВЬ МОЮ
БОЕЦ ПОЛУЧИТ,
И НА ФРОНТ
ВЕРНУВШИСЬ
ВНОВЬ,
ЛАВОЙ
ОГНЕННОЙ, МОГУЧЕЙ
СТАЛЬ
ОБРУШИТ
НА ВРАГОВ.

*"Give blood now." That was the message of the wartime crash collection efforts, whether in the United States, Britain, or the Soviet Union.*

YOUR BLOOD
CAN SAVE HIM

THE PRIME MINISTER HAS SAID:  (*Nov. 9, 1943*)

"The Hazards of Great Battles lie before us"

Here is a warning all must heed. Adequate reserves of fresh blood, plasma and serum, *must* be available for giving transfusions to all 1944 battle casualties that need them. For this reason the Army Blood Transfusion Service calls for many thousands more blood donors of all groups. Will you help by giving a little of your blood? It is simple, painless and harmless, *but the lives of our wounded depend upon it* and thousands more blood donors are wanted.

*Will YOU enrol as a blood donor?*

BRISTOL'S
BLOOD TRANSFUSION CAMPAIGN
Feb. 12ᵀᴴ to 26ᵀᴴ
A.R.P. HEADQUARTERS, 55 BROADMEAD

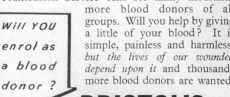

ARMY BLOOD TRANSFUSION SERVICE

rack, storage boxes, a gas-powered autoclave for sterilization, and a kerosene-powered portable refrigerator. He filled it with glassware "of all varieties and kinds—vacuum bottles, blood flasks, drip bottles, etc. . . . In all, our equipment consists of 1,375 separate pieces," he wrote to his Canadian sponsors. "We have enough chemicals to make up solutions for intravenous injections of physiological serums, glucose and sodium citrate to last us for three months. . . ."

Bethune named his operation Instituto Hispano-Canadiense de Transfusion de Sangre, or the Spanish-Canadian Blood Transfusion Institute. He and four Spanish physicians based it in a fifteen-room apartment in Madrid that had been abandoned by a Nazi diplomat, just below the offices of the Communist Party's emergency health service. From there they would broadcast appeals for donors, perform bleedings, and store the blood on the premises. "We collect ½ to ¾ gallon daily, mix it with Sodium Citrate (3.8%) and keep it just above freezing in a refrigerator in sterile milk and wine bottles," he wrote. "This blood will keep for about a week." He lacked the equipment to test for syphilis or malaria, so he took the amateurish precaution of asking the donors to swear on their honor that they had never previously contracted the diseases.

A call from any of the fifty-six hospitals in Madrid would send Bethune and his men sprinting to their van with a backpack full of blood bottles, regardless of the hour. From there they would race to the site of the most urgent need. He described the job in a letter to a Canadian friend: "Our night work is very eerie! We get a phone call for blood. Snatch up our packed bag, take 2 bottles each . . . and with our armed guard off we go through the absolutely pitch dark streets and the guns and machine guns and rifle shots sound. . . . Without lights we drive, stop at the hospital and with a search light in our hands find our way into the cellar principally. All the operating rooms in the hospital have been moved into the basement to avoid falling shrapnel, bricks and stones coming through the operating room ceiling. . . ."

Once at the hospital, they would determine the patient's blood type ("This is done by a prick of the finger, glass stick and serum and takes 2 minutes") and, using their own transfusion equipment, give him the blood.

Bethune had found the action he wanted, and reveled in the sense of purpose and solidarity. "Well, this is a grand country and great people," he wrote, after transfusing a French volunteer who, having lost an arm, raised the other in a clenched fist proclaiming, *"Vive la révolution,"* and a Spanish student-soldier who, shot through the liver, said, *"Nada,"* "It is nothing." Since it was a *mobile* blood service that

Bethune envisioned, he continually delivered blood farther into the field. He ran blood to the fighting men in the Sierra de Guadarrama, racing along the battle lines and stopping to chill the blood bottles in mountain streams. "Plans are underway to supply the entire Spanish anti-fascist army with preserved blood," he wrote proudly to his Canadian supporters. "Your institute is now operating along a 1,000 kilometer front." With a swaggering inattention to danger and an easy sense of macho camaraderie, Bethune could have been a character invented by Hemingway. Indeed, he portrayed himself as such in his vivid articles for the *Daily Clarion*, a Toronto Communist newspaper. In one scene he describes his arrival at a provincial medical center shortly after a skirmish:

> The sure sign of an engagement were [*sic*] the long rows of blood-soaked stretchers, propped up on end, leaning against the walls, waiting to be washed. . . . So up we go to the operating room. Here three tables are at work, the close air heavy with the fumes of ether. Casting a glance, a nod, a *salud* to the chief surgeon as we cross the room to the white enameled refrigerator standing against the wall. The row of empty blood bottles on top tell the story—three, five, seven empties and inside, only three unused.
>
> Out of the door [we go], down the long corridor filled with stretchermen, doctors, nurses and walking-wounded to Dr. Jolly's apartment. His fine, open New Zealand face breaks into a smile as he sees us. We are old friends from early days in Madrid.
>
> "Where's the refrigerator?"
>
> "We have it in the car outside."
>
> "Good, bring it in, we need it. There's a rush on."
>
> The room is packed with wounded. They sit on the floor with bloodstained bandages on head, arms, and legs, waiting to be dressed.
>
> "Sorry, I must go now. Just operated on an Italian captain, poor fellow. Shot through the stomach. Hope he will live. Andre wants to see you. He's used up all his blood." . . . So we bring the refrigerator in and set it down and plug in. Inside we put our remaining four bottles. . . .
>
> This is great! Isn't it grand to be needed, to be wanted!

Bethune was a moral and sensitive man (he came from a family of Presbyterian ministers and showed early talent as a painter), but disguised those traits with his casual drinking and insouciant bravery. An assistant recalled, "He would keep telling me, 'Don't ever get involved.

You can't afford it on this job,' " yet everyone knew the doctor was lying. He was deeply, personally, painfully involved—and that, ultimately, became his undoing. In February 1937, he was delivering blood to Málaga, a city on the Mediterranean which had come under a Nationalist attack. Knowing what happened to captured civilians, the entire city was attempting to escape, creating a human tidal wave along the coast road. "The incessant stream of people became so dense that we could barely force the car through them. . . . The farther we went the more pitiful the sights became. Thousands of children . . . were slung over their mothers' shoulders or clung to their hands. . . . It was difficult to choose which to take. Our car was besieged by a mob of frantic mothers and fathers who with tired outstretched arms held up to us their children. . . . 'Take this one.' 'See this child.' 'This one is wounded.' "

For three days and nights, Bethune used the van as an emergency shuttle, carrying refugees to Almería, 120 miles away. Thousands jammed the town square, standing in bread lines and sleeping in the open air. Then, late one night, when all were asleep, German warplanes dive-bombed the town. Bethune raced amid orange flame and flying debris, picking up wounded and dead children. "What was the crime that these unarmed civilians had committed to be murdered in this bloody manner?" he wrote.

Bethune was wearing out, not only from the emotional burden, but from the physical strain of running blood to the front, twenty-four hours a day. He lost control of his drinking, and his temper flared wildly. Finally, outraged at an imagined insult from the Republican government, he asked his sponsors to bring him back home. Later he joined the Communists in China as they battled the invading Japanese. One day, while performing emergency surgery in a remote peasant village, he slipped and cut his finger. An infection took hold, and two weeks later he died.

Bethune left a legacy of compassion and sacrifice—both China and Canada built memorials to the man—but his transfusion work could never serve as a model. It was too rough, too improvised. His screening and storage techniques were approximate at best. A British doctor who later worked at the institute reported that during a seven-week test period at least 60 percent of blood recipients died (although the majority probably would have died from their wounds anyway). At least half the surviving patients experienced life-threatening post-transfusion reactions.

While Bethune was working from his base in Madrid, a less colorful but more careful physician named Federico Duran-Jorda ran a larger and more sophisticated organization in Barcelona. Whereas Bethune operated out of an apartment and a van, Duran-Jorda filled a multi-story building with laboratories, patient rooms, even a cafeteria (in a city whose residents were going hungry, the promise of a free meal proved a valuable incentive for prospective donors). A trained hematologist, Duran-Jorda believed that the best way to amass blood was not to operate as a fast-moving maverick, but to methodically build a large and stable organization that could deliver the resource even in chaotic times.

Chaos certainly was what he confronted. Barcelona suffered terribly during the conflict. Early on, street fighting broke out among the various factions of the left. Later, when the city united under Republican authority, it became a prime target for Franco, who, through his proxies in the Italian and German air forces, bombarded the town, destroying its oil depots, warehouses, and docks. Municipal services such as electricity and streetcars broke down. Materials ran out. Food supplies dwindled, with no meat, sugar, butter, or milk. "In the whole of Barcelona you cannot buy a piece of soap," wrote an American observer.

Amid such conditions, Duran-Jorda created a service that stood, for its time, as the most advanced in the world. Having studied Yudin's methods and rejected them as impractical (it would be impossible to collect and bleed the bodies during air raids, when most fatalities occurred), he decided to collect blood from the living, while adapting the techniques of mass production. Like the Russians, he refrigerated the blood in bottles with small amounts of citrate and glucose solution. Unlike the Russians, though, he collected only O blood, which, as the universal donor type, could be given to everyone, eliminating the need to blood-test recipients. He pooled the donations in lots of six, diluting any antibodies or antigens that might be particular to one unit or another, producing a more standardized product. Finally, he pressurized the bottles by pumping in filtered air, in order to keep the blood highly oxygenated.

Processing blood as a mass-produced commodity did not mean that he would handle it any less carefully—on the contrary, what distinguished Duran-Jorda's service was the exceedingly high standards he enforced. Unlike Bethune, with his casual appraisals and word-of-honor assurance about communicable diseases, Duran-Jorda's staff rigorously assessed every donor. They performed two separate syphilis tests, for blood type and red-cell concentration, a physical exam, and a written questionnaire about the donor's medical history and habits. The stored blood was used for up to two weeks and no more, leaving a safety margin. Finally, in order to avoid contamination, he enclosed the entire system, from donor's arm to the transfusion bottle, within a sterile network of glass and rubber tubing, manufactured on the premises and of his own design. The system functioned so smoothly and professionally that, even at the height of the fighting, the Blood Transfusion Center of Barcelona (Centro de Transfusion de Sangre Barcelona) was processing up to seventy-five blood donations an hour.

News of his work spread to other countries. In France, Arnault Tzanck, long a believer in donors-on-the-hoof, adopted the Spanish method for Paris. In Britain, the journal *The Lancet* later praised the new methods as "a great advance on any system that has been advocated for this country. The great advantage, especially in time of war, is that large quantities of blood . . . can be withdrawn and prepared for use in a short time under sterile conditions."

Yet, if Duran-Jorda felt pride in his accomplishment, it is doubtful that he found any time to savor it. The bombing continued into the winter of 1938, even as millions of refugees from other parts of Spain flooded in. As the destruction continued and the battle lines drew near, a sense of demoralization took hold. Life in the streets had come to a halt. Even Las Ramblas, that most festive of marketplaces, had fallen silent. The city became ghostly, desperate, and tense. Finally, in late January 1939, as Nationalist forces massed in the outskirts of Barcelona, the city's residents prepared to evacuate. A quarter-million people jammed the streets carrying suitcases, hefting bundles, clutching their children. It was an eerie, heartbreaking scene—the populace milling anxiously, with columns of smoke rising ominously in the background. Now and then, small patrols of Catalan guards would stride by in their red-lined blue cloaks, maintaining the calm with words of encouragement. Sometime before dawn, the procession began. Accompanied by cars, ambulances, mule wagons, and oxcarts, the citizens of Barcelona began a somber migration north to the Pyrenees. Duran-Jorda was among them. Having collected, processed, and distributed an estimated nine thousand liters of blood at the Blood Transfusion Center of Barcelona, he had no choice but to abandon the enterprise. The center was closed, but his methods were not lost: Soon he would teach them to a young woman who would give him safe haven in London.

London was at peace in the spring and summer of 1939, although hardly a relaxed one. The theaters remained open, but the crowds that filled them seemed more tentative than gay. Yet city dwellers did not entirely abandon their revelry as they jammed the pubs for conversation and beer. They danced the "Boomps a' Daisy" and sang along with the latest hit, "Boop Boop Dittem Dottem Whattem CHOO!" They thronged the shopping districts of the upscale West End, where patriotic banners proclaimed "Be proud of our glorious empire." Still, no one could deny that Britain would probably go to war with Germany, and if so the great city would be bombed. If war broke out the German air force could be expected to drop twelve hundred tons of bombs on the city in the first twenty-four hours alone, according to government intelligence estimates, and thereafter a sustained bombing of six hundred tons per day. Six hundred thousand citizens would be killed, and well over a million wounded. (Fortunately these projections would prove high.)

The government prepared for the worst. Thousands upon thousands of gas masks were distributed—" 'Itlers," people called them, in the best Cockney tradition of wry defiance. Tens of thousands of civilians were drafted to serve as bomb spotters, air raid wardens, and firefighters, or in Rescue Units and Decontamination Squads. Officials set up small Quonset hut–type structures called Anderson shelters that, half buried underground, protected their occupants against anything except a direct hit. Hospital emergency wards were moved to the basements. Plans were made to evacuate a million children to the countryside. Public swimming pools were emptied to store makeshift coffins of papier-mâché, and great pits were dug to serve as mass graves.

For all their preparations, however, the British remained surprisingly complacent about blood. London still relied on donors-on-the-hoof, a system that, barely adequate in peacetime, would be swamped in the first days of any conflict. The British medical establishment, like most in the world, knew little about the possibility of storing blood in bottles. Even as the threat of war grew stronger, their attitude remained quaintly old-fashioned. In 1937, during hearings on wartime preparedness, a member of Parliament asked the secretary of war what he proposed to do about the blood supply. The secretary responded that the government's policy was "not to store blood for large-scale treatment as the period for which this can be done is very limited." Asked whether he was aware that the Russians, for example, were stockpiling blood in bottles for extended periods of time, he reiterated his faith in donors-on-the-hoof: "It was more satisfactory to store our blood in our people." Nor was the London medical establishment any more prescient. In 1938, with the Health Ministry predicting thirty-seven thousand casualties during the first week of a war, the staff at four hospitals created an emergency stockpile of blood for the city: a combined total of eight pints.

It would take a considerable force to shatter such complacency, and it came in the person of Janet Vaughan, a young pathologist at the Royal Postgraduate Medical School and Hammersmith Hospital in London. Tall, handsome, zestful, and idealistic (her second cousin Virginia Woolf described her as "an attractive woman; competent, disinterested, taking blood tests all day to solve some abstract problem"), she made a practice of shattering preconceptions. Dismissed by her school headmistress as "too stupid to justify further education," she went on to receive a medical degree from Oxford. Later, after winning a grant from the Rockefeller Foundation to perform research at Harvard, she found that her superiors would not allow her, as a female, to interact with their patients. She pursued her research with laboratory pigeons instead, and made important contributions to the study of anemia. When she returned to London, she wrote the first British textbook on blood chemistry, and eventually secured the position at Hammersmith.

Part of what motivated Vaughan was the desire to use medicine to ameliorate poverty and social injustice. Having worked among the poor during her residency at Camden Town, she had seen the inequities of the British class system and vowed to correct them. For a short time she joined the Communist Party (then seen as the alternative to fascism), only to find it too doctrinaire, and took part in a British physicians' group that supported the Republicans in the Spanish Civil War. In doing so, she became familiar with Duran-Jorda's work, and became convinced that London, like Barcelona, must have stored blood if it ever went to war.

She found the opportunity to act on her convictions in the spring and summer of 1938. Hitler had annexed Austria by then, and decided he needed Czechoslovakia as well. Several times he threatened to invade, each time triggering mobilizations throughout Europe. As one feint gave way to another, the British became certain that war would begin. The tension escalated until the end of September, when the French and English prime ministers signed the Munich Pact, abandoning Czechoslovakia. During this period, Vaughan and a colleague at Hammersmith collected blood from volunteers and stored it in bottles of citrate solution. They had gathered fifty bottles this way—the largest blood storage in London at the time—and in the weeks after the crisis abated, successfully used the stored blood in patients. She later joked: "Everybody used to say that the only blood that was shed [during] Munich was that collected by Janet at the Hammersmith."

Vaughan may have joked about her tiny "blood depot," as she called it, but she developed the unshakable conviction that it would be folly to go to war without blood banks. A city under aerial bombardment would find it impossible to call up donors-on-the-hoof. Doctors, wrist-deep in emergency surgery, would not be able to break away and bleed them. "They must administer blood and not spend time withdrawing blood," she argued in a lecture to the Postgraduate Medical School in London. Yet no one in authority seemed to be listening. When six months had passed and the Health Ministry still took no action, Vaughan took the matter upon herself. In the spring of 1939, she convened a series of meetings of like-minded young doctors to decide what kind of preparations the city should make. They gathered in her flat, above her husband's travel agency in Bloomsbury. There they read papers translated from Spanish and Russian and sifted through piles of apparatus. "The children used to grumble about the old bottles they found about the house in the morning because we were in the process of deciding what sort of bottles we were to use," she said later. The doctors agreed that milk bottles would best serve their purpose and that ice-cream trucks would provide the best transport. (The Medical Research Council, advisory body to the Ministry of Health, later adopted for national use a modified milk bottle with a narrow waist for easy handling; it was variously called the "MRC Bottle" or the "Janet Vaughan.")

By now Duran-Jorda had arrived, having gotten to know Vaughan through the Physicians' Republican Committee in London. Staying with Vaughan's family and working in her lab, he helped the group shape their proposal by teaching them the techniques he had developed in Spain. Together they envisioned a system of four centers positioned around London, each staffed with a hematologist, stocked with modern transfusion gear, and linked administratively to local hospitals. They sent their proposal, "The Supply of Blood for Transfusion," unsolicited to the Medical Research Council. The response was immediate—but not from the council. "It was after that the [department chairman] came to see us and said I was pretty naughty. What was I doing sending memoranda to influential places?" recalled Vaughan. "So I said I was very sorry and forgot about it." Then came a phone call from the MRC, saying they wanted some cost estimates as soon as possible. "So we costed the wonderful scheme; cotton wool so much, rubber tubing so much, and so on, and the estimate went back. . . ."

The proposal had cleared a critical hurdle, yet still encountered weeks of delay. The Treasury Ministry wanted reassurance: "[We] cannot help feeling that the proposals in your letter are rather expensive and we should like to be assured that they represent the absolute minimum cost," read one letter. Health Ministry officials, now sold on the project, assured them that with an investment of merely £20,000 they could save an estimated fifty-five hundred lives each day. Besides, they added, money would only be sent to cover London; hospitals elsewhere in England probably could cover their own costs (although the Treasury Ministry officials did not explain how, or why they should be expected to). The Treasury, astonishingly, continued to balk. Finally, in a masterful twist of bureaucratic reasoning, the Health Ministry officials explained that the quantity of blood provided by the four depots would be "considerably under" the amount necessary for the casualties they projected. Rather than show that the project was inadequate, the missive was seen as a sign of fiscal conservatism, and the Treasury gave the go-ahead on June 5, 1939.

It was now less than three months before Britain would go to war. Hitler, unsatisfied with the concessions of Munich, proceeded to conquer the rest of Czechoslovakia, and occupied Moravia and Bohemia as well. Vaughan and her colleagues rushed to put their depots in place, renting buildings, buying equipment, and hiring personnel from the Red Cross, hospitals, and medical schools. The next step would be to start recruiting and bleeding the donors. On August 22, Stalin and Hitler signed a cynical nonaggression pact, clearing the way for the Germans to invade Poland. War seemed a certainty now, as troops were mobilized and anxious talk filled the air. Vaughan's group had just about gotten their equipment in place when they received a telegram from the Medical Research Council on September 1, 1939. The day they had planned for and dreaded had come. The telegram said simply: "Start bleeding."

# CHAPTER 6

# WAR BEGINS

The same day the telegram arrived, the bleeding began in earnest, although not only in the sense that the message implied. Poland fell in just over two weeks. Half a year later, the Germans had conquered Norway, Denmark, and the Low Countries. In May 1940, they broke into France, to defeat the French army in a mere six weeks.

As governments went into exile or collapsed, so did their medical establishments. In France, Arnault Tzanck, being Jewish, was spirited by friends to safe haven in Chile. The Polish medical service all but disintegrated, but later reconstructed itself at the University of Edinburgh. The Dutch were setting up a medical service closely allied with the Resistance; their transfusion service, although tiny and developed wholly under occupation, would rival any on the continent for the quality of its product and its staff's expertise.

Only the British seemed adequately prepared. During the summer, doctors had worked feverishly to establish the four depots in the outskirts of London—complete with bleeding rooms, laboratories, and industrial-sized refrigerators—close to concentrations of hospitals, yet removed from the center city, where bombs would likely fall. The doctors had recruited some eighty thousand type O donors who stood ready to give when the inventory ran low. These centers, despite their professionalism, had a distinctly improvised feel, wedged as they were into pre-existing buildings. The Northeast London depot, in the industrial town of Luton, was fitted into an abandoned wing of a local hospital; the Southeast London depot, in rural Maidstone, occupied two converted houses. The Northwest London depot, in the manufacturing town of Slough, was located in a social club. Janet Vaughan had rented the facility from a man named Mr. Mobbs. "He didn't think there was going to be any war, but he agreed to give me some rooms in the centre where I could have my refrigerators set up. . . . Now we were very fortunate in our location. My big hall where I did my bleeding was next door to Mr. Mobbs' bar. Everyone used to say, 'How like Janet to set up in a bar!' " The Southwest London depot, in the bedroom community of Sutton, occupied the more sedate milieu of an adult school. "The gym was turned into a laboratory; the library was turned into a room for bleeding donors," remembered Dr. Patrick L. Mollison, who performed research in the center during the war. (Mollison later would write the definitive textbook on blood, now used in medical schools throughout the world.) Unschooled in the art of recruiting donors, Mollison and his colleagues hired a theatrical impresario, who plastered several neighborhoods with posters. "He took us all to a shop in the town, where we sat at tables as people were encouraged to come up and enroll," said Mollison. "Nobody had ever done this before; it was all very new." At the same time, the army set up its own transfusion depot in a hospital southwest of London, in rooms that had been a hospital maternity ward. "It was gratifying to see how easily this organisation, planned without any practical experience of a large-scale transfusion service, swung into action," said a Medical Research Council summary of the project after the war. All was in readiness—centers, hospitals, donors throughout London—for when the bombs would begin to fall.

London remained peaceful during the first year of the war, although British blood had begun spilling overseas. The first great lessons came from the evacuation from Dunkirk after more than sixty-eight thousand men had been killed, wounded, or taken prisoner in northern France. No one really knew how bottled blood would behave under combat conditions—how long it would last, and whether jostling would destroy the red cells—and Dunkirk gave them reason for hope. Doctors used four hundred pints of blood and plasma in northern France and "without exception . . . were satisfied with the results," reported Dr. William d'A. Maycock, a captain in the Royal Army Medical Corps. (Maycock later became a high official in the British Transfusion Service.) The blood not only remained potent for several weeks, but resisted rough handling as well. One surgeon told Maycock that "a transfusion could be given in absolutely any circumstances except in a vehicle"—a procedure that in fact would become utterly routine. The

British doctors also learned that, in this new, kinetic warfare, the medical services too would have to become mobile. Surgeons could no longer wait at base hospitals for the wounded. The injuries from the new high-velocity weapons were so severe that the men would have to be stabilized much closer to the front. That meant moving the blood forward, which Maycock did in portable refrigerators. Just as Bethune had insisted during the Spanish Civil War, the army that could move its blood farther forward would have the greater chance of seeing its wounded survive.

The British were considering this new information when the front came suddenly thundering back to them. Late in the afternoon of Saturday, September 7, 1940, a warm, sunny day in London, waves of German airplanes roared over the city. They pulverized the neighborhoods along the Thames, setting miles of the city ablaze, especially in the poorer East End. "It seemed that all London was burning," a witness recalled. By morning, more than four hundred Londoners had been killed and sixteen hundred wounded.

The attack marked the beginning of the London Blitz, an ordeal of night bombings that lasted for months and caused staggering levels of death and destruction. Hundreds of fires burned every night, sometimes taking whole neighborhoods with them. With water lines interrupted and gas lines ablaze, fires grew to cataclysmic proportions, creating their own weather systems as fresh air got sucked in to feed them. Tens of thousands were killed or wounded in the bombing; tens of thousands more lost their homes and found themselves living on the streets or seeking nightly shelter in the underground tube stations.

Medical people stayed above, caring for the wounded even as the hospitals shuddered under the attack. The British had divided the city into ten emergency medical sectors so the wounded could efficiently be routed to hospitals. They further grouped the medical sectors into the four transfusion zones, each served by a depot. Within minutes of the initiation of a bombing, the hospitals in the sector would telephone the depot with an estimate of their needs. The depots would dispatch delivery vans with the blood. (Vaughan maintained her own group of drivers, mostly women, operating out of Mr. Mobbs's bar.) The volunteer drivers would sprint to their vans and careen through the streets in pitch-darkness, avoiding bomb craters and rubble as buildings collapsed around them. They drove so fast that often they arrived at the hospitals before the casualties. "The deliveries all arrived on time . . . ," according to a postwar report by the Medical Research Council, "which reflected the quality of the drivers employed. Accurate and intimate knowledge of the roads under black-out conditions was essential but had to be coupled with a willingness to drive while a raid was in actual progress, and a determination to get to the hospital at all costs, however difficult the blocking of roads or other circumstances might make the journey."

Doctors labored under near-combat conditions, for bombs had destroyed the hospitals' heat, light, and water supplies. Frequently the only illumination came from the miner's lamps they wore on their heads. Despite these arduous conditions, doctors had almost unlimited access to blood—probably more than before the war—and they used it as never before. Early in the war, they assumed that transfusions would only be useful in treating a small proportion of patients—namely, those who suffered life-threatening hemorrhages. But as the number and variety of injuries increased, so, they found, did the use of transfusions. Never had they encountered so many crush injuries, for example, in which a limb might be smashed under falling debris. Such patients— initially stable, but suddenly and without warning declining—could effectively be treated with blood. Never had doctors dealt with so much traumatic shock, in which a patient's circulatory system collapsed before he could reach surgery; these patients also needed blood. Never had doctors seen so many compound fractures, massive lacerations, or second- and third-degree burns, all of which required blood. Later, after the bombing had abated and Vaughan had time to summarize her findings, she wrote: "It is now widely held that any severely injured patient should be transfused, irrespective of the clinical condition and the level of blood pressure before being sent to the operating theater; in fact, transfusion should begin as soon as possible after the injury."

Not only did they give blood to greater numbers of people than had been anticipated, but they injected a greater quantity of blood into each individual. Using the slow-drip method developed by British doctors some years before, in which the blood was infused slowly, the doctors who treated patients during the blitz learned to inject up to several quarts. The value of blood was becoming so clear to them that they considered it among the top priorities in treating the seriously wounded. At one point, Vaughan came upon a little girl so badly burned that there was no place on her body into which a transfusion needle could be inserted. Vaughan was about to give up on the child when she recalled something she had read about the possibility of giving blood through the bone marrow.

> I went back to have a second look at her . . . took the biggest needle I had in my pack and stuck it into her sternum and hung up a bottle and told my little V.A.D. [Voluntary Aid worker] . . . to pump in as

much as possible. That was the great thing about medicine in wartime; you could take risks. If people died they were no worse off because of what you did. . . . But when I came back a couple of hours later the V.A.D. said, "I've got two pints in." Now this was very exciting and the little girl lived.

[Thereafter] . . . we arranged to have special needles made with flanges on them because the needle I stuck in might have gone right through into her chest. . . . You see, on a rocking boat getting into a vein is very likely to be difficult, whereas you could hope to get into a sternum. So needles with flanges were made and even untrained people could stick them in without too much trouble.

London was consuming vast quantities of blood. No accounting exists of blood consumption during the blitz, although the Medical Research Council estimated that over the course of the entire war the four London depots collected and distributed more than 68,500 gallons. More than 10 percent of the wounded required transfusions; in some neighborhoods this rose to 33 percent. More blood was needed as the bombing spread to other ports and industrial cities, such as Liverpool, Plymouth, and Swansea. As the fall turned to winter and the bombing continued, an unsettling realization became clear: The British had created a remarkable system and were collecting and distributing unprecedented quantities of blood. But it now looked as if they were beginning to run short.

Across the Atlantic, as the United States inched toward involvement, noted transfusionists wondered what they could do. In the spring of 1940, Alexis Carrel, who had performed the historic transfusion on the baby girl in New York, traveled to France to assess the medical situation. Carrel had been busy since that first transfusion, advancing in several areas of science. In the 1930s, he became a friend of Charles Lindbergh, an amateur engineer as well as an aviator, and the two collaborated on an artificial heart. In time they produced a small, exquisitely crafted glass pump that sustained organs for weeks after they had been removed from experimental animals, thus presaging the modern heart-lung machine. Carrel also made a name in the field of metaphysics, with his best-selling book *Man, the Unknown*. In it he discussed the mind-body connection in human health (a philosophy that would come into vogue half a century later), and stressed the healing power of prayer. He also expressed less agreeable views about the proper and subservient role for women, the value of eugenics, and the necessity for widespread capital punishment "in small euthanistic institutions supplied with proper gases." Yet, despite his occasionally retrograde notions (not completely outrageous for his time), he became widely known as a creative thinker, sought by scientists and philosophers alike. He frequently enjoyed long chats with Karl Landsteiner, among others, who was fascinated by Carrel's experiments in clairvoyance. Lindbergh recalled looking up from his work in the laboratory one day to see Carrel walk in with Albert Einstein, engrossed in a discussion of extrasensory perception. Meanwhile, Carrel had made his peace with the French, who had showered him with honors and respect. He purchased a tiny island off the coast of Brittany, where he and his wife spent quiet summers with the Lindberghs, who owned an island nearby.

Now, returning from a trip to his native land, Carrel convened an emergency meeting of the Blood Betterment Association to discuss shipping large quantities of transfusion fluids across the Atlantic as soon as possible. By the spring of 1940, however, with the Germans occupying his homeland, shipping blood to France would only serve the enemy. And so the association turned its attention to Great Britain. Members knew that whole blood could not survive the journey— routine transatlantic flights were unknown at the time, and the blood might not arrive for a week, a significant portion of the liquid's useful life—so they focused instead on a component of blood that was known to exhibit many of the same properties.

If blood is left standing for several hours, it separates into three discernible layers: a layer of oxygen-carrying red cells on the bottom; a thin white, or "buffy," coat of white cells and platelets (white cells fight infection; the disk-shaped platelets in the buffy coat play a role in coagulation); and an uppermost layer of clear amber liquid called "plasma." Plasma is more than just a medium to carry red cells: As a mixture of water, salts, and proteins, it performs some functions that the red cells do not, such as the maintenance of blood pressure. The topic of blood pressure had become crucial to surgeons during and after World War I in relation to shock, which killed thousands. Wound shock, or traumatic shock (as opposed to allergic or psychological shock), occurred when a soldier suffered a wound, burn, or crushing injury. After losing some blood, he would suffer a cascade of alarming symptoms: pallor, clammy skin, thirst, chills, rapid but weak pulse, gasping, and, finally, death. What mystified doctors about shock was that the victims, strictly speaking, did not bleed enough to expire. A person could lose nearly half his body's blood before he died from lack of oxygen carried by the red cells. Many shock victims lost only about 20 percent. Something other than direct blood loss was killing them.

The answer came after many years of examining shock victims. Arteries carry blood under pressure. When filled, they stay open; when emptied, they collapse. Doctors found that a soldier who hemorrhaged could rapidly lose enough blood to allow his arteries to collapse, shutting off the blood supply to his vital organs and producing shock. Many times, soldiers without massive hemorrhages also suffered arterial collapse. A soldier who suffered a crush injury or hidden, oozing wounds, for example, could go into shock even though he initially seemed healthy, because fluids gradually leaked from the broken and shredded blood vessels faster than his body could replace them. Sometimes a patient suffered a systemic arterial collapse, in which blood did not just leak through broken vessels surrounding the wound but drained through the vessel walls over large parts of the body, even where the vessels seemed to be undamaged. Body chemistry changes following a traumatic injury; chemicals released in the blood vessels make them swell and become porous, allowing the liquids that maintained blood pressure to escape. Thus, it was not the loss of red cells that caused shock, but the physical draining of liquid from circulation. A new model of the circulatory system was emerging, as not only a chemical and biological system, but a hydraulic one as well. Volume and pressure mattered almost as much as clotting and blood-type compatibility.

This evolving vision of the function of blood expanded doctors' notions of what they could transfuse. During World War I, British doctors experimentally transfused soldiers with gum acacia, a colloidal liquid with the same consistency as blood. (The liquid only temporarily raised the blood pressure.) One doctor, a Captain Gordon R. Ward of the Royal Army Medical Corps, suggested that field surgeons inject plasma into their patients, although, apparently, no one followed his advice. In the 1930s, Soviet, American, and British researchers began to use plasma on an experimental basis. But it was not until a man named John Elliott made plasma a personal crusade that the Allies began to see it as a resource for the war.

Elliott, laboratory chief at Rowan Hospital in Salisbury, North Carolina, had been experimenting with methods of separating plasma from blood and had collected several bottles of plasma when a young man was admitted to the emergency room. The victim had been stabbed in the heart and obviously had only minutes to live. With no time to take a blood sample from the patient and test it for a matching blood type in the hospital's blood bank, Elliott did the only thing he could: He grabbed a couple of bottles of plasma and infused them. The

response was dramatic. Even before the transfusion was over, the patient regained consciousness. The next day, Elliott gave him more plasma, which sufficiently strengthened the patient for an operation. Three weeks later, the young man went home. "[The] response of the patient to the infusion of plasma was as brilliant as we ever have witnessed with whole blood," wrote Elliott. What's more, the plasma he used had been stored for six weeks, much longer than the useful life of blood.

That and other clinical experiences convinced Elliott that plasma was the best transfusion liquid available, especially in a national emergency. After all, plasma could be stored much longer than blood—for months, as far as he could tell. Plasma could survive rougher handling than blood, since it had no fragile red cells. Most important, plasma did not have to be typed, for it never caused an incompatibility reaction. When a donor gives incompatible blood to a patient, the agglutinins in the recipient's plasma attack the incoming red cells, causing them to clump. But no obvious reaction occurs if the donor gives mismatched *plasma*. Incompatible plasma contains such a small amount of agglutinins in relation to all the recipient's red cells that it has little effect. In other words, anyone who gives plasma is a universal donor.

Elliott became so convinced of the usefulness of plasma that he sought out people who could help him promote its use on a national scale. Having read that the American Red Cross had undertaken a few pilot blood-collections, he wrote to its national medical director, Dr. William DeKleine. He also met with the British Embassy (he gave them some plasma to take home for tests), the blood bank of Baltimore, and the American College of Surgeons, to have them adopt plasma as the new standard. Elliott was relentless, his enthusiasm contagious. DeKleine was so won over by Elliott that he began pushing for a plasma-collection scheme with the army's surgeon general. Meanwhile, Elliott had given a sample to John Scudder, a professor of surgery at Columbia University, who also sat on the board of the Blood Transfusion Betterment Association. Scudder was convinced that if the association was to ship anything to Britain it should be plasma, not blood.

A potent array of forces now aligned itself in favor of shipping plasma to the British. The American military needed information about the usefulness of plasma, and the American Red Cross and Blood Betterment Association wanted to test their ability to process plasma on a national scale. So, in August 1940, the two organizations launched a program known as "Plasma for Britain." They designed it as a cooperative arrangement. The Red Cross would attract the donors, the Blood

Betterment Association would run the collection and processing, and the Red Cross would take over again in packing the bottles and shipping them overseas.

The name most associated with Plasma for Britain was that of its medical director, Dr. Charles Drew. Drew, who studied under Scudder, was the first African-American to earn an M.D. at Columbia. He won a place in American medical history not only because of the work he accomplished, but because his life was so exemplary, his leadership so crucial, his ending so tragic, and his circumstances were so commonly misunderstood.

Drew grew up in a middle-class family in Washington, D.C. The eldest of five, he was always remembered for his confidence, sense of humor, and inspiring, yet unforced, leadership abilities. A gifted athlete, he won a scholarship to Amherst College, where he played football and ran track. Afterward he taught biology and chemistry at a black college in Baltimore and worked with the sports teams, where by all accounts he was an inspiring coach. He attended medical school at McGill University in Montreal (where he still found the time to win national track championships) and pursued a postgraduate fellowship at Columbia. During that period, surgery for blacks tended to be crude, with little preoperative or follow-up care, and black patients often experienced shock. It was only natural, then, that Drew developed an interest in blood and other antishock liquids. He wrote a thesis entitled "Banked Blood," the first draft of which was the size of a New York phone directory; many considered it the most authoritative work to date on the science of blood storage. He and his professor, John Scudder, opened a facility at New York's Presbyterian Hospital, where they experimented with stored blood and, after John Elliott had sent them a sample, liquid plasma.

These were hard times for African-Americans, regardless of their status or station. Segregation was the rule in the South, and in many ways dominated Northern professional circles as well. Blacks were barred from the American College of Surgeons and from many local chapters of the American Medical Association. African-American doctors had no choice but to establish their own professional group, the National Medical Association, along with their own medical journal. Drew had no illusions about what he faced and, in his own determined yet good-natured way, decided to defeat it. He accepted a position at Howard University, the black institution in Washington, D.C., where he planned to train a generation of black surgeons. He wrote to a friend: "In American surgery there are no Negro represen-

tatives; in so far as the men who count know, all Negro doctors are just country practitioners, capable of sitting with the poor and sick of their race but not given to too much intellectual activity and not particularly interested in advancing medicine. This attitude I should like to change."

Soon after Drew started teaching at Howard, his old professor called him with a plea: Plasma for Britain was now under way, and Drew was the only person, as far as Scudder was concerned, who was qualified to lead it. Could he possibly take a leave of absence from Howard and come to New York to manage the program? And so Drew became medical director of the first international program for the shipment of blood products.

Under Drew, Plasma for Britain quickly became a sophisticated operation. Administratively, it comprised eight hospitals throughout New York City, linked to an association office on Manhattan's Upper East Side. There, a dozen telephone operators, equipped with maps of the city and charts of the locations of telephone exchanges, would sit around a doughnut-shaped table taking calls from volunteers. Each time a call came in, the operator would fill out an appointment card and route the caller to the appropriate hospital. The caller would then report to the hospital, undergo blood tests and physical examinations, and donate a pint. The hospital's laboratory would centrifuge the blood to remove the plasma, send samples for testing at a central laboratory, and ship the bottles to a refrigerated warehouse, from where the Red Cross and the military would ship it overseas.

If the program's logistics seemed manageable enough, the technical procedures were anything but. Plasma proved difficult to work with. As a rich mixture of proteins, it was extremely susceptible to bacterial growth. A sample that tested "clean" once, twice, even three times, could later turn out to be dangerously contaminated. One batch that tested clean in New York in August, for example, was found to be infected in London in November. Even the slightest contamination was intolerable, because injecting tainted plasma would most certainly cause death. (For that reason the British often referred to plasma as "liquid dynamite.") "When we began this work we [thought] . . . that preparing plasma would be no more difficult than mixing a cocktail," wrote DeWitt Stetten, chairman of the board of the Blood Transfusion Betterment Association, "but we soon learned, much to our distress, that this was not the case."

Drew approached the problem by leaving nothing open to improvisation or chance. He designed detailed procedures for hospitals to follow at every step of the processing path, from the patient's arm to the shipping bottle. Like Duran-Jorda and others before him, he completely isolated the liquid from the air, enclosing it in a system of tubes, bottles, and vacuum pumps. Each lot of plasma would be accompanied by a sample from the pool, which was kept in a test tube. As a means of monitoring the safety of the lot, the sample would be tested at regular intervals for a period of three weeks, thus screening for bacteria at any stage of growth. Finally, he ordered that all tests must be conducted at the central laboratory at Presbyterian Hospital, where he could closely supervise quality control. His procedures, which set new standards for the nascent industry, eliminated the contamination that had briefly appeared in one of the shipments to London. "Since Drew . . . has been in charge, our major troubles have vanished," wrote Stetten.

The program expanded and did the city proud. Mass donation was new to New Yorkers—it had never been attempted in this country before—yet thousands lined up at the hospitals to give. The Red Cross had mounted a publicity campaign including posters and celebrity endorsements, persuading the public that donating blood was both painless and civic-minded. Nearly fifteen thousand people gave blood, which, after processing, amounted to fifty-five hundred vials of plasma. Plasma, which had never been produced on such a scale, was now becoming part of the American arsenal.

Everyone agreed that Drew had *made* Plasma for Britain; without his diligence and expertise, the program could not have succeeded. Yet, given the esteem in which the association members held him, it is disappointing in retrospect that they did not rise above the ethos of the times regarding the issue of "Negro" and "white" blood. During their organizational meetings, members of the association had debated whether to accept donations from African-Americans for Plasma for Britain. Everyone knew that black blood was no different from white. Their only concern was that, because of the state of race relations at the time and the novelty of transfusion, accepting blood of both races might prove politically untenable. Health centers as renowned as the Johns Hopkins Hospital in Baltimore separated blood on the basis of race. Others did not accept "Negro" blood at all. So the founders of Plasma for Britain made a political decision: They accepted blood from Negro donors, but labeled the plasma so the users would know the race of its origins.

By January 1941, the British had expanded their depot network and built their own plasma facility. Meanwhile, in New York, impressed by the utility of mass-produced plasma, the Red Cross set up a new pilot program at Presbyterian Hospital, this time for the United States Army and Navy, and asked Drew to be its assistant director. Drew had just embarked on the new program when the military announced its policy on blood and race. The armed forces were rigidly segregated at the time, and its leaders thought it best for morale not to collect African-American blood. The Red Cross, whose scientists publicly stated that the policy had no biological basis, acquiesced.

There is no formal record of how Drew reacted to this policy, although his colleagues certainly knew how he felt. In March, just as the program was getting under way, Drew resigned and returned to teach at Howard. There he devoted himself to his main goal of developing a "Negro school of thought" in American medicine. He pursued an eminent career as a surgical instructor, served as a trustee to several national medical bodies, and was a member of a postwar medical team appointed by the surgeon general to tour medical installations in occupied Europe. Settling in Washington with his wife and four children, he prepared for what should have been a long, satisfying life of scholarship, surgery, and education.

On April 1, 1950, Drew and three colleagues were driving through the night to a gathering in Tuskegee, Alabama—the annual free clinic for Deep South blacks in which the rural poor came to be treated by black physicians from all over the nation. It was always an affirmative experience, with old colleagues getting together and doing some palpable good, and Drew and the three other black physicians in the car— Samuel Bullock, John Ford, and Walter Johnson—were glad to be on their way. The previous day had been typically hectic. Drew had performed several operations and then attended a Howard Student Council meeting, and had only slept a couple of hours, so he rested while Bullock drove. Drew took the wheel just south of Richmond. Bullock dozed next to him, while Ford and Johnson slept in the rear seat. Just before 8 A.M., along State Route 49 in North Carolina, Bullock felt a violent bump as the front wheels drifted onto the right shoulder. "Hey, Charlie!" he shouted. Drew, who had dozed off, jerked the steering wheel sharply to the left. The car roared off the highway, plowed into a field, rolled over three times, and came to a stop in an upright position.

Bullock, Ford, and Johnson escaped with minor injuries. They found Drew lying partially out of the car, his right foot wedged under the brake pedal. Deep in shock, he seemed to have suffered extensive crush injuries, the car having rolled over him. A witness to the accident called for an ambulance, which took the men to the Almanance General Hospital in Burlington, North Carolina. Like many hospitals in the South at the time, the Almanance was racially segregated, but the emergency room served both blacks and whites. The on-duty physician, a white doctor named Harold B. Kernodle, examined the patient and looked up in astonishment. "Is that Dr. Drew?" "Yes, we had an accident on the highway," Johnson replied.

Kernodle and two colleagues immediately went to work, applying tourniquets and attempting a transfusion. When their own measures appeared to be failing, they called Duke University Medical Center to consult with specialists. Finally, after nearly two hours of exertions, Kernodle slouched over to Drew's friends in the waiting room. "We tried," he said sadly. "We did the best we could." Charles Drew was dead, at the age of forty-five.

A body of mythology has arisen about Drew, owing perhaps to the scale of his accomplishments, the strength of his character, and the untimeliness of his death. Authors and public figures repeatedly have claimed that Drew invented the process of plasma separation. Indeed, President Harry Truman, at Howard University's commencement in 1952, went so far as to assert that Drew "made possible the very first blood bank in the whole world." What Drew actually did was adapt a laboratory procedure for industrial use—no small accomplishment in itself.

Some have claimed that Drew was more outspoken about civil rights than he seems to have been, that he held a press conference in which he eloquently denounced the military's racial policies on blood. The story is consistent with his character and views, but no record of the press conference has been found.

The most poignant myths involve the circumstances of his death. Often it is said that the doctor who did so much for transfusion bled to death in a Southern hospital's emergency room because the doctors refused to give him "white" blood. It is difficult to trace the source of the rumor, which some say they began hearing within a couple of years of his death. Dick Gregory, the black comedian and activist, told the story to audiences in the 1960s and '70s. Whitney Young, executive director of the National Urban League, repeated it in his nationally syndicated newspaper column. Mainstream publications such as *Time* magazine and *The New York Times* gave credence to the rumor. Even the popular television show "M.A.S.H." recounted the story of Drew's death, elevating it to the level of national folklore.

The story is untrue. The doctors at the hospital in North Carolina recognized Dr. Drew and did not hesitate to give him whatever blood they could. But his injuries were too massive: With his neck broken, his chest crushed, and his vena cava severed, there was no possibility that Drew could have survived. According to Ford, who was with him at the time, "All the blood in the world could not have saved him."

# CHAPTER 7

# BLOOD CRACKS LIKE OIL

With the war under way, a chemist at Harvard named Edwin J. Cohn was growing increasingly concerned. Cohn felt strong attachments to Europe, since he had visited frequently and educated himself in the European tradition. Born to a family of privilege (his father was a wealthy tobacco importer), he attended the finest American academies, including the University of Chicago, Harvard, and Yale, and conducted postgraduate research in England, Denmark, and Sweden. He especially admired European architecture. Cohn, it was said, could take a napkin at a casual luncheon and compose a list on it of the major Romanesque churches in France, discussing their features in exquisite detail. He spoke in pear-shaped, quasi-British tones.

Cohn specialized in the study of proteins, a fairly new field of chemistry at the time. In years past, he had prepared a concentrated liver extract that became a standard treatment for anemia; during the wheat shortages of World War I, he had investigated the protein chemistry of bread in an effort to find alternative ingredients. But now a new thought began to possess him. The Nazis were poised to destroy a great and glorious civilization, one that he personally treasured. He *must* contribute to the effort to stop them.

In the spring of 1940, with the Fascist curtain descending over Europe, Cohn initiated a new direction in protein research. He had heard about the work being done with liquid plasma, and knew of its propensity to become contaminated. He also knew that it comprised several proteins, one of which might be associated with the liquid's antishock characteristics. He felt that if he could isolate that protein he would be able to produce a transfusion fluid that was easier to use than liquid plasma and free of bacteria. The research would be difficult—proteins are complex and highly unstable—yet it seemed worthwhile and intrigued him.

Cohn's work would prove not only intriguing, but absolutely vital, both in the war effort and in the decades to come. The process he developed would become an essential part of the global health industry. It would save millions of lives and yet, in a tragic counterpoint to the technology's blessings, would also cost the lives of tens of thousands. Cohn sensed the potential of his project; that much is clear from his early writings. For the time being, however, he steadied his focus on the immediate and intricate problem of producing a blood substitute from the proteins in plasma. He had just begun to glimpse a solution when a representative of the government called.

Cohn was about to be recruited into the greatest mobilization of scientists to date. The nation's leaders knew that this would be a war of competing technologies and that the country with the fastest, most flexible research establishment would gain a military edge. So, as millions of citizens were drafted for the war, the government recruited scientists through grants and contracts from scores of university, corporate, and military labs. Two new bureaucracies—the National Research Defense Committee and the Office of Scientific Research and Development—surveyed the nation's academic resources and designated magnet facilities such as the California Institute of Technology's center for rocketry and the University of Michigan's center for explosives, among others. Life at these campuses became urgent and exciting, as dormitories filled with military trainees and researchers who had traveled great distances to work with the best minds in their fields.

One of the government's efforts before the war was to mobilize blood for the battlefield, and a scholarly Committee on Transfusion had been established. At the time, they had two options—whole blood and liquid plasma—neither of which would quite do the job. Whole blood could last a week or more in a refrigerator but, considering the time involved in crossing the ocean, would probably be useless by the time it reached the battlefield. Liquid plasma could be stored for much longer, but the committee worried about its tendency to became contaminated. At an early committee meeting, on December 2, 1940, Dr. Max Strumia of Bryn Mawr University described the work he had been doing with the Sharp & Dohme Company of Philadelphia to produce a freeze-dried form of plasma—a fine golden-yellow powder that remained sterile for years yet became fully reconstituted with the addition of water. Then, almost as an aside, Dr. Walter Cannon of Harvard, the nation's pre-eminent expert on shock, said a few words about Cohn. Cannon suggested that, if plasma was to become the medium of transfusion, maybe they ought to learn more about its constituents. It would be "in the interest of clear thinking," he said, "if the protein chemists . . . were brought into the picture."

Based on the committee's recommendations, several blood-related programs got under way in the year before America entered the war. The Red Cross began setting up a system of collection centers; pharmaceutical companies began plasma freeze-drying; and Cohn embarked on a military-funded research program to find derivatives of, or a substitute for, plasma.

He was certainly the right man for the job. Cohn was not only a creative protein chemist, but one of the premier scientific organizers of his day. Colleagues knew him as someone who could assemble a team and produce breakthroughs as if on a time clock. They saw him as a curious mixture of the old and the new. He always sported a wool suit, spats, and a homburg, even on the hottest days, and affected a punctilious formality (photographs suggest a thinner Alfred Hitchcock). Yet he ran his lab as a no-nonsense, production-oriented business, almost akin to today's entrepreneurial biotechnology firms. After recruiting the best and most ambitious men in his field, he would hold weekly strategy sessions, setting definite and clear short-term goals and driving the men with an unrelenting single-mindedness. At a time when universities saw themselves as distinct from corporations, Cohn saw no barrier. Anytime he needed some data, he felt free to pick up the phone and call any expert in the nation. Within minutes, that expert would find himself promising to run experiments for Cohn, regardless of his own deadlines, and would call back as soon as possible with results.

Cohn assembled a team and set up a lab in the ground floor and basement of Building E, one of the great marble buildings that line the quadrangle of Harvard Medical School, with a couple of large rooms and a cold room for plasma. Each week, one of Cohn's assistants would drive a beat-up old Ford to the Red Cross Center in downtown Boston and pick up some fresh blood, place the bottles on the tattered back seat, and drive out to the state toxicology labs in suburban Jamaica Plain, where technicians would spin the blood in a high-speed centrifuge. Then he would return to Harvard Medical School with the jars of tea-colored plasma.

Plasma is a clear but complicated solution, about 93 percent water, about 1 percent salts, and about 6 percent assorted proteins. Cohn knew that adding ethyl alcohol to liquid plasma would cause the proteins to separate and eventually settle to the bottom of the test tube. Unfortunately, all the proteins would rain out together, yielding an unusable mix; he needed to separate them one at a time. He approached the problem by adding the ethyl alcohol several times in succession, but each time under slightly different chemical conditions—such as varying the salt content, temperature, and pH—which favored the precipitation of one protein over the rest. This process, analogous to the distillation and cracking of petroleum to yield oil, gas, and other products, came to be known as "fractionation."

Working through the summer, Cohn's team isolated several fractions from the plasma. They would pour the plasma into a test tube, add some ethyl alcohol, and spin it in a centrifuge for half an hour or so. A pellet of protein would form on the bottom. This protein, which rained out at the first fractionation, or Fraction I, consisted mostly of fibrinogen—the protein that forms the matrix of a clot. They would remove the pellet and fractionate the remaining liquid. They found that Fractions II and III contained globulins, a class of immune agents. Fraction IV contained a mixture of immune agents and cholesterol. But it was Fraction V that attracted their attention: Produced after five rounds of adding ethyl alcohol and centrifuging, this pure white powder consisted mainly of albumin, a highly stable protein that absorbed liquids like a sponge. As such, it seemed a promising component to add to the bloodstream. It was stable over long periods of time and varied temperatures and, unlike salt solution or gum acacia, tended to remain within the blood-vessel walls. It had unusual osmotic strength (the ability to draw in fluids from surrounding tissue, thus keeping the blood vessels fully dilated), yet, as a single protein, took up very little space. In short, it had the antishock capabilities of a package of dried plasma five times its size.

The research moved quickly. By late 1940, Cohn was able to write: "The preparation of large amounts of relatively pure albumin is now possible, and there is practically no upper limit to the amounts of this material . . . that can readily be made available." That was the last he published on the subject until after the war. Albumin had now become classified information.

As Cohn continued experimenting with albumin, several young doctors on his staff began to test the material on humans. They started by injecting a liquid concentrate of albumin into eleven student volunteers from Harvard Medical School whom they had relieved of nearly a liter of blood each. In each case the albumin restored the volume in their blood vessels without causing any ill effects. Then the doctors began to experiment with a larger, more diverse population. "We made connections with emergency rooms throughout the city and they'd call us if they had a particularly bad accident," recalled one of the doctors, Sam Gibson. They would inject the albumin, take note of the patient's reactions, and remove some blood for chemical analysis. Meanwhile, the chemists kept producing more albumin.

They had accumulated a small inventory of the liquid by the first Sunday in December 1941. The staff had been to a lunch at Cohn's house in Cambridge—a typically heavy production, with plenty of cake and custard for desert. Gibson recalled that he was wearily driving home when he heard a news bulletin on his car radio: Pearl Harbor had come under attack. Instantly he wheeled around and screeched to the lab, where his colleagues already were gathering bottles. The navy had ordered Dr. Isidor S. Ravdin, a prominent surgeon from the University of Pennsylvania, to fly to Hawaii to assess the situation, and he had telephoned Cohn to send him every bottle of albumin on hand. They sent him a satchel of fifty bottles, which at that point represented the world's entire supply.

Ravdin took off at midnight, soon after Cohn's satchel arrived, and after hip-hopping across the country and the Pacific, landed in Pearl Harbor four days after the bombing. Nearly twenty-five hundred servicemen had been killed. More than a thousand people were suffering from bullet and shrapnel wounds; many from oil burns that seared the skin so completely it slipped off in rescuers' hands. So many wounded had poured in so quickly that the navy hospital overflowed into barracks, mess halls, and officers' clubs.

For the first couple of days after the attack, doctors could do little more than stabilize the wounded, give them shots of morphine, and mark their heads with mercurochrome to make sure they would not be given an overdose. By the time Ravdin arrived, the doctors were setting bones, treating burns, restitching gaping holes. Ravdin chose seven of the most severe burn patients he could find and injected them with albumin to bring up their blood volume. "All of these patients showed general clinical improvement," he reported. One of the patients was in such critical condition that other doctors questioned whether it was worth transfusing him at all. "He was given a quantity of albumin in the morning . . . and by that afternoon he was delirious but talking. The next morning he was given an equal amount and the following morning [he] was able to eat breakfast."

A few weeks later, Ravdin came to Boston to make his report. By then, eighty-seven people had received albumin in various tests, with only four minor reactions—promising enough for the government, which already was gearing up the production of dried plasma, to embark on a crash program for albumin as well. The military commissioned seven pharmaceutical companies to produce the albumin. Lacking faith in the companies' expertise, Cohn insisted that they send their chemists to his lab so he and his staff could personally train them. James Lesh, a chemist who worked for the Armour Pharmaceutical Company in Chicago, recalled: "My boss called me into his office and said, 'Pack your bags, you're leaving for Boston.' He said, 'Take an extra shirt, because you may be there for some time.' "

It would take vast quantities of blood to produce albumin and dried plasma, not only because of the military's demand, but because of the nature of the blood substitutes themselves. Each of the new technologies required a larger proportion of the blood then the simpler transfusion medium that came before. One unit of dried plasma, for example, required two pints of blood, since about half the volume consisted of nonplasma components. A unit of albumin required 3.6 donations. And so, as the war progressed and more soldiers were wounded, and were treated with more and more blood and its concentrated derivatives, the demand for raw material rose to unprecedented levels.

The Red Cross was ready, or at least they thought they were, having successfully recruited donors in Plasma for Britain. They had opened a donor center at Presbyterian Hospital in New York, and by the end of 1941 had opened six more, all in major Eastern cities. Each center included comfortably appointed reception rooms, refreshment areas, and rooms equipped with cots for the blood drawing. At a time when blood donation was considered exotic and probably painful, the staff were instructed to be courteous and friendly, the donation centers made inviting and reassuring. To encourage donations, the organization hired a publicity staff to mount newspaper and radio campaigns. Posters went up everywhere—one pictured a compassionate nurse leaning over young men, with fighter planes flying by in the background; another showed a wounded soldier with the caption, "He gave his blood. Will you give yours?" Individual chapters mounted incentive programs, bestowing commemorative pins, and forming "gallon clubs" to honor those who gave repeatedly. Military men offered testimonials about plasma, urging their fellow citizens to give. "I'm here to tell you that plasma is practically miraculous," said Donald J. Sutton, a nineteen-year-old pharmacist's mate as he walked into the donor center in Baltimore to "pay back" the plasma he had received in North Africa.

They bring in a guy who's gone into shock. He looks like he's dead. And they pump plasma into him and back he comes.

I remember one [guy], he was literally full of machine gun bullets from air strafing. It looked like he was bleeding to death right there on the beach. But we shot plasma into him and fixed him up so he could be moved to a ship for hospitalization and he'll live.

Often the Red Cross would recruit veteran GIs to report about the "miracle" of plasma infusion. One, a pharmacist's mate named Harry L. Goldman, addressing a donor drive in New York, talked about the morale-building effects of plasma at the front:

When [soldiers] know it's there, they know their chances of living, even if they are hit, are multiplied. They actually love the plasma. . . . They think it's magic. . . . Sometimes they shout, "Doc, hurry up! Bring some plasma," when they see a comrade badly hit. The lads know what it does for them. . . .

Understand, they give plasma right on the firing line; put a bayonet on a rifle and stick it in the ground, hang the bottle on the trigger guard and let it flow. Our pharmacist's mates are trained to handle their equipment in the dark, to rig it up and put it together just the way a machine gunner is trained to take his gun apart and put it together in the dark. Then all that is necessary is to throw a blanket over your head to conceal a light. You put on a small flashlight to find a vein and inject the needle.

In Florida, a radio show called "Women at War!" highlighted the work of Mrs. Grace Jackson, a white-haired lady who headed the volunteer blood bank in Tampa. The unabashed patriotic strains of its rhetoric managed to strike several of the chords that the United States, like other warrior-nations, found so compelling: duty, an idealized sense of life on the home front, the virtuousness of women, and the sanctity of blood. Set in a hospital in which doctors are working to save wounded boys, the drama provides pertinent facts about plasma and where one could donate, and reminds listeners that two-thirds of the nation's blood donors are women. At one point the narrator describes a missing soldier's wife and daughter donating blood to commemorate his birthday. As the broadcast moves forward, one stirring segment after another, it builds toward an inspiring conclusion: "Your blood as a gift, is unlike money, unlike time or work—it is a gift most literally from your heart, straight to the heart of another—to an American soldier or sailor, who may live to help save all you count precious in the world, because you took one simple, generous step to help save him. . . ."

Such efforts, echoed across the airwaves, gave potential donors a powerful incentive, and blood flowed like the mass commodity it suddenly had become. Not all promotions proved so effective, however. Soon after Pearl Harbor, the wife of the surgeon general of the navy showed up at a donor center in Washington, D.C. With such a prominent citizen donating, Captain Lloyd Newhouser, who directed the navy's blood service, decided to conduct the bleeding himself. The woman was rather heavy, especially in the arms, and this obscured the location of her veins. Newhouser had a reputation for being able to find the most elusive of veins, but even his skills were not up to the task: Each time he inserted the needle in her arm, he missed the blood vessel and came up dry. Finally, wishing to avoid an embarrassing situation, he surreptitiously called for a previously filled bottle, which, with a deft sleight-of-hand, he was able to hold high as he graciously thanked the lady.

A more serious impediment was the continuing controversy over "colored" versus "white" blood. During some of the early blood collections, the Red Cross followed the wishes of the military and turned away African-American donors. But in the weeks following Pearl Harbor, the Red Cross persuaded the military to liberalize, accepting blood from black citizens, and then labeling and processing it separately. This way, "those receiving transfusions may be given plasma from the blood of their own race."

The policy proved offensive to many Americans—the country was, after all, fighting a racist enemy—and provoked a storm of correspondence and protest, not just from black and civil rights groups, but from churches and scientific organizations as well. *The New York Times* editorialized: "The prejudice against Negro blood for transfusions is all the more difficult to understand because many a Southerner was nursed at the breast of a Negro nanny. . . . We cannot explain the prejudice that the Red Cross is keeping alive [and assume it] is a survival of the superstition and mysticism associated with blood. . . . Sometimes we wonder whether this is really an age of science."

The New York chapter was not alone in fumbling the issues of blood and race. The Baltimore chapter of the American Red Cross set up a collection center exclusively for blacks, and the New Orleans center set aside one day a week for black donors—both with unsatisfying results. The situation created such continuing protest that in the spring of 1942 the Red Cross held a meeting in Washington to reconsider its position. The organization freely admitted that no difference existed in the blood of the races, yet found it impossible to overcome the assumption that "most men of the white race objected to blood of Negroes injected into their veins." That no one had examined the truth of this belief did not seem to matter—after all, how many white soldiers hemorrhaging from a gaping wound on the battlefield and sinking into a coma would stop a medic from giving them the "wrong color" plasma? Nor did the Red Cross consider the possibility that black soldiers might not want plasma from Caucasians. The bottom line, as far as the Red Cross was concerned, was that 90 percent of the soldiers were white and that policy must conform to "the wishes of the great majority."

It is easy to condemn this policy in retrospect, yet one must account for conditions at the time. The war was a sensitive period for the nation in both issues, blood donation and race. (It was also a paranoid time, when the government saw fit to confine Japanese-Americans to internment camps, regardless of their deeds or affiliation.) Mass donation was a new and somewhat frightening proposal, and segregation was still practiced in the military and in large parts of the country. The leaders of the Red Cross felt that, in order to protect their fragile new blood-collection network, they must maintain the policy of racial separation. Finally, it must be said that some Red Cross employees ignored official policy and distributed whatever plasma they had on hand—regardless of its "racial" origin.

As albumin production got under way, it became clear that the proportion of blood to finished product made it impossible to produce separate batches of albumin from the blood of blacks and whites. Whereas plasma was manufactured in pools of fifty donations at a time, albumin was made from pools of hundreds, even thousands. There was no practical method to ensure separate stocks of "black" and "white" albumin. So the Red Cross banned "black" blood from the preparation of albumin, and adhered to that policy for the duration of the war.

Though these controversies rumbled in the background, they did not interfere with the willingness of Americans to give blood. People came forward by the thousands—individually and in clusters, as families or even whole companies at a time. Businesses and unions signed up to meet weekly quotas: Everyone volunteered, from line workers to executives. At the Schlitz Brewing Company in Milwaukee, five hundred workers donated *en masse*. The American Institute of Banking in Cleveland inaugurated a "Bankers' Hour" with the goal of donating six pints a day for the duration of the war. Streetcar employees in St. Paul, Minnesota, started a blood-giving rivalry, various crews competing against each other to see who could give the most. In Hollywood, Dracula, in the person of Bela Lugosi, donated blood to return his "ill-gotten gains." Collections were made from prison volunteers: San Quentin held first place in donations, with Attica running close behind. Blood donation became an emotional outlet for a populace yearning to sacrifice yet distant from the front. In Dallas, a woman burst into a donor center saying, "Please, I want to give blood. In a newsreel they showed a wounded Marine being given blood plasma. The Marine was my husband." A businessman in New York called his local center to cancel an appointment, only to show up a few hours later. "I changed my mind," he said, a telegram from the army clutched in his hand—a notification that his son had been killed.

To meet the influx of donors, the Red Cross had opened their centers six days a week, and then had set up more centers, spreading their network from the densely populated Eastern cities to the West and Midwest. Eventually they would collect blood in thirty-five cities, with an additional sixty-three mobile units collecting blood from the suburbs. Yet even that could not keep pace with the military's requirements. Each time the Red Cross achieved one collection quota, the military, pressed by the demands of the war, imposed another, several times larger. The demand for 1942 was a quarter-million pints; no one had *conceived* of such volumes before. Yet the Red Cross pushed ahead, and for a brief period actually exceeded the military's quota. The following year, the quota rose to more than a million pints.

Within the first half-year of America's entry into the war, a chain of manufacturing had been developed to produce dried plasma on a national scale. The Red Cross would collect whole blood, chill it with ice, seal it in huge ice chests, and ship it via Railway Express to any of nine pharmaceutical laboratories throughout the nation, distributed so no laboratory was more than twenty-four hours away from its source. The labs would then spin off the plasma, mix it into fifty-bottle pools, freeze-dry it, and distribute the powder in single-dose bottles. Then they would package the dried plasma into field kits—rugged cardboard boxes containing one bottle of plasma and one bottle of sterile water sealed inside a pair of tin cans. The kits were rigorously tested for durability. According to a report from A. E. Willis, assistant sales director of Sharp & Dohme, army personnel "threw the package out a third story window onto concrete and played football with it in one of the large rooms and opened the package to note the effect." The tin cans were battered but did not crack or break. The navy subjected the kits to their own torture tests, placing them on battleships during firing practice—

directly under the twelve-inch guns. Anything that could survive that kind of punishment could stand up to just about anything.

For all its advantages, however, dried plasma was not perfect. The plasma kit, weighing several pounds and the size of a lunch box, could be awkward to carry and cumbersome to use. A medic who wanted to perform a transfusion would have to fight his way into the box, unseal the two tins with the can opener welded to the bottom of one of them, fish out the enclosed rubber tube with the needle in each end, stick the needles through the rubber tops on each bottle, drain the water into the plasma, shake it, and finally administer the liquid. The process could take fifteen minutes or more—an eternity under fire. Albumin, on the other hand, could be packaged as a ready-to-use liquid, since one of the steps in its preparation effectively killed all bacteria. Because it was concentrated, albumin could be bottled in an ampoule the size of a man's fist. Its ease of use, lightness, and compactness would be especially valuable on aircraft and ships; so, even as dried plasma was shipped to the front, the albumin work continued apace.

Cohn's laboratory now had taken on the appearance of a small factory, with dozens of chemists working round-the-clock shifts. Whereas once they worked with test tubes, they now worked with vats, constructing a pilot plant in which they could test-run the procedures to prescribe for industrial facilities. It took about a week to produce each batch of albumin, after which they would heat it at sixty degrees centigrade for ten hours to pasteurize it, and then carefully analyze its osmotic capabilities, viscosity, light-scattering characteristics (a measure of the clarity of the solution), and the purity of its protein. They would prepare it as a liquid and seal it in glass ampoules.

Nor did Cohn's influence stop at Building E. As director of the national albumin effort, he managed an empire of contracts with clinicians, immunologists, and biochemists at universities and companies throughout the United States. He acted as an adviser to the Red Cross and the military, and ran the national system of quality control for all the products he developed. Every time a company produced a batch of albumin, it sent a sample to Cohn's lab, where he tested it. This degree of cooperation between universities and industry was unprecedented. Everything in the vast, sprawling enterprise rested squarely on Cohn's shoulders. "The concentrated attention which he devoted to every detail of every process was astonishing," wrote a colleague; "plans for improvements and modifications poured forth from him in a never-ending stream. . . ." John Ashworth, a veteran of the blood-products industry who as a young man worked at the lab, recalled, "Anybody who didn't want to work twelve hours a day was invited to leave. Hell, we didn't think about the history we were making. There was just Dr. Cohn beating on us every day, probably because the military was beating on him."

Cohn drove his men with a mixture of enthusiasm, example, and intimidation. One never knew when his famous temper would erupt, although it tended to flare over careless mistakes. "A pH off by a tenth could send him into a tizzy," Ashworth remembered. Sometimes Cohn would telephone through the night to check on the progress of an experiment, his ideas for new procedures racing ahead. Once he called a lab assistant at two in the morning. "The young man being told Dr. Cohn wanted to speak to him, hastened to button his collar, tighten his tie and put on his coat before lifting the phone," wrote Dr. Louis K. Diamond, a veteran of the lab. Another staff member recalled how Cohn would stroll into the cold room without donning a parka, holding forth at length while the colleagues who trailed him tried vainly not to shiver. "The good doctor never twitched."

Cohn could certainly leave an impression. "In any scientific gathering, Dr. Cohn is always the man who looks least like a scientist," wrote a magazine correspondent who visited the facility. "The perpetually new-scrubbed pink of his upholstered cheeks, the immaculate white of the fringe around his pate, the perfect fit of the fashionable cloth across his shoulders, and the well-brushed sound of British phrasing in his speech all hint the presence of a brisk and debonair investment-banker-clubman." Colleagues variously described Cohn as a tyrant or mentor; yet, for all his archness and pomposity, Cohn maintained such a high sense of personal and scientific integrity that those who worked for him considered it the most rewarding period of their careers.

As the chemists continued to work in Cohn's lab, their home companies built plants on the industrial scale—facilities that resembled small dairies, with stainless-steel vats, humming refrigerators, and insulated centrifuges, scrupulously clean and patrolled by twenty-four-hour guards. Their representatives met weekly with Cohn to keep abreast of the latest developments. These meetings became something of a local scientific event, attracting some of the most influential people in the blood world, including military advisers, manufacturers' representatives, and researchers from other universities. Everyone would crowd around the oak table in Cohn's office, piled with sandwiches and chocolate cake that, in the heat of the discussion, often went uneaten. The proceedings were preserved in three bound volumes of more than sixteen hundred pages marked RESTRICTED on some pages, CONFIDENTIAL on others. The transcripts show Cohn briskly moving the discussion along to get to the essence of what each person has to say. Long soliloquies are absent from the discussions; there is only the staccato round-robin of people working with a great sense of urgency.

A characteristic meeting took place on June 6, 1942. Things were not going well for the Allies. Germany had conquered most of Western Europe and had thrust deep inside the Soviet Union; Japan was making a sweep of the Pacific. The carnage of battle was on everyone's minds when Cohn gave the floor to Captain C. S. Stephenson of the navy, who announced that the navy would require seven times more albumin than they previously had thought. "There is an urgent request for this material," said Stephenson. "It is up to the manufacturers to get started and produce as much as possible."

The manufacturers were thunderstruck. Here they were breaking all records for industrial development and resource collection, and the military was asking for more! Did the navy comprehend the scope of the project? "The problem of collecting blood is an enormous assembly line, production job," said Dr. Elliott Robinson, representing the Red Cross. "This expansion causes me great concern." The representative from Lederle Laboratories complained, "We had no idea that we would ever be called upon to process 2,000 units of albumin per week." The representative from E. R. Squibb & Sons argued that scaling up so quickly would mean a "a tremendously large job," and the chemist from Cutter Laboratories in California complained that they did not even have the proper equipment.

Stephenson flared. "It appears to me that there is considerable pessimism about this entire albumin program. If the materials cannot be obtained, if the work cannot be done then we must ask ourselves, shall we give up . . . or shall we go ahead and try to get on with the work?"

Yet the manufacturers had raised valid complaints. Just as with the production of liquid plasma, making pure albumin proved devilishly hard, perhaps more so because of the additional processing. One company lost a quarter of their product in a single incident of contamination; another threatened to pull out of the program because they could not achieve Cohn's standards of quality control; still another reported that the albumin they produced was killing the guinea pigs into which it was injected. Albumin required so many steps, so much handling, that some found it impossible to prevent bacteria from invading. Cohn pointed out that the companies could not be expected to make progress if they could not even get the equipment they required. In particular, they needed large, high-speed centrifuges, an unusual piece of equipment made by only one manufacturer—the Sharples Company of Philadelphia. But Sharples had shipped its entire inventory to the navy, which used them to separate oil from bilgewater on ships. Such were the contingencies of wartime production. After several hours of discussion, Stephenson agreed they would get the equipment they needed, and assured them that the centrifuges would arrive.

And so the producers lurched forward, from one crisis to the next. They delivered their first batches of albumin to the navy in November 1942. By the following March, they were collectively producing thousands of bottles a month. It was a commendable beginning, but not enough for Cohn. "We are doing something more than 10,000 units a month now, but we ought to be doing over 25,000," he would say later, in July. "Let us all try to make available to the armed forces the material they need. . . ."

Cohn and his colleagues knew they were not alone in trying to develop blood substitutes, although they had progressed beyond anyone else. They knew, for example, that the Germans were developing substances with antishock capabilities. One, Totofusin, was a simple salt solution that leaked from the blood vessels before it had time to do any good. Another, Periston, produced by the Bayer company, remained something of a mystery until July 1943, when Allied troops captured a bottle in North Africa. They sent it to Cohn's colleagues at Harvard, who subjected it to chemical and biological tests and determined that it was a synthetic antishock agent—an osmotically active derivative of vinyl, consisting of large molecules that would not leak through the blood-vessel walls. It had the negative characteristic of lingering in the livers and spleens of experimental mice. Subsequent efforts to find and study German soldiers who had received Periston never came to fruition.

The Germans also experimented with a blood-clotting agent called Polygal, which they injected into some prisoners at the Dachau concentration camp, as it was later revealed. At the Nuremberg war-crimes trials, the uncle of one of the doctors at Dachau testified that, while visiting his nephew, he read a paper describing the "shooting of four people for the purposes of experimenting with . . . 'Polygal 10.' "

As far as I remember they were a Russian Commissar and a cretin, I do not remember who the other two were. The Russian was shot in the right shoulder from above by an SS man who stood in a chair. The bullet emerged near the spleen. It was described how the Russian twitched convulsively, then sat down in a chair and died after about 20 minutes. In the dissection protocol the rupture of the pulmonary vessels and the aorta was described. It was further described that the ruptures were tamponed by hard blood clots [apparently augmented by the previously injected Polygal]. That could have been the only explanation for the comparatively long span of life after the shot.

Such experimentation did German troops little good, for Polygal found no use in combat. Indeed, having killed or expelled their most able physicians, and rejecting "foreign" medical advances, the Germans relied on medical technologies at least a generation out of date. They made little use of blood-typing laboratory tests, and seemed to have almost no knowledge of blood banking. Instead, they continued to use donors-on-the-hoof, appointing selected soldiers with the "sanitary units" to travel to the front and provide fresh blood to the wounded. A soldier's blood type would be entered on his identity card (often incorrectly, as Allied intelligence later revealed), or tattooed under his left arm if he belonged to the SS. In the event of an injury, he would receive blood from an assigned donor or from a lightly wounded comrade. Whereas the British and Americans had learned to give large quantities of blood through slow-drip transfusions, the Germans seemed completely ignorant of the technique and gave tiny amounts of blood, fearful of overtaxing the heart. The Germans further hampered themselves by accepting only Aryan blood, which eventually frustrated even their own physicians. Dr. Paul Schultze, a German field surgeon captured by the Russians in 1942, complained about the Party's "senseless race theories" and requirement for "pure blood," which made it impossible to collect an adequate supply. So lacking was German transfusion technology that American doctors who came upon enemy wounded often felt compelled to administer plasma. As Dr. Ira Ferguson, an American surgeon who inspected a captured German field hospital reported, "War surgery can be very grim in the hands of untrained surgeons."

Japan similarly crippled herself with xenophobic ideas about foreign science, and employed only the most primitive arm-to-arm transfusions near the front, and rarely at that. After the war, Allied soldiers returning from Japanese prisoner-of-war camps reported gruesome experiments in which Japanese researchers injected them with blood from malaria patients or drained off a prisoner's blood and replaced it with horse blood.

German blindness to the utility of blood not only cost them thousands of lives, but prevented them from seeing innovations developed virtually under their feet. In occupied Holland, the Dutch set up an underground blood collection and processing system that became one of the more sophisticated in Europe. They based it in the Binnengasthuis (Inner Hospital) in Amsterdam (another center, in Rotterdam, had been destroyed during the bombing). They started their collections with donors-on-the-hoof, recruiting hospital staff and then citizen volunteers, then developed ways to store the blood for a few days at a time in sterilized beer bottles.

These were terrible times for the Dutch. Not only were the Germans occupying their soil, killing the partisans, and deporting the Jews, but they had also taken all the machinery, gas, oil, and coal; anything made of metal, even church bells, which they would melt down for copper; almost all the medicine and food. Yet, even thus deprived, the Dutch found ways to modernize their blood system. Dutch doctors had monitored BBC reports about the British army's use of freeze-dried blood plasma in North Africa, and decided to build a facility of their own. The Red Cross recruited Gorard Mastenbroek, a chemist at the Amstel brewery, who, with technical papers smuggled from Sweden and Switzerland, secretly assembled a plasma-drying facility on the company's top floor. He used beer freezers to chill the plasma, a vat from the Dewar's Scotch company to pool it, and centrifuge parts smuggled in from Sweden and a manifold pipe that someone had stolen from a German Messerschmitt to distribute the plasma in multiple bottles. Running twenty-four hours a day, the facility produced more than six thousand units of dried plasma during the course of the war. "The Germans had no interest in any of this," recalled Dr. Johannes J. van Loghem, who later became head of the Dutch transfusion service. "They just left it all alone, probably because they didn't trust our blood."

Thus, during the darkest hours of the German occupation, Amsterdam became a surreptitious blood-processing network, in which donating blood became an act of resistance. Doctors pedaled to and fro carrying bottles. Trucks rumbled into the countryside with hidden crates of plasma for rural folk and the underground. In 1944, during the Battle of Arnhem, in which the Germans trapped thousands of Canadian and British troops in northern Holland, the Dutch smuggled hundreds of pints of dried plasma to soldiers who without it undoubtedly would have died.

The Soviets used massive quantities of blood and its derivatives, including the cadaver blood collected at the Sklifosovsky Institute. The Americans knew about the Soviets' technology, and dicussed it at meetings of Cohn's Blood Substitutes group. Recent intelligence, however, had indicated that the Soviets found the process too cumbersome to yield the volume they needed. As one member of Cohn's research team wrote: "It seems fair to assume that the Russians have found that it is far easier and more satisfactory to secure two liters of blood from four living donors than it is to bleed one selective cadaver by a rather complicated procedure. The living donor may return in two months to repeat the donation; the cadavers can serve their country only once."

Cadaver blood was only part of a system that the Soviets had harnessed as a colossal industrial machine. Having pioneered blood banking and transfusion, the Russians relied on their system of dozens of research institutes and fifteen hundred centers under the Health Ministry. "When war broke out, many of these institutes and stations swiftly and efficiently readapted their work," wrote Professor A. Bagdasarov, director of the Central Blood Transfusion Institute in Moscow. Bagdasarov and his colleagues reorganized the system along four main axes of combat and transportation: a front line from Leningrad in the north to Odessa in the south, on the Black Sea; a second line a few hundred miles to the east, from Archangel in the north through Moscow to Rostov in the south; a third string of collection centers that probed the interior of Russia along the Volga River; and a fourth grouping scattered throughout Siberia. Each division was organized to take maximum advantage of existing air routes, rail lines, and large population centers for harvesting blood. "During [the first] twelve months of the war literally hundreds of tons of blood have thus been applied to wounded men in all stages of evacuation, and with truly tremendous success," wrote Bagdasarov, not without some patriotic pride.

Doctors would collect all types of blood, using type O for universal whole-blood replacement and the other types to produce plasma. The quantities were staggering, considering the hardship under which the people lived—one thousand metric tons of whole blood alone, and plasma in such quantities that doctors spoke in terms of "tank cars" of plasma rather than pints. (One major drawback of the Soviet program was a generally high level of bacterial contamination, possibly because they used plasma in the liquid form. At one point, wrote Bagdasarov, 40 percent of the recipients of large plasma doses were experiencing reactions.) The Soviets also produced an array of blood substitutes, including "Petrov's solution," a suspension of salts in sterile water; "Seltsovski's solution," a liquid designed to stimulate the blood-producing bone marrow; and the "Federov and Vasiliev solution," a gelatinous liquid designed to maintain blood pressure. All prevented shock for up to a few hours, but none had the long-lasting benefits of plasma or albumin.

Whatever directions the Soviets followed in their research, there was no denying their commitment and patriotism. Each day more than two thousand Muscovites lined up to give blood, a level equaling that of the affluent and well-fed New Yorkers. "From early morning till late at night these [collection centers] seethe with activity," wrote Bagdasarov. Donors received food and money for their blood, although many apparently returned the cash payment. Virtually all the donors were women, most of the men having been called to the front. "The use of the woman's name on the bottle of transfusion liquid sometimes leads to correspondence and friendship between the donor and recipient, an occurrence which no doubt increases enthusiasm among neighboring donors," wrote Dr. Wilder Penfield, a Canadian field surgeon who toured the Soviet front. "One soldier, wounded for the second time and in a critical condition, is said to have demanded blood from the same woman who had saved his life before!"

Penfield, who was visiting as a member of an Allied medical mission, found Soviet war medicine "well organized, efficient, modern." He marveled at the field hospitals—rustic wooden structures "of the type of construction seen in a Canadian lumber camp" that the medical staff built and camouflaged themselves. "On one occasion our automobile halted in what seemed an uninhabited place, although we were actually between two special hospitals on either side of the road. The presence of one of them was betrayed by the sound of an accordion." Everywhere were examples of rustic improvisation: a special wooden washing house where patients were cleansed and sorted; large whitewashed wards with hundreds of beds; and a tented sterilization room where doctors could sterilize their surgical instruments in an autoclave mounted on a wood-burning stove. "The blood bank was a little cellar dug into the ground where a large cake of ice kept blood and serum cold."

Nothing would test the Russians' bravery, resourcefulness, and ability to marshal blood and transfusion fluids as the Siege of Leningrad, one of the most brutal chapters of the war. As German forces invaded the Soviet Union, Hitler directed his Army Group North to encircle the city, destroy it, exterminate the population, and then sweep south to participate in the obliteration of Moscow. (Having decided that his armies would not be able to defeat the Soviets in a strictly "knightly fashion," Hitler ordered an Eastern-front war of mass extermination.) The Soviets fought the Germans to a standstill, retreating across a series of defensive belts. Finally, in September 1941, the Germans laid siege, cutting all rail lines and roads to the city, and terror-bombing daily. For the next three years, Leningrad became a scene of the most miserable deprivation, with no electricity or gas, almost no heat or water, and with snow sweeping through the bomb-shattered windows. The populace relied on a daily food ration of a moldy piece of bread the size of a fist. Many were compelled to eat dogs, cats, rats, and a soup they concocted from the glue scraped from furniture joints and wallpaper. At the height of the blockade, more than thirty-five hundred people were dying per day. By the end of the nine-hundred-day siege, more than 630,000 people had died—more than the combined American and British losses during the entire course of the war.

Amid the appalling suffering and destruction, the staff at the Leningrad Institute of Hematology and Blood Transfusion carried on. The institute had no electricity or running water, and few windows remained to block the wind and snow. Fragile glass equipment had been wrecked; even the guinea pigs used for lab tests had died. The doctors substituted scavenged materials—vodka bottles, copper tubes, anything they could find. Mounting autoclaves on woodstoves in the style of the army, they would sterilize water from melted snow to make Petrov's solution to mix with blood to stretch their supply.

Outside the institute and the blood-collection centers, donors formed ragged lines in subzero temperatures. Patriotism ran high: The city was plastered with thousands of posters depicting donors as "Heroes of the Soviet Union," and the party had a way of issuing non-refusable "invitations" to factories, universities, even to schools. Over the course of 1941, despite the deteriorating conditions, the number of donors in Leningrad doubled. The institute staff also experimented with techniques to obtain more blood from each individual, taking a series of small allotments over several days, for example, rather than the large withdrawals favored by the well-fed Americans. Meanwhile, the institute staff trained over a thousand transfusion specialists to serve at the front, sending them forward with kits of blood and plasma. It was surely one of the most heroic civilian actions of the war. Not only had the citizens of Leningrad survived a savage blockade, but while doing so they provided the Soviet armed forces with more than one hundred metric tons of blood and plasma.

Cohn too was thinking in terms of volume. Having shown that blood substitutes could be produced on an industrial scale, he now found himself limited by the amount of raw material—human blood. The public had responded well to blood drives, but no one wanted to depend on them forever. Donations had also shown an unnerving tendency to slack off during calm periods, exactly when the inventory should be expanded for the next crisis. And so Cohn began searching for a replacement for blood, and all the collection and storage problems it entailed. The ideas paraded through his lab, as universities and industry vied to participate. Knox Gelatin Company volunteered gelatins with the same consistency as blood and with apparent anti-shock capabilities; the California Fruit Growers Exchange offered pectin, a liquid carbohydrate derived from the white inner portion of citrus-fruit rind. None of the liquids fulfilled the requirements of remaining long enough within the vessel walls to maintain blood pressure without accumulating in the liver or kidneys.

Cohn actually came close to finding a nonhuman source of albumin in his experiments with plasma from cows. It could have been an ideal blood substitute. Cow plasma is chemically similar to that of humans, and slaughterhouses could have provided a limitless supply. Furthermore, Cohn found that, by fractionating cow plasma, he could produce an albumin virtually indistinguishable from that of humans. Scientists in Boston and Minneapolis began testing the bovine albumin with increasingly encouraging results. In April 1941, Dr. Charles Janeway of Harvard reported injecting up to twenty-four grams of bovine albumin into more than thirty patients with only one minor reaction. Another team later found "no significant reaction except one chill due to a contaminated bottle." They expanded the program, injecting tiny amounts into the skin of nearly three thousand volunteers from prisons and medical schools, with virtually no adverse allergic reactions. Meanwhile, Armour and Company in Chicago, under Cohn's direction, began producing bovine albumin in quantity, gearing up for military production and providing test samples for clinical research.

The prospects for bovine albumin looked excellent for a time. Then, in July 1942, a sixty-two-year-old man reacted ten days after an injection, developing fever, anemia, discoloration, and welts. Cohn wondered what hidden property caused the bovine albumin to trigger an immune response on a time delay—whether it was contamination or an instability in the protein that allowed it to break down eventually. In order to address those possibilities, Cohn modified the process to make the product more pure and twenty times more stable than the albumin he had started with. He seemed to be making strides with stability when a colleague came in with another discouraging report. One of the team's doctors had been working with prisoners from the Norfolk Prison Colony in Massachusetts who, in exchange for volunteering, won sentence reductions. Of the sixty-six men he injected, twenty-one experienced the same delayed reaction of muscular pain, anemia, and welts. One of the prisoners reacted quite severely. Nineteen days after his injection, he suffered fever, muscle aches, and pains in the hip. On the eighth day after the onset of his symptoms, the prisoner awoke and "suddenly and unpredictably" died. An autopsy revealed that he died of a heart attack brought about by a swelling of the heart tissue—a sign of an allergic immune reaction.

The scientists discussed the experiment in a meeting on October 19, 1942. Cohn, as the leader of the project, attempted to take full responsibility, something his colleagues would not permit. Indeed, "the whole group had urged that the project go ahead," according to the navy's Captain Stephenson, who added for the record that he felt the prisoner's sacrifice must rank as "one of the finest example of heroism in the history of medicine." The group passed a resolution in favor of continuing the research, but Cohn suspended all work with bovine albumin on human volunteers. This prisoner had been the only fatality from Cohn's experiments.

As the war escalated in savagery and intensity, the nation's blood-processing machinery pounded on. Seven industrial laboratories were producing albumin, nine producing dried plasma, and the Red Cross supplied them all with millions of pints of raw material. Cohn's lab developed new products from fractions they had previously discarded. From one discarded fraction his scientists produced gamma globulins, a class of immune agents especially effective against measles and mumps, a scourge of military bases at the time; from another they developed laboratory reagents for determining blood groups; from yet another they produced fibrin film and fibrin foam, which surgeons used as patches during brain surgery.

Blood was being used in greater volume and in more permutations than ever before. By the close of 1943, the military, which at the beginning of hostilities had estimated it would need one hundred thousand pints of transfusion fluids, had received more than two and a half million packages of dried plasma and nearly 125,000 ampoules of albumin. The quantity was staggering, yet, as battlefield surgeons would soon report, even that would prove insufficient.

# CHAPTER 8

# BLOOD AT THE FRONT

As Edwin Cohn was fractionating plasma, his colleague Dr. Edward "Pete" Churchill was observing its effects on soldiers overseas. Churchill, tall, bald, and reserved to the point of seeming enigmatic, was a professor of surgery at Harvard and the Massachusetts General Hospital, recruited as one of the army's traveling "consultants" in the spring of 1943 to assess the situation among American troops in North Africa. And what he saw did not greatly please him.

One of the strengths of the American research effort during the war was that the government did not merely dispatch equipment to the field, but sent researchers forward to assess its performance. With each new weapon or technology came field engineers to detect any bugs or glitches so technicians at home could modify the design. Similarly, the military dispatched a cadre of medical consultants to analyze medical and surgical techniques. These consultants, recruited from some of the finest universities and hospitals, represented a variety of fields, including chest surgery, plastic surgery, neurosurgery, ophthalmology, oto-laryngology, and several specific diseases. Figuratively peering over the field surgeons' shoulders, the consultants would observe and critique, prepare written circulars to introduce and standardize procedures, organize training seminars to discuss advances in medical care—and not infrequently jump in to lend a hand. Their work did not always proceed smoothly—they had no direct authority *per se,* and their suggestions occasionally rankled those who had spent their careers in the military—yet the fact that their opinions saved lives won them respect and eventually admiration.

Churchill belonged to that peripatetic band, and had personal reasons for joining. Born in Chenoa, Illinois, a farming community near Chicago, he was educated at public schools and at Northwestern University before completing his medical degree at Harvard. He developed several new procedures in thoracic surgery and had achieved an international reputation. He had also reorganized the surgery department at Mass. General into a streamlined modern prototype. Yet, despite his contributions in civilian life, he yearned to enter the military. During World War I, he had remained in medical school on a student exemption, signified by the bronze caduceus pin the medical students wore. One night he was walking up the aisle of a theater with some friends when several audience members who had not seen the pin, assuming them to be draft dodgers, began to hiss. He never forgot the humiliation of that moment. When the war came, he volunteered for the first action he could find, and in the spring of 1943 became colonel and surgical consultant to the North African–Mediterranean Theater of Operations.

Churchill had become a participant in Operation Torch, the first major American action of the war. Led by a talented lieutenant general named Dwight D. Eisenhower, the invasion was to be an enormous pincer movement from the Nile to the Atlantic Ocean undertaken by the American and British armies. On arriving at the U.S. Army headquarters in Algiers, Churchill realized that he faced a considerable task. Military surgery had long been regarded as the uncultured stepbrother of its civilian counterpart. Although certain techniques had been developed for the battlefront—during the Spanish Civil War, for example, José Truetta, a colleague of Federico Duran-Jorda's, invented the concept of triage, in which surgeons systematically winnow through the wounded—surgeons who went to war were generally supposed to adapt what they knew from civilian medicine. Yet this war, more than any before it, imposed situations qualitatively different from those of the nonmilitary world. Conditions that had been tamed in the homeland, such as streptococcus infections or the fungal infestation called "trench foot," became pandemic at the battlefront. The war created wounds unlike anything the civilian doctors had treated—not the knife slash or bullet hole of a street fight, but mass-produced gore, with death in the form of a mangling explosion or in a storm of machine-gun fire cutting across the torso. "A surgeon would say: 'But

I've worked for years in the Detroit Receiving Accident Hospital. I know how to handle wounds,' " wrote Churchill. "But he still would have no conception of the destructive force of high-velocity missiles or the timing factor so important in treatment, and the many other elements that make war-time surgery a specialty in itself. . . . Military surgery is not to be regarded as a crude departure from accepted surgical standards . . . [but as] the surgery of trauma encountered in epidemic proportions."

His first task was to update physicians on some of the newer surgical techniques. Previously, for example, when a soldier came in to a surgical field station, doctors would clean a wound, dress it, and then wrap it in a closed plaster cast, so the soldier could be shipped to a hospital without the stitches' ripping open. After studying the procedure, Churchill reported that the casts did more harm than good, since they cut off circulation as the tissues swelled. He introduced the doctors to the split plaster cast, which enabled tissues to expand without danger. Another change involved the long-standing practice of letting a wounded soldier sit for hours before the operation, in the mistaken belief that he would somehow "stabilize." Churchill and the other consultants showed that the man did not stabilize, but merely weakened, and persuaded the surgeons to attend to the wounded as quickly as possible.

One of the areas that most concerned Churchill was the use of plasma for transfusions, which had given him doubts from the beginning. He appreciated the strength of plasma's reputation, but wondered whether the benefits had been exaggerated. As he traveled from one field hospital to the next, his concerns about plasma and albumin grew stronger. Their effects seemed only temporary. A soldier sustained on plasma for more than a few hours began to suffer "air hunger," as field surgeons called it, with a rapid pulse and a characteristic gasping for breath. His organs became enfeebled by the lack of oxygen; in this weakened condition he might not survive surgery, which in itself was a formidable challenge to the body. As one medical observer wrote: "At a certain stage an adequate circulation cannot be restored by merely filling the system as one does an automobile radiator." The problem was that plasma and albumin lacked the oxygen-bearing red cells. As such, they could forestall immediate shock by restoring blood volume, but they could not address the body's lasting need for oxygen. This gap became apparent during surgery, when many patients sustained on plasma slipped into a secondary shock, which could only be addressed by the infusion of whole blood.

The field surgeons also revealed that, since no blood was being shipped from America, they were forced to harvest blood from whoever happened to be around, regressing to the old method of donors-on-the-hoof. An American nurse in North Africa wrote about a desperate search for blood after a battle:

> The corpsmen had donated all the blood they could spare. Nurses offered but were refused. The doctors couldn't afford to give blood—they had to keep going, all of them. Colonel Wiley hit upon a solution. He got in contact with Ordnance and Quartermasters [the divisions that handle explosives and supplies] and asked them for blood donors. Within two hours thirty men had been sent to our area. They were to rest in an empty tent, and as blood was needed they were called.

Sometimes the doctors had no donors but themselves. Dr. Kenneth Lowry was a surgeon from Ohio who, with his brother Forrest ("Frosty"), served in a surgical unit in Gafsa, Tunisia. When Churchill visited him at his surgical-aid station, Lowry showed him the diary he had been keeping about medical conditions at the battlefront. In an entry for February 2, 1943, Lowry described his attempts to get blood for the wounded.

> We have had all types of surgery; sucking chest wounds, abdominal wounds, compound fractures, and amputations of arms, legs, feet, and thighs. . . . To date we have lost only one case here, a lower one-third thigh amputation with multiple wounds of the left leg and thigh. He was in profound shock is spite of plasma, 500 cc. of blood and lots of glucose. The operation did not increase this shock but neither did he improve. More blood might have helped. Blood is so precious, so urgently needed! What we do give is being obtained from our own personnel who are most willing but they really need it themselves after putting in long hours without rest or sleep. We could not find a donor for a splendid chap from Maine last night. He was in severe shock and needed something in addition to plasma and glucose so Frosty gave his blood, took a short rest and went back to operating again.

The next day, he wrote in a letter:

> I cannot help but add one remark which I have observed in our work. Dried human plasma is saving hundreds of lives that would surely otherwise be lost. Of course whole blood is better but difficult to obtain.

During his early weeks in North Africa, Churchill visited British medical installations in Tunisia, where, unlike the U.S. Army practice, official policy had sanctioned whole blood from the beginning. The British had set up a layered organization to move the blood in stages to the front. Led by Brigadier Lionel Whitby, an erudite and unflappable physician who had lost a leg as a machine gunner in World War I, the system comprised a series of storage depots where blood would be received, inspected, refrigerated, and moved forward. The chain began in Britain; refrigerated army vans would fan out across the country, collecting blood from volunteers and delivering it to the military medical unit that had been set up at Southmead Hospital in Bristol, England. Here—at the "home depot," as Whitby put it—the blood was chilled, processed, and readied for shipment. From there it was airshipped in insulated crates to "base transfusion units"—large, selfcontained blood banks in the major theaters of operations. It was the job of the base units to estimate the need and receive the blood and transfusion equipment from England. From the base units the blood would travel forward in refrigerated vans to "field transfusion units"—mobile transfusion stations that could quickly be moved to where they were most needed. At every point in the chain of transportation, the blood remained under the watchful eyes of technicians whose only job was to monitor and protect it. During the early campaigns, thousands of bottles traveled to base transfusion units in Norway and France. (The British, considering it medically risky to ship the blood all the way to North Africa, instead sent equipment to the base transfusion unit in Cairo, so doctors could set up a depot of their own, collecting blood from noncombatant troops and shipping it forward in refrigerated trucks.) The entire chain, from the home depot in Bristol to the forwardmost unit, constituted an integral transfusion service, administratively separate from other parts of the Royal Army Medical Corps.

As much as the British valued whole blood, even they at times underestimated the need. In the weeks preceding the battle of El Alamein, doctors at the British base in Cairo assembled thousands of bottles of blood and plasma. But in their protracted desert combat they ran short of critical supplies. "When I visited the British base transfusion unit in Cairo that had supplied blood to the battle of El Alamein Major Buttle, who was running this unit, told me that he had to send his men out in the streets of Cairo to pick up old beer bottles, bring them back, wash them and draw blood into these bottles to ship up to the front line," wrote Churchill. ". . . I saw Egyptian civilians sitting on the floor, placing bottles of blood into containers and packing straw around them. It seemed unbelievably primitive and yet in the opinion of the doctors who needed this blood, Buttle's accomplishments warranted the Victoria Cross."

Churchill was reaching the inescapable conclusion that, whatever the difficulties of transport and storage, American soldiers must have whole blood. Other officers in battle zones agreed. On March 24, 1943, Churchill sent a memorandum to Washington suggesting that the army begin shipping whole blood and transfusion equipment directly to North Africa. "With the liberal use of plasma certain casualties recover from shock, but are not in condition for a major surgical operation," he explained. His request was received politely, but ignored. A few weeks later, he sent a second request, stressing that whole blood was needed for a "significant proportion of the wounded." Again, no response. In June he made a third request for blood and transfusion sets.

Despite these entreaties, the one man who could order the shipment refused. Army Surgeon General Norman T. Kirk, a diminutive, silver-haired, and splenetic individual (his temper earned him the nickname "T.N.T. Kirk"), was an orthopedist, or bone man, by training, and lacked patience for laboratory science—blood, proteins, plasma, and such. As far as he was concerned, the question of transfusion had been answered with plasma, and he was loath to reopen it. Beyond that, whole blood presented logistical problems that the army would rather not have to confront. Unlike plasma, with its room-temperature shelf life, blood had to be refrigerated and handled with care, and the army had neither the cargo space nor the refrigerators to ship it. It also seemed to him that the army, having invested so much time and money in blood substitutes, did not need to consider an alternative. As Churchill later wrote: "A huge vested interest had been built up starting from assumptions and erroneous thinking. . . . Civilians were busy helping with the war effort and many had their prestige at stake in the collection and use of plasma. Publicity had been launched to provide plasma for the wounded soldiers, and the Red Cross as well as the N.R.C. [National Research Council] was behind it. Edwin Cohn was working to improve plasma and was trying to get albumin solution into production. . . . It gained size and momentum like a rolling snowball."

Churchill's strongest ally in the attempt to procure whole blood was Lieutenant Colonel Douglas B. Kendrick, M.D., a career army medical officer and subordinate of Kirk's, who directed the army's transfusion program. A former fullback at Emory University, Kendrick was as large and amiable as Kirk was small and obstinate, and hounded him endlessly over the issue of whole blood. On one occasion Kendrick, a participant in Edwin Cohn's Subcommittee on Blood Substitutes, helped draft a resolution urging the use of whole blood as well as albumin and insisted on hand-carrying it to Kirk. "He turned me down cold," Kendrick later recalled. "He said they already had plasma and couldn't spare the planes." Kendrick kept returning with proposals, memos, and recommendations. At one point, after presenting Kirk with several pages of evidence, Kendrick could not help exclaiming: "You've just *got* to start sending whole blood." Kirk, his famous temper rising, glared up and snapped, "Goddamnit, Kendrick, if you bring this up again, I'll throw you out of my office!"

There was no mistaking the surgeon general's intention: The army would fight the war on plasma and albumin. If Churchill and Kendrick wanted whole blood for the wounded, they would have to find a way to get it themselves.

With no help from Washington, Churchill and his men improvised in North Africa. Borrowing glassware from the British, purchasing citrate from the local French pharmacies, scavenging plane wrecks for glass and metal tubing, they established a small blood bank in Algiers, in which they took blood donations from noncombatant troops and the walking wounded. To encourage donations, they paid $10 a pint. The laboratory facilities were so rudimentary that the blood received was not even tested for syphilis or malaria. "Better alive with malaria than dead of wounds," they would say, and continue transfusing.

They adapted the system for the invasion of Sicily in the summer of 1943, shipping transfusion equipment with the troops in addition to the normal allotment of plasma. During the attack, doctors established a network of *ad hoc* blood banks a few miles back from the fighting, collecting from combat troops and lightly wounded soldiers. Meanwhile, Churchill continued to hector his superiors, to the increasing discomfort of everyone above him. At one point the army theater surgeon in charge of North Africa asked him to stop communicating directly with Washington and restrict himself to going through channels in Algiers. Churchill was not used to being ignored, nor could he fathom Kirk's refusal to accept the medical evidence. When a combat correspondent for *The New York Times* came through the area, Churchill took him aside and said, "You must break the story that plasma is not adequate for the treatment of wounded soldiers." Soon afterward an article appeared entitled " 'Live-Blood' Banks Save Soldiers' Lives in Sicily When Plasma Proves Inadequate." Later Churchill bypassed his superiors even more flagrantly when he directly ordered transfusion equipment from the Baxter Laboratories of Chicago. ("Churchill had balls," a colleague recalled. "Besides, he was from *Harvard:* In his mind that gave him enough rank to face down the United

States Army.") Kendrick supplied the necessary paperwork. "It is quite apparent from your report, and others received from the various Theaters, that stored whole blood must be made available in the forward surgical hospitals," he wrote on official letterhead. Contravening Kirk's wishes, he added, "The Surgeon General's Office is perfectly aware of the need for whole blood in the Theaters and is anxious to supply the necessary equipment to make it available."

"We worked *around* Kirk," Kendrick later confessed. "I had worked with Baxter labs and I authorized the shipment. Kirk could have had me court-martialed but, given the results, who would have testified?"

By late fall, the equipment began arriving—thousands of glass vacuum bottles for closed-system collecting, miles of rubber tubing, several refrigerators, and hundreds of pounds of laboratory reagents. The army, having routed the Axis forces in Sicily and invaded the Italian peninsula, set up a new headquarters in Naples. With the city half demolished by the Germans—the university burned, the hospitals looted—the Americans constructed a makeshift medical laboratory and blood bank on the outskirts of town. (A second blood bank was set up in a whitewashed cow barn that had belonged to Mussolini's son-in-law.) There, against the backdrop of the mountains and the bay, off-duty soldiers with type O blood would line up to donate and collect their fee.

The war moved ahead. In January 1944, the Americans and British invaded Anzio, a beach resort south of Rome. The landing went well for the first couple of days, but as the Allies paused to consolidate their forces, the hills erupted with artillery and rifle fire. For months the Allies found themselves in one of the war's bloodiest and most fruitless engagements. German warplanes attacked the hospital ships that lay off the harbor, despite their prominent Red Cross insignias, sinking one with seventy-five patients and doctors. Bombs flattened the 95th Evacuation Hospital, sitting in plain view on the open beachhead. The scene was chaotic. Patients were fleeing from one hospital to the next, "sometimes with attendants continuing transfusions en route, one holding the bottle while another steadied the needle," according to an Army doctor at the scene.

The staff at the blood bank in Naples had been working around the clock to set up the equipment, collect the blood, and pack it for shipping, but they were not ready in time for the invasion. With American soldiers bleeding on the beachhead, the medics at Anzio, even though stocked with albumin and plasma, borrowed thousands of bottles of blood from the British. It was not until six weeks after the start of the invasion that the first blood from Naples arrived. Later, as the Americans pushed inland and northward into Italy, they set up a more effective distribution scheme. Now the blood began to move with due speed—collected at donor centers in Rome, Pisa, and Florence, flown out on a designated "blood plane," and met at the landing strips by special couriers who shuttled it to hospitals throughout the peninsula. Finally, surgeons could administer both the whole blood and plasma that the wounded men needed.

Considering the official resistance they encountered, Churchill and his allies accomplished something remarkable, starting with virtually nothing in North Africa, and setting up a system that spanned the Mediterranean. The blood bank in Naples, born as a "temporary structure of rough hewn boards and beams," had quickly grown into a state-of-the-art medical installation supplying more than three hundred bottles of preserved blood per day within a radius of several hundred miles. It was a stunning achievement, especially since Washington did so little to help. Indeed, of the nearly eighty thousand pints of blood transfused into the soldiers who fought in the Mediterranean during the course of the war, not one had come from the United States.

One of the services the Naples blood bank provided was the shipment of blood that the Free French collected in North Africa to their forces fighting in Corsica and southern France. The Nazi occupation of their country had robbed the system of its sense of idealism; donations in Paris plummeted to such a degree that hospitals hired professional donors, who, on at least one occasion, threatened to strike. But the ideals of transfusion still burned in Algeria, where a group of French military doctors created a service for the army. Their director, Dr. Edmond Benhamou, was a World War I veteran who had performed arm-to-arm transfusions at Verdun, and had been operating a clandestine blood bank in the basement of the Mustapha Hospital. "They made me 'Dictator of Blood' [but] I was to be a dictator without a dictatorship," he said in a speech after the war. "[We] had no bottles, no glass ampules, no needles, no rubber tubing, no blood donors and no transfusionists." But he did have his government's unequivocal support, which put the resources of French Africa at his command. When no stainless steel could be found in Algiers, he commissioned the city's jewelers to forge needles of pure silver; when he could not find gum rubber for tubing, the government froze all the rubber-carrying ships leaving the port of Dakar, Senegal, nearly two thousand miles away, and commandeered the cargo.

The French designed their new system as an autonomous body with its own processing laboratories, vehicles, and propaganda specialists to encourage donation. A central facility, or "Mother House," was established in Algiers. There doctors gathered and bottled blood and plasma and shipped it to a "Moving Wing" traveling with the French Expeditionary Corps fighting in Africa, Corsica, and southern France. The Mother House gave birth to "daughter" facilities in Tunisia and Morocco that also fed blood to the Moving Wings. The daughter facility in Fez, Morocco, was directed by Dr. Jean Julliard, who later became the surgeon general of France. Isolated in the mountainous town, Julliard developed innovations with materials at hand, including a system to force-filter plasma through layers of cotton (he used elevator motors to power the compressors) and a "donation chamber" that became popular throughout France. Donors would sit outside these sterile chambers, with their arms protruding in through holes. Inside, white-clad technicians would bend over the disembodied arms, drawing blood under germ-free conditions. The donors, accompanied by soothing attendants, saw nothing; they sensed only the prick of a silver needle as it slid into their arm.

The French doctors called their organization ORT, for Organisme de Réanimation-Transfusion, or the Transfusion-Rehabilitation System, and saw it not just as medical technology but as something more transcendent and meaningful. To them it represented a philosophy of medical care, embodying all that was both modern and humane, especially in contrast to the values of the fascist enemy. Blood donation was benevolent, voluntary, and welcomed from all, French and Arab alike. Blood thus became more than a pharmaceutical; it symbolized a new social contract, bringing fulfillment to both donor and recipient. Years later, when the AIDS virus would taint the French blood supply, the public would wonder why their transfusionists had reacted so slowly, why they continued to accept blood from populations with high levels of AIDS, such as prisoners, long after others had stopped. Only the most thoughtful observers would realize that the historical French reverence for blood—for its symbolism as an uplifting social contract—would blind many doctors to the virus that tainted it.

With such problems unforeseen at the time, the doctors of the ORT spread their philosophy with missionary zeal. A small plane had been placed at Benhamou's disposal, and he flew from one North African outpost to the next, talking in hospitals, cinemas, wherever a crowd could gather, "explaining the purpose of our mission and why we must have blood and more blood."

There was one brilliant transfusionist in France whose talent went unused throughout the war. Alexis Carrel, winner of the Nobel Prize and pioneer in the first surgical transfusions, had retired from the

Rockefeller Institute and returned to live in quiet solitude on an island off Brittany. But Carrel was not destined for a peaceful retirement, or for a permanent reconciliation with France. After the French government fell, he reported to Paris, badgering the new leaders to put his experience to use. An essentially apolitical man, he was not discouraged by the Vichy government's politics: Frenchmen were dying and he wanted to help. For a while he worked to develop antishock agents. Later he established a nonpolitical Institute for the Study of Human Problems funded by the Vichy leader Henri Philippe Pétain—a fact that did not sit well with the Resistance, whose leaders spread the word that Carrel was a Nazi sympathizer. The charges were false, as could have been seen in Carrel's writings. Yet, in those dramatic and polarized times, the mere fact that he accepted Vichy support made him a collaborator.

Carrel did not live well during the war years. He and his wife refused to accept firewood from the Vichy government or a greater-than-normal allotment of food. Cold and hungry, with his boat and car confiscated by the Germans, he spent his days in declining health and isolation. As the Nazis withdrew and the Free French moved in, the French press denounced him as a racist and Nazi apologist—charges that left him in abject despair. After liberation, the police repeatedly invaded his home "to insure that he had not fled the country." The final tragedy of this misunderstood man occurred on a November afternoon in 1944, when the government radio station reported that Carrel had fled France to avoid standing trial as a Nazi collaborator. In truth, earlier that day he had died of a heart attack.

Salisbury, England, is a town of uncommon serenity. Set amid verdant hills about eighty-five miles southwest of London and a few miles south of the megaliths of Stonehenge, it is a place of medieval squares and great Gothic churches. Outside the city, checkerboard patterns of meadows and fields evoke the image of a "green and pleasant land." But the region seemed otherwise in the spring of 1944. With jeeps and tanks carving up the countryside, billeted troops overrunning the towns, and planes and ships cluttering the airstrips and harbors, southern England had become a vast military depot, buried under millions of tons of weapons, thousands of tanks, and thousands of airplanes. Warehouses brimmed with uniforms, packs, K-rations, and helmets, and such domestic items as toothbrushes and toilet paper. Whole villages were evacuated as troops bearing a kaleidoscope of uniforms moved in and rehearsed their maneuvers. It was all in preparation for the largest amphibious assault in history—Operation Overlord, the Allied invasion of Normandy.

A medical stockpile had been growing as well, as the Medical Corps amassed tens of thousands of tons of supplies including bandages, plasma, albumin, morphine, surgical instruments, bedpans, oxygen tents, and X-ray machines—everything from complete setups for medical facilities large enough to accommodate a thousand beds, to hundreds of millions of finger-sized ampoules of penicillin. The items ranged from the mundane to the macabre: from tens of thousands of pairs of spectacles to thousands of artificial eyes in four sizes and five colors. These materials and more were stockpiled in sixteen sprawling depots throughout the United Kingdom. They moved in so many directions—shelved, reshelved, and ticked off on countless lists—that one supply officer described the preparations as an ongoing state of "organized confusion."

The man responsible for managing this materiel was Brigadier General Paul R. Hawley, chief surgeon for the European Theater of Operations. A hale, hearty, and forthright career man, he spent a year wrestling with the bureaucracy to ensure the swift and steady movement of supplies. Now, as the clock ticked down to the invasion of Normandy, he worried about a material that was *not* being forwarded from the United States. All the western Allies—the British, French, and Canadians—were shipping whole blood by now, and Hawley knew that, if the army wanted to avoid an appalling mortality rate, American soldiers must have access to the liquid. The surgeon general had made it clear that the Americans would fight the war on plasma and albumin, so, just as Churchill had done in North Africa, Hawley formulated a plan of his own. Starting in the summer of 1943, he convened a working group of doctors and consultants at the American medical base in Salisbury to find a way to supply whole blood to the soldiers.

The group, the Whole Blood Service Committee, represented a formidable array of talent and personalities. There was Elliott C. Cutler, tall, hawk-nosed, brilliant, and amiable, professor of surgery at Harvard Medical School and surgeon-in-chief of Peter Bent Brigham Hospital in Boston who, having volunteered for service, rose to the rank of brigadier general, the highest of any surgical consultant. There was Colonel James C. Kimbrough, a rough-hewn Kentuckian—"a great shouter," as a colleague described him—a urologist and career army man who, as director of professional services, coordinated the work of the consultants and the army. There was Lieutenant Colonel James B. Mason, another career army doctor, who oversaw the planning, and

Captain Robert C. Hardin, a hematologist from the University of Iowa, who would manage the blood bank if it ever took form. These men and several other consultants took on the formidable job of determining how much blood the army would need and how the medical service could locally provide it.

In order to estimate the need for supplies, military planners worked with a standard parameter of D+90—the amount of munitions, rations, and fuel needed to sustain the invasion for ninety days. Hawley's group performed a similar calculation with blood. Military intelligence told them to expect 1,875 casualties for each day of the invasion. The experience of the British in North Africa had taught them they should expect to provide blood in the ratio of one pint for every ten wounded men. Applying that ratio to the estimated casualty rate gave them a projected need of about two hundred pints a day.

Then came the question of how and where they could obtain the blood. Since the surgeon general had ruled out importing it from America, the doctors looked to the next-largest source—the million and a half GIs stationed in Britain. That number may seem large, but it shrinks rapidly when one considers what proportion of those men could reliably be expected to give. Most were scheduled to be shipped to the continent; of those who stayed behind to work in support units, only about 40 percent would have the type O blood that the blood bankers sought. That number had to be whittled again, since past experience showed that even among the most motivated donors only about two-thirds would regularly give. The committee thus peeled away one layer after another of unavailable, unqualified, or unwilling donors, and in the end projected that they could collect about 650 pints of blood a day—well above what they thought they would require.

If the committee members felt hopeful about what they might accomplish, one of them held his optimism in reserve. Elliott Cutler was a bright and generally diplomatic man whose standards and integrity nonetheless forced him to make starkly honest judgments. And his judgment at this point was that there were problems ahead. An operation of this size had never been attempted on foreign territory before. What would happen, he wondered, if, as the casualties mounted on the continent, the troops based in England were drawn off to replace them? That would leave fewer donors, perhaps too few to ensure the supply. Or what if the base hospitals in Britain were swamped by the wounded returning from Normandy? "I am worried what might happen in this theater if a big attack started and great quantities of wounded people were brought to this island and our blood bank was not working," he wrote in a memo. The others assured him that the collections could succeed, and hurried with the plan.

Through the fall and winter of 1943, the blood committee outlined the procedures and processed the paperwork to acquire the necessary equipment. They envisaged a system similar to that of the British: Bleeding teams with refrigerated trucks would collect blood from soldiers at various military bases and return it for processing to the home depot, from where it would be flown and shipped to advanced blood depots in France. Once the blood had arrived on the continent, refrigerated trucks would take it on "milk runs" to the evacuation and field hospitals closer to the front. The only difference was that American blood would not be handled by a separate branch of the medical corps, but would be shipped along with other items of medical supply.

It was now January 1944, five or six months before the expected invasion (the actual date was a tightly held secret), and the pace quickened in training and supply. The Americans fitted big-volumed LSTs ("Landing Ship, Tank") with surgical theaters and bunks to carry the wounded back to England. Cutler and a colleague visited the ships and mapped out routes for the stretcher bearers to follow so they could effectively negotiate the narrow doorways, sharp turns, and steep ladders. They also designated cold rooms for the blood. Hawley and his staff made elaborate plans to bring back the wounded aboard ships and airplanes, treat them at field stations right at the docks and airstrips, and then speed them back to any of seventy-nine hospitals located throughout England. The most gravely wounded patients would travel directly to a "southern belt" of hospitals; those with less critical injuries would go to more distant facilities, in East Anglia. The whole system would have to be "fluid," commanded Hawley: It was essential that patients should not be immobilized in these hospitals and beds thus "frozen." Cutler captured the urgency of those weeks in a hurried diary entry: "Catching up; things moving. Hospitals arriving daily. Hard to keep up with the work."

Work on the blood bank continued apace, although bottlenecks developed. In one instance, crucial shipments of trucks and refrigerators never arrived: Having been duly processed and cleared, they sat somewhere on a dock in New York, overlooked in the preinvasion stockpiling. Hawley's group begged a few dozen trucks from an ordnance unit and borrowed bottles, tubes, and other equipment from the British. They kept moving forward. By early spring, having received, swapped, or borrowed most of what they needed, the blood bankers felt they were well on their way. They had not yet started collecting the blood—that would come later, given the liquid's limited life span. Then, in late March, a medical officer arrived with distressing news about blood use in North Africa and Italy. In the early days of the campaign, he reported, one pint of whole blood had been used for each eight casualties; now, with more patients needing longer-term care, the consumption of whole blood had quintupled. Then came the disturbing report from the commander of American forces in England that the number of donors had fallen far short of expectations. The committee met to reassess their projections. Accounting for the fact that more blood would be needed yet less would be available, they determined that, instead of supporting troops for D-Day plus ninety days, the blood bank could only help them for D+45. It was an "alarming reduction," as Mason later described it. The committee recommended "that immediate consideration be given to acquisition of whole blood . . . from the U.S." Yet Hawley knew that Kirk would not bend. The panel grimly lowered their preparedness estimates, trained more personnel, and began collecting blood.

It was now just over a month before D-Day, and the Allied forces seemed a coiled spring waiting to be released. Some three million men had been assembled in southern England, with six million tons of equipment to support them. Thousands of ships bobbed in the harbors, packed with tanks and amphibious vehicles. Orders were sealed, communications monitored. With the invasion fully planned and stocked, secrecy and deception became the order of the day. During these final weeks, a security breach occurred. An article in *Time* magazine stated that the opening of the blood bank would signify that the invasion would begin in less than three weeks. Hawley hit the roof. In a memo entitled "Subject: Violation of Security," he demanded to know who leaked the information. Hardin, who was managing the blood bank, sheepishly explained that, when the reporter had visited the facility several months earlier and asked how long the blood could be stored, Hardin had told him about twenty-one days—an innocent but compromising reply. The reporter, making an "uncomfortably correct" conjecture, linked the blood collections to the invasion. Hardin was ordered to ameliorate the damage. So, just as the military had set up a phantom Army as a feint from the actual site of the invasion, Hardin deployed phantom transfusion teams, sending them "hither and yon" to disguise the collection date. Pretending to take blood behind ostentatiously locked doors, they shuffled the bottles in such a way that "there was no evidence . . . of how little blood was actually being collected." Finally, after a week of phony collections, the general summoned Hardin to headquarters: "I was taken to the middle of a large room . . . and the date of D-day was whispered in my ear. . . . I was told that this was a planning date and that the actual invasion would occur within a 48-hour span of this date." The collections began.

D-Day approached and the inventories grew, although their shelf life and size were known only to select officers. On June 2, Hawley addressed his staff: "I don't know when D-day is, and if I did I couldn't tell you anyhow. But it is logical to assume that it is not too far off now; if we have anything left undone, the time is falling short. We must be ready to go." Hundreds of tons of medical supplies had been loaded, everything from penicillin to portable hospitals. Plasma and albumin, with their long shelf life, had already been lifted aboard. Finally, Hardin gave the order to load up the blood bottles, iced and packed in insulated cans.

On the morning of June 6, the largest amphibious landing force ever assembled threw itself against Hitler's Atlantic Wall. The carnage was terrific as the men ran full-face into German fire. As morning gave way to afternoon and the invaders inexorably moved up the beachhead, they witnessed a new but increasingly familiar tableau: medics kneeling over the wounded, holding aloft bottles for transfusion. In the heat of the battle, surgeons did not deliberate over the merits of plasma, whole blood, or albumin, but grabbed what they had on hand. They used 250 bottles of whole blood that day—a sizable quantity, but actually somewhat less than expected.

Back on the English side, Cutler and Hawley raced from one hospital to the next, from Winchester to Netley to Stockbridge back to Salisbury, checking the surgical teams and supplies. "The continental invasion is on at last," Cutler wrote in his diary. "All are excited; too much so. . . . I wouldn't have believed it possible two weeks ago." As patients came in, he scrawled hurried case reports, noting any weak points in treatment and transport. Two weeks later, Cutler flew to Normandy. The fighting had moved off the beaches by now, and the army medical units at the beachhead were efficiently distributing transfusion fluids. One of the detachments had buried a refrigerator near the Omaha Beach airstrip to secure the incoming blood, and delivery trucks shuttled this way and that. "The tremendous demand for blood completely justifies the establishment of the blood bank," wrote Cutler, "and from reports and observations it is clear that we must have saved life. . . ." Yet he also renewed his concerns about supply. All indications were that the surgeons were using even more blood than their counterparts in Italy. "I could not judge from statistical data the exact relation between blood and plasma," wrote Cutler, "but I had the impression it was being used almost as frequently as plasma; that is, in the ratio of

1:1. . . ." Several days later, Hardin arrived and, after closely studying the progress of the wounded, concurred that they needed more blood than had been predicted.

Now the fighting was moving inland, into the high, impenetrable hedgerows of Normandy, where the combat proved long and bloody. As the American Third Army, led by General George S. Patton, swung into action, the army's need for whole blood increased to more than a thousand pints a day. Cutler knew that the blood-bank supply was shrinking. It was only a matter of time before the army ran short. In a staff meeting on July 28, he implored Hawley "to make one more try to obtain blood from the United States. . . ." Hawley doubted it would do any good. Cutler insisted: "We all believe in this. We think it can be gotten from the U.S. It should be given precedence."

Hawley looked at the men who surrounded him—Cutler, Kimbrough, Hardin, Mason. These were seasoned military doctors, even those who had entered the war as civilians, and they did not seem inclined to hastiness or hysteria. In a few short months, they had organized and constructed a blood system capable of meeting the needs of a modern army. If they now maintained that the blood was running out, that the blood bank they organized could no longer keep pace, then he had better believe them. He sighed deeply and said, "The Surgeon General is definitely opposed to it, but I am willing to put it up to him." He sent Kirk a radiogram: "Burden is being imposed that the ETO blood bank cannot meet in the demand for whole blood in forces fighting in France. That blood is necessary and is saving lives, all are convinced. It is believed necessary that daily air shipments of 1,000 pints be sent. . . ." A few days later, he wrote Kirk a letter, and sent his medical officers to Washington to make the case personally.

Kirk was a stubborn man, but certainly not a fool. In July, he had toured the combat zone in Italy and, seeing surgical theaters firsthand, realized that plasma products could not fill the need. When he received Hawley's message, he convened a meeting to calculate how much blood could be diverted from the plasma program and how quickly it could be sent. He wired Hawley that he could soon begin shipping about 250 pints a day, with larger quantities to follow. Then Cutler arrived, and Kirk agreed to scale up the shipments to a thousand pints daily by September. "The Surgeon General openly expressed the opinion," wrote Cutler in a letter to Hawley, ". . . that if the surgeons of the E.T.O. wish for blood they should have it. . . ." It was now late August 1944. At virtually the same time that the Allied armies marched into Paris, to the embraces and cheers of a jubilant populace, the first plane bearing blood from America touched down on European soil.

By now all the Allies were pouring blood into the battle. Arnault Tzanck, the great French transfusion pioneer, had returned to Paris from exile in Chile and established a National Center of Blood Transfusion, which came to be known as "ORT 2." Filled with enthusiasm and patriotic fervor, he established national "Days of Blood," in which masses of civilians turned out to donate for the wounded, and the "Road of Blood," a route of train, ambulance, and jeep transport to carry French blood to the forward hospitals. The exiled doctors of the Polish Faculty of Medicine had set up an Institute of Blood Transfusion in Edinburgh, and with Britain's help air-dropped a thousand bottles of blood into Warsaw during the city's uprising. The British, in addition to their sophisticated depots and collection centers, had constructed two plasma freeze-drying plants, and were flying tons of the material to the continent. The Soviets had their "tank cars" of plasma and anti-shock liquids; the Canadians, Australians, and New Zealanders had blood and plasma—a world in which transfusion was a novelty just a few years before now seemed awash with blood and its derivatives.

In the fall of 1944, the American Red Cross began to collect blood for the Pacific, the most distant front for the Americans in the war. This would be the ultimate challenge, to ship whole blood to the other side of the world, where primitive conditions and withering heat would threaten its usefulness. Responding to the surgeon general's request, the Red Cross began collecting type O blood at its centers in San Francisco, Oakland, and Los Angeles, later expanding the network to include Portland, Chicago, and, finally, the East Coast. The bottles were gathered at the naval depot in Oakland, packed into insulated plywood boxes, and then air-shipped to Hawaii and then to the naval station in Guam, where they were re-iced and forwarded to the isles throughout the South and Central Pacific. The transit time proved a rapid five to seven days. Meanwhile, the episodic nature of the fighting, combined with the lag time in gearing up collections, caused disorder on the home front, as the military submitted wildly changing quotas.

Nor was the confusion limited to home. Even though the blood traveled forward efficiently, the military, by an oversight, had sent the authorization and instructions for its proper handling and use via ordinary surface mail. The blood arrived weeks before the paperwork, leaving the doctors who received it irritated and confused. Such was the case in the invasion of Leyte, a rugged tropical isle that the military saw as the gateway to the Philippines. (The naval engagement preceding the attack saw the first appearance of kamikaze or suicide planes, whose toll in shock and burn patients made plasma and albumin indispensable.) The soldiers who participated in the invasion went ashore with plasma and albumin; they also had small amounts of blood collected in Australia and New Guinea—which they used sparingly, because it tended to be tainted with malaria. When large quantities of clean blood arrived, however, they literally did not know how to use it. Uninformed about its proper handling, they removed the bottles from their insulated containers, tossed them in the back of ordinary supply trucks, and drove them several hours over rugged roads in tropical heat. The army blood chief, Kendrick, who arrived to supervise the blood use, was appalled. "No command in either the Central or Southwest Pacific had been advised officially of the whole blood program by either Army or Navy sources," he wrote to his superiors. Indeed, since Kendrick himself lacked the necessary paperwork, the doctors he encountered often resented him. "[I] was told at one installation, where confusion was rampant, that it was not necessary for a War Department representative to come out and tell them how to run their transfusion service." At one point, his "activities were restricted, and [I] was prohibited from interfering . . . on the ground that officers in charge of the program were competent to handle it."

Not all the new arrivals met with hostility. Navy Lieutenant Henry R. Blake, working with the Red Cross, came to the island with the first shipment of blood from San Francisco. Walking through the tent flap of a medical field station, Blake encountered a young soldier on a cot bleeding from a severed leg artery. Doctors had sustained him on plasma for a time but, lacking red cells, could see they were losing him. Nearby lay a comrade who obviously had given blood once too often, yet insisted on attempting to donate again, perhaps at the cost of his own life. The surgeon was wondering what he should do when Blake lumbered in with his crate of bottled blood, which he proceeded to administer to several in the tent. Such cases would have been more common on Leyte if not for the mishandling and lack of information. As it was, about half the blood that had been shipped to Leyte was spoiled.

Kendrick continued to meet with the area's medical officers, indoctrinating them not only on the technical points of handling fresh blood but on the astonishing proposition that they could have a virtually unlimited supply, provided they send word on what they needed. The result was a more efficient operation on Luzon, the big northern island of the Philippines on which Manila is located. It became common on Luzon to transfuse a wounded man with up to six pints of blood— about half his body's normal complement.

The first successful widespread use of blood and its multiple derivatives during the Pacific war took place during the Battle of Iwo Jima, the most costly invasion in American military history. The volcanic island topped by Mount Suribachi represented a stepping-stone in the conquest of Japan. Less than eight hundred miles off the coast, it could serve as a support base for B-29 Superfortresses bombing Tokyo. The Japanese understood the place's strategic importance and reinforced it with more than twenty thousand troops installed in a heavily fortified network of tunnels.

The preinvasion planning had gone on for months. Each Marine was backed up by 1,322 pounds of materiel. The medical supplies included two white hospital ships to be anchored off the coast, four big-bellied LSTs outfitted as evacuation hospitals, several tented field hospitals, and a field depot complete with thirty days' worth of medical supplies including plasma, albumin, and refrigerated whole blood. The troops were inoculated against typhus, scrub typhus, and plague; their uniforms were dusted with DDT. Pharmacists' mates carried seabags stuffed with medical equipment, including battle dressings, morphine, sulfa, surgical knives and scissors, plasma, and albumin. It was not without reason that they made such preparations: Military planners predicted a casualty rate nearly twenty times higher than during the invasion of Normandy.

And then it began. Four medical parties landed in the first wave, quickly setting up aid stations in shell craters. A pharmacist's mate was shot in the jaw before he could finish descending from a landing craft. A group stashed their seabags on the beach as they scrambled forward to establish an aid station. By the time they returned, a few minutes later, a shell burst had pulverized their equipment. All the boxes of dried plasma were smashed. Looking out to sea for the boat that was carrying their other equipment, they saw it hit on the way in.

With casualties running to a thousand a day, the medics rose to heroic levels of performance. As bullets flew around one corpsman, he crawled to a soldier who was hemorrhaging from a bullet wound in the neck. Dragging the wounded man back a few yards, he buried the lower part of the soldier's body in the sand so it would offer a smaller target, located a vein in the exposed part of the body, and administered a transfusion under fire. (Despite his efforts, the soldier later died.) Vulnerable to enemy gunfire as they crept forward to treat the wounded, the medics suffered unspeakable casualty rates, in some battalions exceeding 60 percent.

As the Marines inched inland, the Medical Corps set up a chain of aid stations to pass the wounded from the fighting to the rear. The hospital ships *Samaritan* and *Solace* were anchored twenty miles away. Each dawn, one or both would steam to within two miles of shore to receive the wounded from amphibious vehicles. Meanwhile, blood products moved in a countercurrent toward the front. Whole blood, now carried in portable refrigerators, traveled as far forward as the clearing stations; dried plasma and albumin moved farther, carried by the medics right up to the fighting. Never before had blood and its products been so completely utilized in a battle, a fact that immeasurably heartened the soldiers. "The most precious cargo on this island of agony," was how Marine Corps correspondent Ralph W. Myers described the materials.

> The primary item in each Medical Corps man's front-line survival kit, the arrival of whole blood . . . brought a new smile to the lips and bloodshot eyes of overworked Navy doctors in the forward areas. . . .
>
> When the front line was the edge of the beach the whole blood was right behind the troops, in hospital ships and the sick bays of dozens of LSTs and assault transports to which the wounded were evacuated. . . .
>
> Then as the fighting inched northward, small portable containers of the invaluable liquid moved right up to the company medical stations, a matter of minutes behind the shell-shattered No Man's Land to the north of Motoyama airfield No. 2.
>
> The first supply rolled into the Third Marine Division's Medical Company B station on a huge truck, [carried in] two big ice "reefers" [refrigerators]. It had a lot of takers—two tents of men lying and crouching behind sandbags under a pair of big tarpaulins. . . .
>
> In a few minutes the wine-colored bottles were held aloft above the prone bodies and the stuff was doing its miraculous work.

As the weeks passed and more blood and plasma were loaded ashore, it became clear that they gave soldiers a psychological boost— not only because they increased the men's chances of survival, but because those fighting knew the enemy had virtually none. Even the cosmetic effects of blood proved uplifting: Soldiers who were used to seeing their wounded comrades look death-white, even with plasma, now beheld a rosier glow.

Pharmacist's mate John H. Willis was clambering over the southwest slope of Mount Suribachi, attending to the wounded with bandages and plasma, when a piece of shrapnel caught him in the shoulder. After allowing himself to be patched up at an aid station, he quickly made his way back to the front. Halfway up the mountain, he saw another Marine in a bomb crater, hemorrhaging. Willis, jamming a rifle bayonet-first into the ground, hung a bottle of plasma on the trigger guard, inserted the needle, and initiated the flow. With one hand he steadied the transfusion tube as eight grenades in succession landed in the crater; he rapidly tossed them out with the other. The ninth grenade exploded before he could release it, his free hand still grappling with the tube. He was posthumously awarded the Medal of Honor.

Soon the Marines would surmount the dead volcano and be immortalized by one of the most famous photographs of the war—four men straining to raise the American flag—which, cast as a sculpture, now stands as a monument in Arlington National Cemetery. Yet this was but one of many heroic spectacles, not only on Iwo Jima but throughout the Pacific. Carl Mydans, a photographer for *Life* magazine, wrote of the iconlike image of "combat medics in bouncing jeeps," who, holding on with one hand, "raised the other high, as one would a torch, holding a bottle of plasma, pouring life back into a broken body. I think I have never seen a soldier kneeling thus who was not in some way shrouded with a godlike grace and who did not seem sculptured and destined for immortality."

Marine Corps Sergeant Myers described the gritty endurance of corpsmen lying on their backs in foxholes, "holding plasma bottles over their heads until their arms grow numb. . . . They sit on careening jeeps and trucks, holding their bottles aloft while wounded leathernecks fight their individual battles for life en route to the rear areas. And there are more bottles of blood, whole and plasma, lashed to poles and upended rifle butts at the grim wayside stations along the beaches."

Kendrick, the army's transfusion chief, described another scene on an unnamed island in the Pacific, in which a wounded man lay in a depression directly under the line of fire of a Japanese machine gun. "To leave him without treatment would have meant his going into irreversible shock. To move him would have meant certain casualties for the litter squad. A staff sergeant, who was later awarded the Silver Star for bravery, crawled out to him, dressed his wounds, splinted a fracture, and then administered three units of plasma to him by lying by his side and elevating the bottle of plasma with one hand."

Something symbolic was happening to blood, although it may not have been clear to the medics at the time. Even as they worked, bullets and shrapnel zipping around them, they were taking part in a transmutation as elemental as any that had come in the centuries before. In every epoch, people had seen blood as more than it really was—more than a liquid or component of their body, but as a reflection of their image and understanding of themselves. To the ancients, powerless in a frightening world, blood became a symbol of the potency they desired. The Greeks, as we have seen, looked at blood as a reflection of the universal order, one of the four bodily humors that so perfectly mirrored the balance of nature that an excess of the substance was presumed to cause illness. Their Christian descendants imbued blood with spirit, infusing it with the qualities of the creatures in which it flowed, from the noblest of men to the most gentle of calves. Medicine changed drastically in the late nineteenth century, shifting from humors to germ theory; yet, despite its increasingly practical use, blood retained its mystical resonance. The early Soviets saw blood as the expression of the collective; the Spaniards, as a covenant between a people and their cause; and the Nazis, perversely, as the literal and figurative embodiment of racial purity.

Now blood would reach another level of allegory, forged in history's most murderous conflict, at once both uplifting and practical. For the Americans were imbuing blood with their own kind of metaphor—not of the magic or potency from above, but of their *own* power, inventiveness, and humanity. Theirs was a typically practical symbology—not the lofty social contract of the French, but a pride in their competence to develop and use a resource. After all, the Americans processed blood into more products than anyone else—as blood, plasma, and albumin, and as the host of pharmaceuticals that emerged from Cohn's lab. Indeed, they had used blood in such a large quantity and in so many forms that strange little coincidences began to occur. In a hospital in Paris, a Corporal Kenneth L. Johnson received plasma that turned out to have been donated by General Dwight D. Eisenhower. "The Corporal Has the Blood of a General," read the caption of the newspaper photograph. In the Pacific, a wounded navy gunner named Harry Starner was receiving some plasma when, glancing at the bottle, he saw his own name; he had donated blood while on leave in Washington, D.C. That infinitestimally small possibility demonstrated just how industrialized—how *mastered*—blood had become. It was the ultimate stage in the liquid's evolution. Blood, once held in religious regard, had been removed from a man's circulatory system, separated, processed, freeze-dried, packaged, shipped, and reconstituted—only to be injected into the same individual in a completely different form several months later and half a world away.

PART THREE

# BLOOD MONEY

# CHAPTER 9

# DR. NAITO

August 1945: The gray metal deck of the SS *Sturgis* heaved beneath the feet of Lieutenant Colonel Murray Sanders as it steamed into Yokohama Harbor in the defeated empire of Japan. Sanders was among the first group of Americans to arrive in Japan, after the atomic attacks, on a reconnaissance mission which only he could conduct. A bacteriologist by training, he had served in a clandestine research facility at Camp Detrick, Maryland, which had been established in response to reports that the enemy was trying to develop agents of biological warfare. Rumors had surfaced that the Japanese were developing weapons that carried bacteria and viruses. Documents found on the bodies of Japanese soldiers hinted at the existence of an "Epidemic Prevention Unit" or "Water Purification Unit," one of whose missions was to spread epidemics by infecting the enemy's water supplies. Japan was even said to have developed a "plague bomb." Sanders's mission was to investigate these capabilities, preferably before the Soviets discovered them.

In briefing Sanders, American intelligence had given him a handful of photographs of several Japanese scientists thought to have participated in the project. One in particular was called to his attention: Dr. Ryoichi Naito—medical doctor, bacteriologist, and the only man in the group who spoke English. Sanders had no idea how to find this informant or what he would say when he finally made contact.

The ship groaned to a halt at the dockside, giving Sanders a moment to look out over the city. Yokohama had been spared from atomic destruction, but one never would have known that from the city's appearance. Once the muscular heart of Japan, with shipping, manufacturing, munitions, and steel, Yokohama, like Tokyo, had been fire-bombed, and now existed as a landscape of ruin—buildings reduced to piles of gray rubble, survivors staggering in deprivation and shock.

Sanders turned and descended the gangplank. A stranger walked up to him—a wiry Japanese man with glasses, thin lips, and straight hair combed back to the right. He carried a photograph of the American (Sanders never learned how he got it, but suspected he may have obtained it from American intelligence). He introduced himself as Dr. Ryoichi Naito, and offered to serve as Sanders's interpreter.

So began one of the strangest episodes in postwar Japan, and one that would influence the nation for decades: For while Naito would inform the Americans about the existence of the biological-warfare experiments, he would initiate a cover-up that would mask their true horror for another forty years. He would shape the nation in other ways as well. In closing out one chapter of Japanese history, Naito was among those who would initiate another—an era in which Japan rebuilt itself into a peaceful modern industrial democracy. In his case, the contrast was especially dramatic. Having spent the war designing heinous biological weapons, he would eventually become a benefactor to millions, as one of the world's foremost collectors and processors of blood.

The story of how Naito arrived at a position to accomplish such evil and good simultaneously is itself a tale of a complex personality, a gifted but temperamental individual who tackled all tasks with a kind of ferocity. A poor but brilliant young man from Kyoto, the young Naito was a spindly youth with a volatile temper and a fierce intellect. Enamored of both languages and science, he learned German and Esperanto, an international language with a following at the time, and read an Esperanto version of Marx's *Communist Manifesto,* even though he was majoring in engineering. Soon after hearing his first radio transmission, he built a receiver that students and professors alike came to hear. When a visiting British professor remarked that the boy did not understand English very well, Naito hurled himself into mastering that language by purchasing a used copy of *Around the World in Eighty Days* and memorizing it. He was an unusually determined young man.

Naito enrolled in medical school at Kyoto University—to support himself, he taught Esperanto and worked as a paid pallbearer—where he came to the attention of an army colonel named Tomosado Masuda. Masuda recognized Naito's brilliance and drive, and tried to persuade him to enlist in the army. These were heady times for Japan—in the years preceding World War II the country was amassing an economic and military strength that would allow it to shrug off the Western imperialists—and the place to be, said Masuda, was the army. There a young man like Naito would find steady work, steady pay, honor, respect, and the chance to serve the nation in its ascendancy. He would also find unparalleled challenges. Masuda explained that he was working with a scientist named Shiro Ishii to form a bacteriological unit. Chemicals had shown their usefulness in the last war, despite the combatants' later revulsion. Ishii and Masuda both felt that the next war would rely on an even more sophisticated arsenal—the harnessing of the power of microbes. They were just now preparing the groundwork, and offered to include Naito in this exciting new development.

Naito enrolled in the Army Medical School, served for a time in Manchuria, and then returned to Kyoto University to manage a microorganism facility doing military research. In 1937, the army sent him to Berlin, ostensibly to study at the Robert Koch Institute; by night, however, after the other scholars had departed, he sifted through the garbage to gather intelligence about German research.

The next year, he traveled to the United States on a similar intelligence mission. There, according to American reports, he unsuccessfully tried to obtain a vial of yellow-fever virus from Rockefeller University. Then he moved to the University of Pennsylvania, where he observed the freeze-drying developed by Sharp & Dohme. The possibilities for the technology fascinated him. On the one hand, it opened the way to preserve plasma, which he knew, even then, would save countless lives; on the other, it fit perfectly with the Japanese army's plan to develop biological weapons. Freeze-drying micro-organisms would make them potentially viable for years, easy to be delivered in a bomb. What a weapon such technology could produce! He must bring home one of the American vacuum pumps, a key part of the freeze-drying process. Lacking the budget, he used his food money for the purchase, living on canned food and, later, scraps he found in the garbage. Finally, after six months in the United States, he returned home with the vacuum pump in hand.

Back at the Army Medical College in Tokyo, Naito's research followed two divergent tracks, as different as the conflicts in his own personality—one unmistakably evil, and another that led to ultimate good. He worked in the epidemic-prevention department, a high-security laboratory behind a double barbed-wire fence, laboring to isolate poisons that could be freeze-dried and placed in a bomb, such as the fugu toxin found in the livers of blowfish. He also gave the freeze-drying technology to his mentor Shiro Ishii, who headed a germ-warfare research facility near the city of Harbin, in Manchuria. Ishii cultivated a variety of pathogens, including plague, anthrax, dysentery, and typhus—at one time he had grown a large enough quantity to kill the world's population several times over—and developed ways to deliver the pathogens in food, water, aerosols, and bombs. But what made Ishii's laboratory notorious was not so much the weapons he produced as the unusual methods he employed to develop them: He tested his pathogens on thousands of captured soldiers and civilians.

Harbin has been described as a Paris of the East, with its broad boulevards and old European buildings. But the facility the Japanese built near the city's outskirts can only be described as the embodiment of hell. Sprawling over three square kilometers, Unit 731, as it was called, comprised 150 buildings, including laboratories, insectariums, an animal house, even amenities such as sports facilities and a religious shrine. At the center of the compound stood a square enclosure known as the *ro* block (the Japanese character *ro* is shaped like a square), and inside the *ro* block were two buildings that Ishii referred to as his "secret of secrets." There he maintained a stock of human guinea pigs. Gleaned from captured territory as the Japanese army devoured Manchuria, these people included resistance fighters and civilians, Chinese and Russians, women and children—shipped into the prison as "special consignments" or "new material." These victims were so deprived of their humanity that his soldiers referred to them as *maruta,* or "logs."

From before the war until the day after the A-bomb fell on Hiroshima, Ishii and his men subjected these people to some of the worst horrors of the war, rivaling the German concentration camps. They used the *maruta* to study frostbite and starvation, allowing the conditions to progress until death. They tested such new weapons as flamethrowers on the prisoners—and new forms of transfusion, such as emptying their circulatory systems and refilling them with horse blood. Most important, however, was the use of these people for the development of microbial weapons. Each time the researchers devised a new way to spread lethal pathogens, they would infect the prisoners with injections or aerosols, with contaminated feathers brushed under their noses or bacteria poured into their food, and then observe the progress of the disease. In order to test plague bombs, they would march the prisoners outside the compound, clad them with metal shielding, and detonate the explosive about twenty yards away. After the contamination, they would march the prisoners back to their cells and take careful notes on the progression of the disease. No one escaped the experiments alive. Those who survived the first contamination would then be exposed to a second, and so on, until they died. Alternatively, Ishii's staff would dissect the subjects of the experiment—whether or not the prisoners were still alive. Thousands of prisoners are thought to have died.

Ishii's activities apparently took place with Naito's scientific and administrative support. After all, Naito supplied the freeze-drying technology that made it possible to load pathogens into bombs, and officially visited the camp more than once. A veteran of the era described Naito as "the very strong right arm of Dr. Ishii . . . the person who made Ishii's work possible." Moreover, as we shall see, it was Naito who orchestrated the cover-up of Ishii's activities, cloaking his ghastly secret and shielding him from prosecution after the war.

Meanwhile, Naito also threw himself into a more benign use of freeze-drying technology. Before the war, almost no one in Japan had given a thought to transfusion, although some had experimented with donors-on-the-hoof. Minds began to change, however, in 1942, when the Swiss Red Cross delivered packages of dried American plasma to the prisoner-of-war camp at Yokohama. With War Ministry approval, Naito constructed a plasma-drying facility at the Army Medical School, with seven collection centers throughout the city. The donor force consisted entirely of women, since so many men were involved in the fighting. In his memoirs, Naito nostalgically recalled the donor centers: "The presence of so many young women gave the centers the atmosphere of a flower garden," he wrote. Naito estimated that he produced three tons of dried plasma in the course of the war, although he never had the opportunity to use it, since most of the shipments were destroyed by the Americans. Naito wrote that Japan's use of dried plasma barely reached "one two-thousandth" of the Americans'. Eventually he had to abandon the endeavor, since so many firebombs were falling on Tokyo that people could not safely line up to give blood.

During the war years, Naito displayed a characteristically fierce attitude toward whatever task had been placed before him. He demanded that his staff work seven days a week, and pushed them mercilessly to achieve results—they referred to him as the "Typhoon." No sooner had a bacteria been isolated than he demanded to see a vaccine; no sooner had his men produced a vaccine than he demanded to set up the means of mass production. "His mind worked so quickly that other people could not follow him," a colleague later remembered. Always temperamental, under stress he exhibited bouts of explosive rage, punching subordinates and shattering glass doors. Sometimes, in an effort to control himself, he would sketch a clenched fist and cross it out. He drifted through the laboratories pale as a ghost, grizzled and unshaven, with a distant, unfocused gaze from lack of sleep.

Now, as the war raced to its conclusion, Ishii and Naito threw themselves into a frenzy of activity, each in his own way and in his own domain. When Ishii heard that the first A-bomb had fallen, he knew that he must leave no evidence of his crimes, and ordered his soldiers to sanitize the area. For three days his soldiers dynamited the buildings, gassed the surviving *maruta,* machine-gunned the local laborers, and released hordes of infected fleas and rats. Then they fled south, passing through China and Korea on a special train, and crossing the Korea Strait to Japan. Ishii went to the Wakamutsi Hotel in Tokyo, opposite the Tokyo Military Hospital. There he hurriedly disposed of the specimen jars of *maruta* organs that had been shipped there for study. Then he slipped away.

Naito acted quickly as well, although in a less ignominious way. When the bombing of Tokyo had intensified, Naito moved his research facility to Nigata, a small, quiet city on Japan's eastern coast, where the staff continued to manufacture serum and vaccines. (Soon afterward the Tokyo headquarters was demolished in an American air attack.) Immediately after the Japanese surrender, he addressed hundreds of technicians in Nigata who had gathered to hear him. It was a different Naito—no longer arrogant, humbled by the trauma of losing the war. He spoke simply and modestly: "Japan is defeated but fortunately will not be divided like Germany. I want to keep this group together. Please do not worry. Please make every effort for the rebuilding of Japan."

At this point the chronology of his life becomes foggy, like so much in immediately postwar Japan. It is known that someone—American intelligence, perhaps—requested that he make his way to Yokohama and assist Murray Sanders.

Naito seemed the ideal informant—humble and obedient, contrite and efficient, briefing Sanders by day and going home to his family at the end of each day. In reality, however, he was serving other masters. Each night, instead of returning home, he would travel to the suburbs, where Ishii and his associates remained in hiding. There he would recount his conversations with the American. Together the group would analyze Sanders's questions and decide which information Naito could reveal. In this way he did more than respond to Sanders's inquiries; he subtly steered them.

After several weeks of this manipulation, Sanders began to catch on. He increasingly pressured Naito to tell him more. Finally, when it became impossible to continue resisting his demands, Naito gave the American what he was looking for—a twelve-page handwritten document outlining the work of Unit 731, describing its organization and activities, and naming names. Written in flawed English, it portrayed Naito as a man who had decided to tell the truth from the deepest motives of conscience and morality. "The purpose of this information is only to rescue our poor, defeated nation, and to avoid the damage, according to your words, that if we offer the truce [truth] as a science, you may help this poor nation with your every effort. . . ." In several long sections of organizational charts and prose, Naito described the structure, function and key personnel of the biological-warfare units. At the end he wrote: "I ask you to understand that I am staking my life doing this information; I shall be killed if any one knows that I have done this information. My only hope is to rescue this poor, defeated nation."

As we now know, Naito had nothing to fear: The men he named in his document were the same ones he was effectively protecting. For, despite all the details he so carefully provided, he said nothing about human experimentation, omitting, in essence, the unit's core activity. Sanders, as thrilled as he was at having finally obtained the information he needed, probably harbored lingering suspicions, because at the end of Naito's document he hand-printed a notation: "I have asked Dr. Naito whether prisoners were ever used as experimental 'guinea pigs.' He 'vows' this has never been the case."

Naito had one more duty to perform after giving his "confession" to the Americans. He negotiated a promise of immunity for Ishii and the others in exchange for all their technical information. The men would never pay for their crimes against humanity.

At that point Naito disappeared. Having fulfilled his obligations of loyalty, he retreated from the avenues of power. He declined a position at Kyoto University, and opted for the life of a simple country doctor, setting up a practice in a small town near Osaka. He seemed to undergo a deeper change as well, converting to Roman Catholicism and abandoning his characteristic ferocity (although some would forever suspect his motives). No longer the unpredictable tyrant, he became known locally as Inginburei, which, roughly translated, means "He Who Is So Polite As to Be Almost Rude." He became a familiar sight in his working-class neighborhood, bicycling from one patient to the next. People would nod kindly at Inginburei, knowing that their doctor was a humble, devoted, and honorable man.

These were years of quiet struggle for the doctor, as they were for the rest of the nation. Japan, now a political and economic ruin, had become a public-health catastrophe as well, with conditions reminiscent of the previous century. Epidemics raged through the populace, fueled by poor sanitation, lack of medicine, and neglect. The Americans responded by inoculating millions. Then they set out to rehabilitate the medical infrastructure. American specialists built hospitals, introduced new medical-school curricula, and redesigned the archaic pharmaceutical industry. They established a National Hygienic Laboratory to assure the quality of pharmaceutical and medical supplies, and a National Institute of Health to coordinate Japan's often warring universities.

They also attempted to help modernize the blood supply. After the war, an improvised blood system had developed that had more in common with the black market than with a health system. Hospitals obtained blood through independent blood brokers who, for a commission, would supply donors-on-the-hoof for "bedside transfusions," as they were called in Japan. Aside from a cursory examination for blood type, hospitals did not test for disease. A terrible accident was bound to occur, and in 1948 a woman contracted syphilis at a government branch of the Tokyo University Hospital. The incident triggered a national scandal. When the news reached the occupying Americans, they instructed the Japanese Ministry of Health to take steps to provide for a safe national blood supply.

Naito too had been thinking about blood. He never forgot how close he had come to producing transfusion fluids during the war. Now working among the poor, he saw how desperately they needed transfusions, and how all too often they could not receive them. He began to make plans. At night, while his family was sleeping, he would formulate solutions to his nation's chronic shortages. He did not envision the kind of nonprofit enterprise the Japanese Red Cross was planning to establish. He favored a corporate entity, after the model of American companies the postwar Japanese had come to admire—a national chain of blood-and-plasma banks. He tuned and tinkered obsessively with the plan, eagerly awaiting the opportunity to use it.

Naito and his fellow doctors in Japan were not alone in seeing the need for a national blood supply. The war had demonstrated the importance of transfusions on the national scale. Those nations that had not already organized collections realized they must do so. But how should they proceed? Blood represented a new kind of resource, and nations were not sure how to manage it. Should donors be paid? Blood was, after all, a valuable raw material; it would seem only fair to compensate donors for their trouble and time. Yet blood was more than a mere commodity. It was a part of the human body; some doctors referred to it as a tissue or an organ, and felt that to pay for it would be obscene. In

1948, the International Red Cross adopted a resolution that "the principle of free blood be universally applied." But it wasn't that simple: In some countries, poor ones particularly, giving blood represented a genuine personal sacrifice. And so, as the world rebuilt itself from the war, each nation embarked on its own course, influenced by a combination of factors—medicine, philosophy, economics, and chance.

In Germany, which used blood very little during the war, blood banks became established haphazardly. One of the first was at the Marburg Surgical and Clinical Hospital, under the direction of an obstetrician named H. Schwalm. Schwalm began with little in the way of resources—Allied bombing raids had destroyed most of the building, and the de-Nazification process had left him critically short of staff. Yet Schwalm felt that a blood service could not wait. Using the only glassware available for storage—Coke bottles he had received from the Americans—he established a little blood bank in the basement. The hospital had no donor room, so he bled donors in the clinic's ambulance. Doctors in localities throughout West Germany acted under the same impulse, so that by the early 1950s more than forty hospitals had established blood banks. Most of them paid for donations with money and food.

At the same time that individual doctors were setting up blood banks, the German Red Cross tried to establish a system on a national scale. With commercial collectors already servicing most of the cities, the Red Cross focused on the rural parts of Germany, building processing centers well away from the major cities so they would remain safe in case of an atomic attack. From there, mobile collection vans would course through the countryside, stopping at factories and small towns. Thus a dual system evolved—paid clinical centers in the cities and the voluntary Red Cross stations in rural areas. Both did a reasonably effective job, but the tension between these two elements in the system would fester for decades.

In East Germany, as in many Communist states, transfusion became a state enterprise, in which blood was collected in factories or in the military and paid for with money, food rations, or vacation days. Despite the rewards, people saw the system as coercive—the ultimate invasion of privacy by the state—and when Communism fell, some years later, blood systems throughout Eastern Europe collapsed.

Poland, although dominated by the Soviet Union, resisted the demands of the state and allowed its Red Cross to manage the blood. This may have come about because of the influence of Ludwig Hirszfeld, the courageous doctor who had performed research on blood and the human races and resisted the Nazis' attempts to distort his work. Hirszfeld had remained indomitable throughout the war. Cast into the Warsaw Ghetto by the Germans (though he was a Catholic, his having been born a Jew was enough to condemn him), he secretly taught medical courses until he escaped with his wife and daughter. During the war he lost his daughter to leukemia. Now he presided over the Scientific Counsel of the Polish Red Cross. Under his guidance, Poland passed Europe's first law specifically regulating blood, outlawing its sale for profit. Hirszfeld became such a revered figure in Poland that after his death the Institute for Medical Microbiology in Wrocław was renamed in his honor.

In Switzerland, the Red Cross gained stewardship of blood partially because of a tax break. During the war the group had organized some arm-to-arm transfusions, and had even stored some blood in mineral-water bottles, but it really amounted to little more than a "shadow organization," according to Dr. Alfred Haessig, who later became director of the group's Central Laboratory. But as the Americans left Europe, they donated thirteen thousand units of dried plasma to the Swiss Federal Office of Hygiene. There was a pause while the office decided what to do, never having received such a gift before. Eventually it donated the powder to the Swiss Red Cross, because if given to a charity, the gift would be tax-free. The Red Cross distributed the plasma to doctors and hospitals, who responded so enthusiastically that the agency constructed an American-style plasma-drying plant. Later it set up another facility, in an immense underground cavern hollowed out of the Bernese Oberland mountains. By the early 1950s, the Red Cross had set up a sophisticated network of a central laboratory and regional transfusion centers. Swiss expertise and the purity of their product would eventually make them a world player in the blood business.

The Dutch, who had cleverly duped the German invaders, continued to astound with their courage and ingenuity. They faced a desperate situation by the end of the war, with their infrastructure wrecked and virtually no food. The British and Americans sent medical "starvation teams," restoring people who were too weak to eat with infusions of blood plasma mixed with proteins derived from milk and beef. Princess Juliana, who had been spirited to Canada, donated eight pints of blood to her citizens. Yet, despite the deprivation, Mastenbroek and his colleagues at the blood laboratory in Amsterdam moved ahead. They had read about the Cohn process in a popular magazine and studied the accompanying pictures, and within a few years of the cessation of hostilities they developed their own fractionation techniques. Their technology grew at such a rapid rate that by the late 1940s they were exporting know-how to other European nations.

The Canadians, who had supplied freeze-dried plasma to the British at Dunkirk, developed a blood system from the partnership of two respected and remarkable entities. Before the war, Dr. Charles Best, the codiscoverer of insulin at the University of Toronto, had formed a plasma freeze-drying facility at the university-owned Connaught Laboratories. In order to accumulate a supply, he arranged for the Canadian Red Cross to recruit donors. After the war, the alliance continued. The Red Cross took the role of Canada's blood manager, recruiting donors, maintaining supplies, and distributing blood and plasma products. Connaught became the nation's main plasma laboratory, churning out derivatives from source material provided by the Red Cross. Canadians pointed with pride to the alliance, a partnership of a humanitarianism and pharmaceutical expertise. Decades later, however, the partnership would crumble, resulting in an epidemic of blood-borne AIDS and hepatitis.

In Italy, an assortment of blood centers developed, some public, some private; some in hospitals, some in private clinics; some voluntary, some paid. One center was established by an American relief committee under the leadership of Dr. Max Strumia, the Italian immigrant who had become one of the foremost hematologists in America. (Strumia performed some of the early plasma freeze-drying research at Bryn Mawr College, near Philadelphia.) The Italian government attempted to consolidate these centers into a network, with all donations free and coordinated by the Red Cross. Despite their best intentions, blood shortages continued for several years. Meanwhile, a black market had developed in Rome that raised the price of blood to about $100 a liter.

Blood centers spread throughout the Third World as well. In China, doctors at the Peking Union Medical College set up a small blood bank and stocked it with material from people who had been paid for their blood. In Hong Kong, the local Red Cross established a program to aid thousands of refugees from the Chinese Revolution. In Thailand, members of the royal family inaugurated the nation's blood service by giving donations. Their blood was drawn with needles made of gold.

In India, the army maintained the wartime service established by the British, collecting from recruits and supplementing the supply with blood from paid donors. Dr. George W. G. Bird, a British physician who worked in India for many years, recalled that the army was often quite persuasive in recruiting volunteers:

When I used to go to army units I . . . sought the assistance of the key figure in the unit, the Subedar Major. With this man on one's side there was no problem at all. For example, a very fierce looking Subedar Major once paraded the whole unit, a very large one, and spoke to them on my behalf. It did not matter at all what he said during most of his speech. It was his last sentence, a blatant form of moral blackmail, which was totally successful. This sentence was simply this: "Now if there is any one of you who does not wish to give blood to save the life of his dying brother take one step forward." Understandably no one moved: every soldier was recruited.

More often, however, blood was difficult to obtain, because it played a heightened role in local superstition. Many Indians believed that giving blood sapped strength from the arm, and donors' arms sometimes went limp from psychosomatic distress. Some donors fell into a trance. Doctors knew never to mix the blood of Muslims and Hindus, or to give the blood of a Harijan (untouchable) to a member of a higher caste. Bird once attempted to transport a group of Indian policemen who had agreed to give blood, but had to give up when the men's wives lay down in front of the trucks. "I learned that it was a local superstition that blood letting led to sexual impotence and the women did not wish to have impotent husbands."

Africans saw blood as sacred, even magic. Many refused to have "their soul put into a bottle," according to an International Red Cross report. Tanganyikans imagined that the Red Cross workers who invaded their territory belonged to a secret society of blood drinkers, as one team from the Red Cross found out: "One night, an African passed by a tent laboratory where blood samples were being examined under a microscope; he saw on a table where the evening meal had been served either a bottle of Chianti or a bottle of tomato ketchup which has the same red colour as blood. The effect of this was the entire village, armed with spears and clubs, attacked the Red Cross medical camp which was obliged to retreat immediately."

Meanwhile, the great blood powers moved forward, adapting for civilian use the systems they had developed for the war. In Britain, the Medical Research Council, which so effectively managed blood during the bombing of London, ceded authority to the newly established National Blood Service, part of the Ministry of Health. The politics of transfusion were simple for the British: Having nationalized their health system after the war, they considered blood, like all medical care, a resource for the people, and placed its management in the hands of the government. William d'A. Maycock, the former army captain who had supervised the blood use at Dunkirk, presided over fourteen transfusion centers in England and Wales (Scotland developed an autonomous system). Each center collected, preserved, and distributed blood to subsidiary blood banks and hospitals in its region at a level sufficient for two weeks' consumption. Any blood that sat longer was returned to the center, which drained off the plasma and sent it to facilities near London to be dried or fractionated and returned to the region. Some of Britain's best scientists participated in the system, including Sir Allen Drury, who had presided over the Medical Research Council's blood services for much of the war, and Patrick Mollison, who had developed ACD, the standard blood preservative, who now directed the government's Blood Transfusion Research Unit. Citizens donated their blood without payment. The system seemed so effective and fair that it shone for many years as an international example. In time, however, territorial jealousies arose between the regional managers and those in London, which nearly pulled the system apart.

For sheer idealism surrounding blood, no country approached France, where blood donation, as we have seen, had become associated with the Resistance. Arnault Tzanck and his colleagues Marcel Bessis and Jean Dausset re-established a donor service at the Hôpital St.-Antoine, where thousands of citizens had given blood during the war. But conditions had changed. With no emergency to compel them, even the liberated French lost interest in giving. In the spring of 1949, however, a cruel but fortuitous accident occurred: An explosion at a small factory in Vincennes brutally burned three people. Transported to Tzanck's hospital in Paris, they survived only after massive plasma infusions. The mayor of Vincennes was so grateful for their recovery that he offered to lead his townspeople to Paris for a mass donation, but the hospital personnel offered instead to come to Vincennes. Arriving with a van loaded with transfusion equipment, nurses, and doctors, they set up a temporary donor center in the town's city hall and mounted a citywide Day of Blood (Journée de Sang). The day proved an instant success, as hundreds lined up to give. From that moment, Days of Blood became a French institution. The charismatic Tzanck traveled from town to town illustrating the ease of donation, and citizens everywhere donated blood. Later a national blood-giving club was formed, the French Federation of Blood Donors, with hundreds of thousands of members, social events, conferences, and significant political influence. The members made Tzanck their first honorary president.

With his nation's idealism rekindled, Tzanck and the others now moved to harness it, to keep the flame burning while managing the blood as an industrial resource. They developed a national blood policy, which became part of French law in 1952. The keystone of the policy was *bénévolat*—a concept that means the blood should be voluntarily given, with no payment to donors and no commercialism of any kind. Under this concept, blood was not merely a resource but part of the social contract that bonded French citizens. Even prisoners were encouraged to give, for participating was seen as a humanizing influence. The law organized the nation's transfusion centers into a loose network of *postes,* or offices, and regional laboratories. The *postes* would collect and distribute whole blood and liquid plasma, then sell excess plasma to the regional laboratory for processing into dried plasma and fractionation products, to be sold back to the *postes.* The prices were set to eliminate all profit. Special status was granted to the regional laboratory in Paris. Renamed the National Blood Transfusion Center (Centre National de Transfusion Sanguine), it carried the additional responsibilities of research and training, thus serving as a pacesetter for the nation. The overall design created a model of efficiency and idealism, a benevolent, altruistic, and autonomous blood system. Until the AIDS scandal, it gave the French people an enduring social bond about which they could feel uplifted and proud.

Japan began with no grand philosophy or design. After the syphilis scandal of 1948, the Americans pressured the Ministry of Health to organize a system of blood banks. The ministry, in turn, authorized the Japanese Red Cross to open a volunteer blood bank in Tokyo. Like virtually all Japanese institutions at the time, the Japanese Red Cross received training and equipment from its American counterpart. In 1952, after years of planning, building, and training, the Japanese Red Cross opened the doors to the nation's first all-volunteer blood center.

The effort was doomed from the start. Giving blood freely was still a foreign concept in postwar Japan. Unlike the French, the Americans, or the British, the Japanese had not mounted national blood drives during the war, and the public never got into the habit of seeing blood giving as community service. Other conditions made their job even more difficult. Japanese people are generally small—each can safely donate less than half as much as his American counterpart—and citizens tended to feel that losing blood would be dangerous. The Japanese health-care system discouraged donation, because the national health-insurance program provided blood at no cost. People saw blood as a commodity automatically provided by an invisible source, and did not feel the need to give. "People would say, 'We paid for health insurance and it buys our blood,' " one blood banker wrote many years later. " 'Why should we pay again in blood and pain?' " Finally, cultural factors interfered. Blood carries meaning in Japanese tradition, more so than in the West. We may say, "Blood is thicker than water," but people in Japan mean it almost literally. Blood equals family. Siblings are said to share the same blood; if a child exhibits characteristics of a grandparent, he will be said to have the older person's blood. Such a commodity, so rich in feeling and tradition, is not casually given away. Thus, in 1954, for example, a year in which France collected more than 312,000 donations, the Japanese Red Cross took in just over five hundred.

Meanwhile, a competing idea had been born. Dr. Ryoichi Naito had almost completed his plan for an industrial-style national blood system when two old friends from the military, who had become quite influential in the pharmaceutical and financial communities, paid him a visit. After several late-night cups of sake, the men asked him to show them what he had been working on. The document was vintage Naito—111 pages of neatly scripted characters in which every aspect of the venture was thoroughly explored. It reviewed the importance of blood and plasma, outlined the production technologies, described the facilities, calculated their cost, and spelled out the requirements for management and staff. It projected the cost of purchasing blood and the profits from selling it, and calculated balance sheets for the corporation's first three years of operation. This impressive and compelling document was a portrait of an idea that clearly had found its time. Eager to promote Naito, his friends put him in touch with the vice-president of the Kobe Bank, a wealthy and influential man with an interest in social welfare, who lined up other business leaders to back him. Naito still needed permission from the Ministry of Health to open a blood bank, and it was here that his work with Sanders served him well. The Americans afforded him considerable credibility, and gave him technical advice and support. His application sailed through the ministry. In March 1951, the Japan Blood Bank opened for business.

The company looked unprepossessing enough, occupying a two-story white wooden building in an industrial neighborhood of Osaka. The second floor contained facilities to produce preserved blood, typing sera, and dried plasma; the first housed a walk-in donor center. For two hundred cubic centimeters of blood, donors would receive 400 yen—nearly twice as much money as a Japanese laborer could make in a day.

Naito's company in Osaka quickly became a focal point for the disenfranchised and poor. Hundreds lined up for the 8 a.m. opening—war widows, unemployed day workers, students, and the homeless.

Teenagers sold blood to help their parents support the family. The enterprise attracted its own spin-off businesses, and vendors gathered up and down the lines, to sell fortifying dishes of *zenzai,* a soup of beans and sugar, and *niku udon,* noodles with beef and eggs. Some in the queue would sell the maximum allowable quantity of blood and then circle back, line up, and sell it again.

The spectacle bothered Naito—"The concept that the blood bank exploits the poor is against our morality," he later wrote—but this did not prevent him from moving ahead. He opened another blood center in Kobe, an industrial city a half-hour away. His centers operated at full capacity, even as the Red Cross waiting rooms sat empty; in a land of the desperately unemployed, the simple fact that he paid made all the difference. Now, as he resumed the working pace of his earlier years, he did so with none of his previous anger, but tempered his behavior with consideration and respect. Instead of calling him "Typhoon," subordinates referred to him as the "Locomotive" for his power and drive; but they did so with affection, even love. (Naito himself came to relish the sobriquet: When he later assembled a collection of his writings, he entitled it *Noises of an Old Locomotive.*)

And a locomotive he was, unstoppable once he had built up some steam. Seeing an expanding market in exports, he sent samples all over the world, along with brochures that he personally typed. In 1953, he made an agreement with Cutter Laboratories of California, one of the original fractionators under Cohn, to ship them raw plasma for the production of pharmaceuticals. The contract inaugurated a long-term relationship that eventually would give him a foothold in America, and fueled his expansion into more Japanese cities and foreign countries. By 1954, less than three years after he established the company, the Japan Blood Bank's sales exceeded $500,000, with exports accounting for more than 60 percent, dwarfing the activities of the Japanese Red Cross. Naito had laid the groundwork for a company that eventually would became a powerhouse in the blood world—a multinational giant worth $1.5 billion, producing a host of blood and fractionation products with subsidiaries in eleven foreign countries.

It had been quite a journey for Naito, from his shadowy military days to a position of accomplishment, wealth, and respect. Colleagues around the world came to admire him, and would freely admit it even after they had learned of his past. But Naito's life journey was not destined to be simple; his most challenging trials still lay ahead.

# CHAPTER 10

# DR. COHN

The audience was abuzz in the lecture hall of the Instituto Superior Technico in Lisbon, where some of the world's most prominent blood and protein scientists had gathered to hear a state-of-the-art review of their field by the revered and redoubtable Dr. Edwin Cohn. Cohn had remained busy after the war, turning his attention to theoretical studies, such as the fundamental nature of protein. He also had begun to examine the cellular elements of blood—the red cells, white cells, and platelets. But, like so many products of Cohn's mind, even those explorations had practical implications: In order to study the cellular components, he had to find a way to separate and preserve them.

Hence the machine on the stage. Variously called the "blood machine" or "biomechanical apparatus," it was a marvel of miniaturization and automation, a stainless-steel-and-plastic box about the size of a modern-day clothes dryer. Blood flowing into the machine ran through a column of chemically impregnated beads that prevented it from clotting, over a heat exchanger that cooled it, and into a series of centrifuge bowls and tubes for further separation. The centrifuge bowls were a marvel in themselves. They replaced the swinging bucket-type centrifuges, which had to be frequently emptied and refilled. Modeled on cream separators, they consisted of two bowls, one inside the other, connected at the lips and turned upside down in the shape of a dome. The blood flowed in through a central tube that ran to the bottom of the space between the bowls. As the centrifuge spun, the red cells, being heavier, were pulled down and outward, forcing the lighter plasma to the top. Eventually the plasma flowed up and out through a central collecting tube. Every once in a while, the machine would be stopped while the red cells were removed.

Then the blood components traveled to other devices within the machine for further separation. The result was that what entered as blood exited as a half-dozen ingredients. This, Cohn felt, was the future of blood—administered not in its whole, undifferentiated form, but as a series of constituents. Component therapy, as he called it, would ensure that patients would receive specifically what they needed. It would also promote the efficient use of blood, since one donor could provide several recipients with red cells, platelets, or the fractions of plasma.

And so the lecture hall hummed with anticipation. Before the audience sat the biomechanical apparatus. Behind it crouched Dr. James Tullis, a young researcher who had traveled to Portugal with Cohn, tinkering with some final adjustments. In strode Cohn, elegantly dressed, with a handkerchief peeking jauntily from his breast pocket. Normally in such a presentation, he would draw blood from an assistant as he lectured so the audience could see it run through the machine. But today he had something special in mind. In order to demonstrate the painlessness and ease of the procedure, he would connect *himself* to the apparatus, processing his own blood as he talked.

Cohn bared his arm, extended it to Tullis, and commenced his lecture as the blood began to flow. The audience leaned forward. A spotlight had been trained on the bowl so they could see the blood separate into red cells, the pale-yellow plasma, and the white "buffy coat" of white cells and platelets. "The event began faultlessly," Tullis would later write.

Tullis took out his pocket watch. Normally it took about four and a half minutes for the cells to spin out and the plasma to emerge through its tube at the top. But, unknown to Tullis and Cohn, an overzealous technician back at Harvard had made an error. Cohn had instructed that all tubes and surfaces that came in contact with the blood be coated with silicone to avoid damaging the cells and platelets. But the technician had sprayed the silicone too liberally, clogging a vital outflow port. The result was that, as Cohn's blood flowed into the spinning centrifuge, pressure inside the machine began to build.

Tullis checked his watch. Three and a half minutes had passed. Cohn lectured on, waving with his free hand to point to charts and to various parts of the machine. Four minutes and ten seconds; four and twenty-five. Tullis sensed that something was wrong, but could not piece together what it might be. Four minutes and forty seconds. The plasma should have emerged by now. Dare he interrupt? Suddenly, at five minutes, something went *pop!* Tullis looked up in time to see the audience scrambling over their seat backs. Instantly he realized what must have happened: The pressure in the machine had built to such a degree that a seal in the high-speed centrifuge burst, showering the people in the first several rows with the lecturer's blood. They regrouped on higher ground as Cohn continued speaking with his usual aplomb.

In the immediately postwar years, Cohn and his colleagues did not find it easy to maintain the nation's interest in blood as a resource—surprisingly so, given all that blood had done for the war effort. A sense of complacency had taken hold in America, and the national blood-collecting machinery had shut down. The Red Cross closed its blood centers, and the government dropped its contracts with the pharmaceutical companies. People were almost happy to get out of the blood business. Blood was an expensive and troublesome material, difficult to handle and vulnerable to disease. Collecting it had exhausted the Red Cross of money and manpower; they were still soliciting contributions to restore the $16 million they had spent on the National Blood Program. Indeed, except for independent community and hospital blood banks, no one was very interested in the resource.

Cohn and a few other doctors and military men argued that this was exactly the wrong time to demobilize. Granted, the immediate emergency was over, but a new conflict loomed with the Soviet Union. The next war, if it came, would dwarf the carnage of World War II. It would be an atomic war, affecting untold millions of civilians. The transfusion needs would be "almost beyond estimate," in the words of General George C. Marshall, requiring "more blood in a single week than all our requirements for a full year during the [previous] war."

Meanwhile, some in the Red Cross were thinking that a national blood program could make their agency more economically stable. After all, they reasoned, before the blood program how many Americans ever really thought about the Red Cross? People had only heard of it occasionally, usually after fires and floods. The sporadic nature of the Red Cross's activities made fund-raising difficult. On the other hand, the blood program had made the organization a national presence, the very symbol of patriotism, serving all the people all the time. The argument became, "Look at the good things we can do come fund-raising time," according to Sam Gibson, one of Cohn's principal researchers, who later became the Red Cross's medical director. After months of discussion, the combined weight of arguments carried the day. At a meeting in Cleveland in June 1947, the Red Cross announced that it would once again become involved in the business of collecting and distributing blood.

The first step in initiating the enterprise would be to round up some powerful and persuasive leaders. The Red Cross had always been able to recruit well-connected people, often generals and other military men. Its president at the time (and during the war) was Basil O'Connor, a former law partner of Franklin Delano Roosevelt. O'Connor had recruited Dr. Ross T. McIntire, Roosevelt's personal physician and former navy surgeon general, to administer the blood program. McIntire, an ear, nose, and throat man, knew little about blood and asked Cohn to recommend a medical director. Cohn suggested one of the rising stars in transfusion, Dr. Louis K. Diamond, a pediatrician and blood-bank director at Children's Hospital in Boston and professor at Harvard Medical School.

Diamond had already made a name for himself by treating a rare, fatal blood disease among newborns, called "erythroblastosis fetalis." This mysterious condition afflicted neonates with swelling, jaundice, and fatal anemia. No one knew what caused the symptoms. But after microscopically examining the babies' blood, Diamond linked them to a shortage of mature, functioning red cells and an excess of immature, nonfunctioning ones.

Other scientists linked this imbalance to a rare incompatibility reaction between the blood groups of the mother and child. The number of blood groups had grown dramatically since Landsteiner had discovered the original four at the turn of the century. Dozens had been added. Unlike the original blood groups, however, these new categories—or subgroups—were only partially incompatible with each other, sometimes causing a dangerous reaction and sometimes not. One of these subgroups involved a protein called Rh (so named because it was first isolated from the blood of rhesus monkeys). Scientists found that, when a person without the Rh protein (Rh-negative) receives Rh-positive blood, nothing seems to happen, but if transfused with the same blood again he may react strongly enough to die. What actually occurs is that the Rh protein triggers a mild immune response which in subsequent exposures becomes more severe.

Diamond confronted a dangerous complication of the Rh reaction, in which an Rh-negative mother gives birth to an Rh-positive child. In theory, the mother's immune system should ignore the foreign protein, since maternal and fetal blood do not mix across the placenta. Reality is messier than theory, however, and the two bloods make contact before or during the normally bloody birth. The mother's immune system initiates a mild antibody response. The first baby usually escapes. But during subsequent pregnancies, the mother's immune system, geared up by the previous exposure, furiously attacks the blood of the new baby, wiping out red cells wherever it can find them. The result is what Diamond had observed under the microscope—a shortage of red cells and an excess of immature ones growing to take their place. Such babies generally died.

It would be years before scientists discovered how to prevent this rare and fatal reaction. Meanwhile, the options were limited. Given that the baby's blood was essentially destroyed, the only action doctors could take was to drain out the old blood and replace it with new. But it was no small matter to put blood into a newborn, as Alexis Carrel had shown years before. The fragile veins and fluttering heart could barely sustain a simple transfusion, much less a total blood replacement.

Diamond came up with an elegant solution. Using the same circulation routes that nature had designed, he untied the baby's umbilical cord and inserted a tube into the vein that ran through it. Then, with a syringe, he alternately withdrew the old blood and injected new blood to replace it, a few teaspoonful at a time. As the new blood circulated throughout the body, he kept withdrawing old blood from the umbilical vein, until virtually all the old blood had been washed out by the new. The treatment became standard for the next fifteen years. Diamond won a national reputation as a transfusion innovator.

Those were exactly the credentials that Cohn had in mind when he telephoned him about the Red Cross position. Diamond had not even known it was available. As he later remembered, Cohn simply said, "You're going to go down to advise Dr. McIntire how to set up blood banks in all these Red Cross units." One did not lightly spurn an offer from Cohn. Diamond went to Washington, with little idea of the task that awaited him. "I was greeted like someone special, red carpets rolled out, and people waiting, 'Yes, Dr. Diamond,'—'No, Dr. Diamond.' " He assumed the consultancy would last about a week. But on the third day he found a sign on his door saying "Chief of Blood Transfusion Service."

"I said, 'Look! I have a job back in Boston. I can't do this!' " Someone told Diamond he was wanted on the phone. It was Cohn, calling from Boston. "I've talked to the Dean," he said in his cultivated tones. "You are relieved of all pediatric responsibility. You're staying in Washington, D.C., and organizing the Red Cross Transfusion Services."

Diamond protested: "I've got a wife and two children, and a home! I've got a pediatric service and a blood bank! I can't leave there!"

Cohn paused a moment. "Well, the Dean said you're leaving. If you wish I'll talk to the President of the University."

"No, no!" said Diamond and the matter was settled.

Despite his misgivings, Diamond could not have landed in a more exciting place. The size of the new program would rival all that had been accomplished during the war. The system would include dozens of fixed and mobile collection units in every region of the nation. An army of technicians would draw blood and store it, separate the plasma, send it to processors, and distribute the finished products. The entire program would be free, supported by voluntary blood donations, fund-raising, and a small charge to hospitals. Furthermore, the Red Cross already had a head start. After the war the army found itself with a surplus of one and a quarter million packages of dried plasma, which they gave to the Red Cross. Acting through state health agencies, the organization dispensed it free to the nation's doctors and hospitals. The goodwill engendered by this act was incalculable, although it also created problems, as we shall see later.

Diamond realized he would have to move carefully. In many communities, such as Boston, Phoenix, San Francisco, and New York, local blood banks predated the Red Cross effort; doctors in those areas might resent the intrusion. So the agency's strategy—at least in the short term—was to establish new blood banks only where the Red Cross was specifically invited.

The first invitation came from Rochester, New York, where an internist at the University of Rochester Medical School, Dr. Herbert R. Brown, had galvanized the medical community to support a regional blood center. Brown had been stationed in Guam during the war. There, under the command of then Navy Surgeon General Ross T. McIntire (now Diamond's superior at the Red Cross), he directed the distribution of whole blood in the Pacific. After the war he returned to the university, and persuaded the local medical community to link up with the National Blood Plan. When Brown called the Red Cross in Washington, McIntire flew to Rochester to lend his office's support.

The grand opening took place in near-blizzard conditions. Brown had acquired a beautiful facility, a historic mansion donated by the Rochester Institute of Technology, which he fitted with the most modern equipment. The walls were painted a soothing light green, with music piped in to comfort the donors. A small crowd of dignitaries arrived, from the mayor of Rochester to the head of the American Medical Association to a cluster of Red Cross officials from Washington. Eight hundred token donors were invited, representing veterans' groups, organized labor, business, religious groups, the police, and student organizations. Even prisoners from Attica were represented: Earlier that week, Red Cross volunteers had gone to the institution and collected several hundred units of blood.

The opening began with a celebratory luncheon at the Rochester Chamber of Commerce. There, amid popping flashbulbs and glaring klieg lights, Dr. George F. Lull, secretary and general manager of the American Medical Association, praised the initiation of a national program. The mayor of Rochester, Samuel Dicker, said his entire city felt "honored and proud" to be chosen for the site of the initiation. The Red Cross's president, Basil O'Connor, declared the day "a milestone in medical history." Then they all trudged through icy streets to the center, where the donors lined up to give blood and relax with fruit juice and snacks.

A poignant moment came when a veteran named Carl Piccarretta arrived. Piccarretta, a member of the Barb Wire Club of Rochester, had spent three years in Japanese prison camps, and credited his physical and emotional survival to the Red Cross food packages that had reached him. Now he had come to pay a debt of gratitude with his blood. As he waited in line, he chatted with reporters about his capture in Corregidor and confinement in a hard-labor camp near Kobe, Japan. Finally, a volunteer called out his name. Piccarretta practically ran to the examining room; but when he got there the doctors said they had to disappoint him. "With all the tenderness at her command, the volunteer behind the desk told Carl Piccarretta that his blood was not acceptable because he had had malaria," wrote one observer. Dr. Brown himself came over to console him, and volunteers gave him coffee and cookies. The ex-POW was the picture of dejection as he slumped back to the reception hall, retrieved his coat and hat, and walked out into the cold.

Rochester was the first of many grand openings. In short order the Red Cross centers opened in Wichita, Kansas; Stockton, California; Atlanta, Georgia; and Kansas City, Missouri. Later would come Portland, Oregon, and Mobile, Alabama. Each was accompanied by busy photographers, visiting dignitaries, lines of donors, and officials delivering inspirational statements.

Behind this picture of goodwill, however, loomed the continuing shadow of racial discrimination. The Red Cross had liberalized its wartime position, in which it had segregated blood according to the donor's race. Now the organization declared, "On the basis of recorded scientific and medical opinion, there is no difference in the blood of humans based upon race or color." But this statement did not end racism in blood collection. In a concession to the Southern states, the Red Cross authorized its chapters to label the blood they collected by race and let the hospitals they sent it to decide how to distribute it. It was, in the words of the policy, a way to "collect and hold blood in such a manner as to give the physician and the patient the right of selection at the time of administration." In other words, though the chapters themselves would not discriminate, they would not interfere with local hospitals' desire to do so. Like the Red Cross's policy during the war, the guidelines represented an attempt at a compromise.

The policy was not destined for a peaceful existence. Early in the Korean War, the Red Cross tried to establish a center where United Nations employees could give blood to the war effort. The employees were so offended by the racial designation on the donor cards (especially since it violated the United Nations Charter) that they threatened to boycott the collection. They relented only after the Red Cross agreed to let the UN print its own donor cards, with no reference to race. (Later the Red Cross omitted the racial designation from its cards as well.) The issue kept festering. It erupted again in the late 1950s, when Louisiana and Arkansas passed laws requiring the segregation of blood. Louisiana went so far as to make it a misdemeanor for physicians to give a white person black blood without asking permission. Asked for his reaction, the Red Cross president at the time, General Alfred M. Gruenther, told *Look* magazine that the matter was of peripheral concern to the Red Cross, since no Red Cross blood centers operated in Arkansas or Louisiana; the blood centers in those states were run independently. He noted, however, that during emergencies the Red Cross did send shipments of blood to those states, and did so "in conformity with applicable state laws."

The uncomfortable compromise continued for decades, with the National Red Cross officially condemning segregation but allowing Southern chapters to wink at the practice when local regulations made it unavoidable. They did, after all, have other priorities. They were trying to blanket the nation with blood centers, providing blood for all Americans, regardless of social standing or color. And if that meant deferring to certain local customs, then so be it—the blood program must move ahead. Not until the late 1960s, after the American civil rights movement, did the segregation of blood disappear.

Though it undoubtedly rankled, the race question did not impede the Red Cross's progress. Nor did it affect the celebratory mood in one grand opening after the next, as politicians, doctors, and Red Cross officials toasted each other and the areas' citizens. The Red Cross officials particularly liked to sound patriotic themes, equating the Red Cross with Americanism in an effort to maintain the public's support.

At the grand opening in Wichita, Basil O'Connor declared, "It is fine to see that the spirit of the pioneer still lives in Kansas. . . . It was the spirit of pioneering which inspired . . . the establishment of the blood service which is being formally dedicated today." In Stockton, California, another Red Cross dignitary said the local Red Cross staff members were pioneers, "as your ancestors were pioneers in this great state one hundred years ago." That's how Red Cross officials saw themselves as they inexorably pushed forward into virgin blood territory. At one point, O'Connor expanded on the pioneering theme, proclaiming a kind of Manifest Destiny, with the Red Cross collecting blood from Americans lined up at donor centers from sea to shining sea. "There can be no recession from the Red Cross movement," he stated. "Far from retreating, the American Red Cross is prepared to go forward . . . The Red Cross National Blood Program may well become one of the greatest single health programs the world has ever known."

Not everyone was inspired by O'Connor's vision. For, if the Red Cross portrayed itself as a group of pioneers, others found themselves cast in the role of Native Americans, already inhabiting a land that intruders wanted to conquer. During and immediately after the war, many physicians had organized their own blood banks in local hospitals. Many had operated successfully as independents, and did not need a monolithic agency taking over. It irked them that the Red Cross intended not to *supplement* the existing blood banks, but to *replace* them—the organization had, after all, declared its intention to become the "total" blood supplier for the nation. It also irked them that a volunteer agency was attempting to trespass in medicine. Prominent physicians such as Dr. John Scudder of Columbia University, who had directed the original Plasma for Britain project, objected to the agency and its meddlesome volunteers. Blood, after all, was the province of physicians. It was one thing for the Red Cross to *recruit* the blood donors; they had done so admirably during the war. But to allow the agency to perform medical functions, like supervising blood tests and processing? Never. "Upon us [the doctors] rests the responsibility to fulfill those tasks."

Scudder's words presaged a conflict that was beginning to grow throughout the nation. As the Red Cross—flush with its success in the war, righteous and excited—set down its roots in one community after the next, local opposition began to grow. Nowhere was this resistance more vehement than in San Francisco, home of a respected independent blood bank. And no one was more forceful in her opposition than the administrator of that blood bank, a woman who became such a determined opponent that it earned her a nickname: the Wicked Witch of the West.

Bernice Hemphill was a tall, gangly newlywed when she walked, almost literally, into the world of blood banking. A native of San Francisco, she married a dentist who served with the navy in Pearl Harbor. They had been living there for a month when the Japanese air force launched the attack. Bernice was driving home from church that Sunday morning when an artillery shell nearly blew her off the road. She navigated home, only to find her husband waving frantically from the balcony, the soap from his shave still clinging to his face. "They've bombed Pearl Harbor! I've got to get over there!" he shouted, and left with the car.

Sitting in the apartment, pinned to the radio, desperately looking for something to do, Hemphill heard the governor make an emergency call for blood. She ran to the street and hitched a ride to Queens Hospital in Honolulu, where a doctor named Forest H. Binkerton had established a small plasma bank. People of all races had shown up to give—white, Hawaiian, Chinese, Japanese—even the elderly former governor and his wife. Hemphill stood in line for an hour, then, frustrated by the waiting, pushed her way into the lab. Amid scenes of "utter confusion," she grabbed some equipment and began helping to run blood-typing tests. She worked tirelessly, grabbing sandwiches and naps whenever she could find them. "I didn't know where my husband was and he didn't know where I was, but there were people to talk to, and you just worked and worked," she later recalled. She did not leave for three days. "At the end of the third day, Binkerton came up to me and said, 'Girlie, I've been watching you very intently.' " He asked her to work for the blood bank full-time.

She managed the lab for another two years, then returned to San Francisco with her husband, where she volunteered her services at the Irwin Memorial Blood Bank, one of the nation's few community blood banks at the time. Founded by a British doctor named John Upton and San Franciscan DeWitt Burnham with support from the county medical society, it occupied a beautiful stone mansion donated by the prominent Irwin family. During the war, the staff collected blood from civilians, processed and redistributed it as plasma to British troops, and as plasma and whole blood to the area's hospitals. Hemphill ran the lab, coordinating recruitment efforts and rising to the position of blood-bank administrator.

It was at Irwin that Hemphill got her first taste of the coming blood-bank competition. Early in the war, the Red Cross had set up a provisional center in town to supplement Irwin's efforts by collecting plasma for the American armed forces. The two had agreed to coexist peacefully—indeed, for the first few months the Red Cross was in town, Irwin let the agency use its facilities. But soon conflicts developed over the relative importance of the two blood banks' missions. Since the Red Cross collected blood for American troops, its officials felt they had a greater right than Irwin to San Francisco's blood donors. The Red Cross asked Irwin to stop advertising, and castigated Irwin's recruiters for "stealing" its donors. The Red Cross festooned the cable cars that ran past the Irwin blood bank with placards bearing the familiar Red Cross logo. The implication was clear: Give to the Red Cross, the *real* blood bank in town. "This went on during the entire war," recalled Hemphill; it ended only when the Red Cross left the city.

After the war, peace returned to blood banking in San Francisco, mainly because Irwin controlled the territory. But in 1947, Red Cross Medical Director McIntire, on a national promotional swing, stopped in at Irwin to tell them about the National Blood Program. "He told us that they planned to cover the country with blood centers, all run and operated by the Red Cross," Hemphill remembered. "He was very arbitrary, very dogmatic, very arrogant. You know, the Red Cross was the biggest charity at that time. But bigness does not mean greatness. So we just took the stance of 'That's what you think.' We had no intention of fading away."

The directors of Irwin chose to resist. Consulting with the California Medical Association, Upton laid the groundwork for a way to preempt the Red Cross with an affiliation of blood banks to serve the entire state. Under his plan, any local medical society that wished to establish a community blood bank could obtain a loan from the California Medical Association. Once established, this local bank could join with others into a regional collection-and-distribution network. The network would be set up like spokes in a wheel. Rural blood banks surrounding urban areas would feed blood into the center for processing and distribution according to the regional needs. With all banks in the system operating as nonprofits, Upton felt they could deliver the most blood at the lowest possible cost.

Upton's plan quickly caught on. Over the next several years, seven blood banks were formed and linked into his regional organization. But the Red Cross was not ready to concede. Having opened a blood center in Stockton, they now moved into San Jose and Los Angeles. Upton criticized the Los Angeles center as inferior. "We have a few problems, they have a thousand," he told a medical convention. He added that replacing the Los Angeles center with one based on his example "would clean up the situation quickly." At his urging, the California Medical Association passed a resolution that "it is not the proper function" of the Red Cross to run blood banks.

California was one of several states where the National Red Cross would attempt to face down local unaffiliated blood banks. Clearly a larger battle was looming. In order to balance the weight of the Red Cross, Hemphill joined with other blood-bank directors to form a national organization of independent blood banks called the American Association of Blood Banks, or AABB. Hundreds attended their opening meeting in Dallas in November 1947. They included innovators such as John Elliott, who had pioneered the use of plasma transfusions and was now the director of a blood bank in Florida; prominent doctors such as John Scudder and Lester J. Unger, inventor of one of the early stopcocks used in the days of direct transfusion and now director of a blood bank in New York. Their ostensible purpose was to trade expertise and technical information, but the political agenda quickly became clear. Their real *raison d'être* was to oppose the Red Cross. Participants described blood banks as their "birthright," one not to be impinged upon by an "outside agency." Scudder called the Red Cross movement "nationalization of the blood banks by the government," even though he knew the Red Cross was not a government agency. Some denounced the National Blood Program as socialistic.

On the final day of the conference, the members passed what seemed, at first glance, to be a generous resolution. They invited the Red Cross to sign on as a member of the AABB. "WHEREAS, the American National Red Cross has expressed an interest in the procurement of Blood, BE IT THEREFORE RESOLVED that the American Red Cross be invited to membership in the Association and to aid in the procurement of Donors according to local needs." Soothingly couched in the terms of inclusion was the true intent of the group's resolution. It is found in the phrase "to aid in the procurement of Donors." To those in the business, the meaning was unmistakable: It was OK for the Red Cross to procure the blood donors, but the *professionals* at the blood banks should handle the rest.

So began a conflict that would simmer for years. From today's vantage point, the argument between the Red Cross and the independents may seem to have involved a petty struggle for territory and power. But it reflected deeply felt differences about the proper use and management of blood. The Red Cross felt lay administrators should direct the blood banks, along with the help of volunteer staff and part-time physicians; the AABB felt physicians should be in charge. The Red Cross believed in centralized control over the local blood banks; the

This mobile blood collection center in Bristol, England, typified many throughout the free world. The Nazis collected blood far less effectively, since they accepted it only from certified Aryan donors.

In Milwaukee, Wisconsin, in 1942 a hundred nuns lined up to donate blood for the war effort. Such outpourings of patriotism were common at blood centers.

A container in which whole blood was shipped to field hospitals in Europe during World War II. The bottles were packed in dry ice for preservation.

*Dried plasma and albumin traveled far and wide during the war, whether by mule in Italy or by tribesmen in New Guinea.*

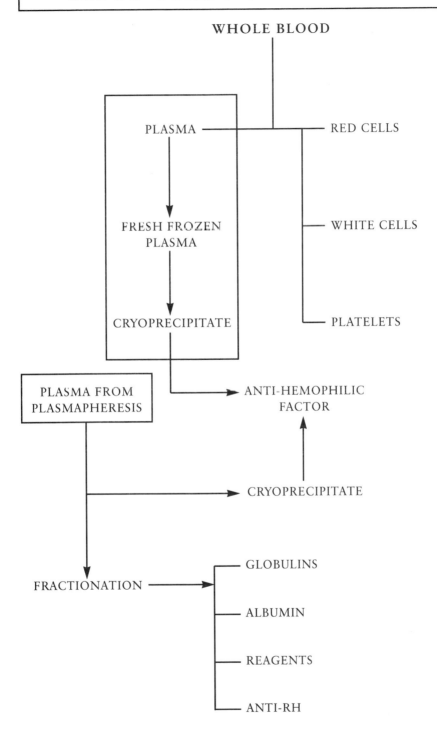

# BLOOD COMPONENT THERAPY

WHOLE BLOOD

PLASMA — RED CELLS

FRESH FROZEN PLASMA

WHITE CELLS

CRYOPRECIPITATE

PLATELETS

PLASMA FROM PLASMAPHERESIS

ANTI-HEMOPHILIC FACTOR

CRYOPRECIPITATE

FRACTIONATION

GLOBULINS

ALBUMIN

REAGENTS

ANTI-RH

*In the years following World War II, blood would be seen as a collection of components, each of which had therapeutic value.*

*In the immediate postwar years countries around the world hurried to develop their national blood supplies, so important had the resource of blood become. This recruitment poster was designed to enlist donors in France during the country's "Days of Blood" campaign.*

*In the 1960s American drug companies began employing plasmapheresis to harvest plasma and other components selectively. The process, although time-consuming, represented a leap in efficiency over the withdrawal and use of whole blood, and led to the further industrialization of the resource. Later the process was automated and made faster, as shown in this modern photograph.*

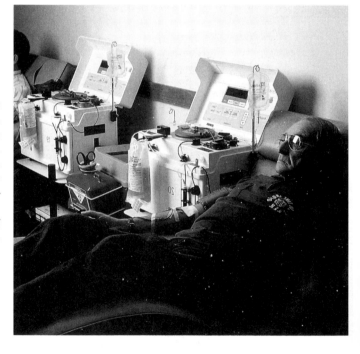

Dr. J. Garrott Allen *warned early and often about the dangers of blood-borne hepatitis and the "plasma mills" that spread it. His warnings would prove prophetic, for both hepatitis and AIDS.*

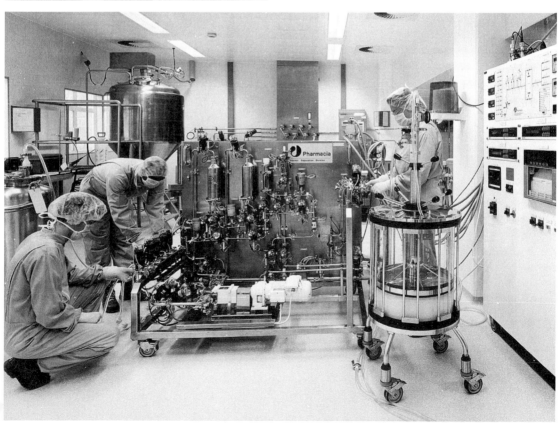

*Blood as a resource: technicians working at a modern fractionation facility.*

*Casualty of AIDS: Susie Quintana, a homemaker and crochet artist from rural Colorado, contracted AIDS from a blood tranfusion. After prolonged litigation a jury awarded her more than $8 million—the day after she died.*

*Michael Rosenberg, one of thousands who contracted AIDS from the clotting factor he took for hemophilia, cofounded the activist group Hemophilia/HIV Peer Association. He struggled to bring the hemophilia holocaust to national attention before succumbing to AIDS in 1994.*

*Corey Dubin, journalist, activist, and a director of the Committee of Ten Thousand, was the first child to receive experimental clotting factor in the mid-1960s. He is now HIV-positive. He poses (left) with a friend at the Tenth International AIDS Conference in Yokohama, Japan.*

*As HIV spread through the blood systems of the world, scandals erupted involving some of those officials who delayed taking preventive measures. Michel Garretta (center) and Jacques Roux, two principals in the French "Affaire du Sang Contaminé," enter a Paris courtroom at the start of their trial in 1992.*

*HIV in blood products triggered a massive scandal in Japan, and demonstrations such as this march in Osaka in December 1995.*

*In a widely broadcast public act of humiliation, Takehiko Kawano, president of the Green Cross, bows and apologizes for negligence.*

AABB promoted local autonomy, with advice and assistance from a national professional organization. The AABB emphasized technical education, and the Red Cross lagged in this respect.

Yet even those differences represented only the surface of a more fundamental disagreement. For the Red Cross and the AABB held warring philosophies about the collection and distribution of blood, differences that extended to issues such as citizenship and social service. In short, their views were as divided as the American character itself.

The Red Cross, which forged its philosophy in times of disaster and war, followed a doctrine of community responsibility. Under this philosophy, blood was a resource for everyone that the community at large should freely provide. No demands were placed on the recipient. No matter how much he consumed, the general population would replace it through a sustained level of voluntary giving.

The AABB, consisting of medical professionals, found the Red Cross's viewpoint absurd, impossible to sustain in nonemergency conditions and nothing like normal medical practice. They followed a mode called individual responsibility, in which the recipient of blood took on the burden of helping to replenish it. A patient who received, say, three pints of blood was expected to round up the donors to replace it. If he could not meet that responsibility, the blood bank would charge him a replacement fee—usually about $25 a pint. The replacement fee was not inflexible; any recipient who truly could not pay was let off the hook. But most people, they hoped, would feel a sense of obligation and spread that feeling among the donors they recruited.

Both organizations shared some fundamental principles: that blood should not be considered a commodity, for example; that the processing fees be kept as low as possible; and that the enterprise should run on a nonprofit basis. As a magazine commentary would say: "If the two organizations could only learn to make no money harmoniously, the public would be better off."

Harmony, however, lay years away. As the Red Cross opened one center after another, Hemphill traveled the country, spreading her own gospel of individual responsibility. She also introduced a new way to trade blood that vastly improved regional cooperation. Called the National Blood Clearinghouse, her system operated like the Federal Reserve Bank. Under it, hospitals and blood banks could trade blood and blood credits, almost like drawing checks from a bank. If a hospital ran short of type A blood, for example, it could, through the clearinghouse, quickly obtain the blood from whoever had it. In return, the hospital would give a credit, like a check, promising repayment in the form of blood or money. At the end of each month, hospitals would settle their accounts by sending money or blood to the clearinghouse—again as with a bank. Formerly, when blood banks ran short they had to borrow directly from each other, with repayment in kind. Hemphill's system made providing blood immeasurably easier, as currency makes material trade easier than barter. It also meant that someone who received blood from an AABB blood bank in one part of the country could recruit friends and relatives to help replace it at their local AABB blood bank, anywhere in the country. Almost immediately, blood banks in California adopted the plan, and Hemphill pushed for its use on a national scale.

Meanwhile, the Red Cross kept expanding its influence. By the end of the decade, the two groups had just about evenly divided the country, with thirty large regional Red Cross blood banks established, and the AABB representing more than a thousand smaller community entities. It was as if two fundamentalist churches were competing—each energetically proselytizing and each proclaiming its righteousness over the other. The conflict over who should give blood and who should receive it, who should manage blood and what philosophy they should follow, would rend the American blood establishment for decades.

As the situation became tense in the world of blood procurement, the atmosphere in Cohn's laboratory became somewhat more relaxed. With the crisis of the war past, Cohn's sense of humor returned, and he allowed himself to show more of his natural warmth. He received much recognition in the form of honorary degrees, awards, and lectureships throughout the United States and Europe, including a Medal of Merit from President Truman, which must have helped salve his ambition. He also became the first scientist at Harvard to receive a University Professorship, a lifetime appointment reserved for "men of distinction" to pursue their research wherever it might lead. The professorship gave him added autonomy, a balm to a personality like Cohn's. He detached his lab from the medical school and recast it as a multidisciplinary, free-standing entity which, with his penchant for pomposity, he named "University Laboratory for Physical Chemistry Related to Medicine and Public Health of Harvard University."

Cohn's attitude may have eased a bit, but it did not affect the intensity of his schedule. He believed in a kind of *noblesse oblige,* that scientists, by their very knowledge and position, had a permanent duty toward the public, regardless of personal cost. And so, despite the war's end, Cohn remained as active as before. He relentlessly shuttled from Boston to Washington, working as a consultant to the National Research Council and the National Institutes of Health, and as special adviser to the American Red Cross, whose new president, General George C. Marshall (he succeeded Basil O'Connor), was a personal friend. He worked with Marshall and others to renew the nation's commitment to blood collection.

Cohn also continued to oversee the nation's plasma products. After the war, most pharmaceutical firms lost interest in fractionation as an expensive and troublesome process. But the technology returned after a disastrous sequence of events. When the military gave the war-surplus plasma to the American Red Cross, they did not present an unblemished gift. The dried plasma had been contaminated with hepatitis, a viral disease that in its most serious forms can cause liver damage and death. After several people died, the Red Cross recalled it, then contracted the Squibb pharmaceutical company to fractionate it into products that did not transmit the disease.

One of those products was gamma globulins—more commonly known as antibodies—which come in several varieties, each resisting a specific disease. During the war, drug companies had isolated the gamma globulins that resist measles and hepatitis, and had distributed them to military personnel. (Scientists thought it was the gamma globulins' disease-fighting properties that made them unlikely transmitters of hepatitis, even when fractionated from contaminated plasma.) Now researchers made an exciting discovery. Working with the material provided by Squibb, they isolated the gamma globulin for polio.

Polio was the most feared epidemic of its day. Caused by a virus that invades the central nervous system, the disease presents itself as fever and stiffness, often followed by paralysis and death. The most tragic aspect of the polio epidemic was that it targeted children, infecting their spinal cords and leaving them unable to walk or even breathe. Images of children in "iron-lung" machines and of "poster children" hobbling bravely on crutches haunted the public mind in the early 1950s, creating a bleak counterpoint to the postwar American dream.

In 1951, Dr. William McDowell Hammond of the University of Pittsburgh showed that gamma globulins could provide children with "passive" resistance. Hammond found that injecting children with the polio antibodies conferred immunity to the disease as long as the gamma globulins remained in their bloodstream—generally for about a month. The ideal solution would be to provide an "active" resistance, in which a dead or harmless form of the virus is injected, stimulating the patient's immune system to produce its own gamma globulins actively whenever necessary. Almost immediately, the National Foundation for Infantile Paralysis (sponsor of the March of Dimes campaign) purchased every drop of gamma globulin in the country and gave it to the government for mass distribution. For two years, the gamma globulins became the nation's best defense against polio, until Jonas Salk's "active" vaccine provided permanent protection. (Dr. Albert Sabin's oral vaccine followed later.)

The market for gamma globulins rejuvenated the nascent plasma industry. Several companies that had dropped the technology decided to re-enter. Soon a host of fractionation products were being made—not only gamma globulins, but albumin and laboratory-testing reagents, all of which became increasingly profitable. But in order to reap their share of those profits, the drug companies would have to go through Cohn. The good doctor never felt comfortable with the companies' quality control, nor did he have great confidence in the government. So he set up a foundation to keep control over the patents. He would license them to drug companies free of charge on the condition that they submit their products to a commission he created for rigorous quality checks. Like many of Cohn's public endeavors, this one never earned him a dime.

Cohn's work with plasma derivatives was leading to a new level of sophistication about blood. Cohn maintained that administering blood in its whole form was a waste, since the liquid contained so many useful constituents. This much he had demonstrated with plasma, when he derived several pharmaceuticals from an undifferentiated liquid. Now he felt that he could do the same with whole blood. By isolating the cellular components of blood—the red cells, white cells, and platelets (disklike structures that serve as the physical foundation of clots)—he could make one unit of whole blood yield several units of usable products. He called this approach "component therapy," or "blood economy."

One of Cohn's colleagues, Charles Janeway, explained in a paper how the blood economy could work. Normally, he wrote, if you collect four units of blood you can treat four individuals (assuming they require only one unit each). But if you first separate the liquid into red cells and plasma, you can then treat six individuals—four with the red cells and two with the plasma. (Each unit of blood comprises several components that are present in different strengths. So it takes two units of whole blood, for example, to yield one dose of plasma.) You can increase the blood's usefulness again if you first fractionate the plasma into albumin, gamma globulin, and another component which Janeway identified as Fraction I. In this way, you can treat a total of twenty-three people—all with the original four units. Thus, even with the technology of the early 1950s the efficiency of each unit of blood could be increased by nearly 600 percent. As Janeway concluded, "The beauty of the work that has been done by Dr. Cohn's group and others is that for the first time real economy in the use of blood becomes possible."

Cohn envisioned a country interlaced with a network of blood centers, collecting and separating blood in all its forms, using the resource to its maximum economy. He had been actively promoting this idea since a seminal conference in 1949. The meeting had been held at the behest of the National Research Council and the military, who were looking for a source of platelets and white cells to stockpile along with plasma and albumin. Doctors who had visited Hiroshima after the surrender had found that gamma rays were lethal to blood. The rays targeted the hemopoietic system, destroying the production sites for red cells, white cells, and platelets. People who survived the blast suffered a dramatic decline in white cells, blood-clotting platelets, and plasma protein content, especially albumin—all of which contributed to the frightening new ailment that came to be known as radiation disease.

The implications of this syndrome were not lost on military planners. They realized that, in the event of an atomic war, blood and its components would be more vital than ever. Indeed, with the specter of atomic war hanging in the air, General Leslie R. Groves, director of the Manhattan Project, wrote that the time had come to accelerate blood research:

> If an atomic bomb is ever used against us, the matter of therapy for the victims will be of utmost importance. The best medical opinion is that the most essential part of this therapy will be the use of whole blood. It is true that during World War II dried plasma was used with remarkable success in treating the injured. However, plasma itself will not be sufficient for an individual who has been injured by ionizing radiation. Whole blood will be given until such a time as the patient's bone marrow has a chance to recover. We are not aware that any process has been developed for the storage of the cellular elements of the blood for long periods. . . .
>
> It seems obvious that some system must be devised within the next few years whereby a large amount of all the elements of blood will be prepared and stored to be ready in the case of an atomic war.
>
> I urge that the solution of this problem be given a high priority by the military establishment.

Similar concerns motivated the nearly 140 scientists who had gathered at the conference at Harvard Medical School, titled "The Preservation of the Formed Elements and the Proteins of the Blood." They presented research on extending the life of red cells, on isolating white cells, and on the mechanical properties of each. They gave papers about platelets and other "Components Concerned with Blood Coagulation." Everyone agreed to push forward with component therapy as the next stage of blood resource development. Yet out of that meeting came a disturbing realization. All blood-collection equipment at the time employed glass, citrate, and rubber tubing. Glass bottles stored blood, citrate kept it from clotting, and rubber tubing transported it from one vessel to another. But each of these materials destroyed the components that scientists wanted to isolate and use. Glass and rubber had wettable surfaces that caused platelets to rupture; citrate killed white cells. Cohn realized that if he wanted to develop true component therapy, he would have to replace the technology on which the industry had come to depend.

Just across the courtyard from Cohn worked a medical innovator with an equally cantankerous personality. Dr. Carl W. Walter of Peter Bent Brigham Hospital had already made a name for himself in the field of asepsis—the development of sterile operating procedures. Trained as a surgeon, he eventually became less concerned with the operation than with the environmental factors that affected the patient. All too often, patients would survive the surgery only to die from an infection contracted in the operating room. Walter developed new methods to sterilize instruments and to remove airborne contaminants. His book *The Aseptic Treatment of Wounds* became a standard medical-school reference throughout the United States.

Walter also had years of experience with blood and transfusion. A protégé of Dr. Elliott C. Cutler, who helped organize the blood supply for the invasion of Normandy, Walter got his introduction to the field as a medical student in the early 1930s, in the early days of transfusion therapy. One day he was assisting in an operating room, managing the tubes in an arm-to-arm transfusion, when the tubes kept getting clogged. To keep the blood flowing, the chief resident would order the nurse to squeeze harder on the simple ball-hand pump. Suddenly one of the tubes exploded, showering everyone in the operating theater with blood. "Jesus Christ!" Walter shouted. "There must be a better way!"

A few years later, Walter set up a blood bank at Brigham. There he stored citrated blood in glass bottles and exchanged it among local facilities. "We got into trouble with the trustees over it," he later recalled. "They thought it was immoral and unethical to traffic in human blood." Later, when Cohn began his fractionation experiments, it was Walter who provided the raw material.

Now Walter too became dissatisfied with glass. He saw rigid glass bottles as repositories of contamination, since the blood made contact with air as it filled them. The same went for rubber tubing. In fact, every stopper, needle, and tube junction represented a possible contamination point. The solution, he felt, would be to replace the glass bottle with a soft plastic container which could be stored flat and airless until the time came to fill it. It might even be possible to weld the plastic blood bag to needles and plastic tubes, producing an entirely sealed and sterilized system.

Walter worked for years to put his concept into practice, experimenting with one variety of plastic after the next. Some burst during sterilization, or ruptured during freezing. Some cracked when handled too roughly. Some leached toxic chemicals to the blood. Finally, after five years of trial-and-error experiments (many of which took place on the family's kitchen table), Walter came up with the right formulation—a pillow-shaped, flexible, one-pint container that was biochemically inert, immune to temperature extremes, and tough enough to survive a two-thousand-foot drop from an airplane (as the government required). Walter knew that Cohn was awaiting a substitute for glass. When he finally found the right polymer, he marched into Cohn's lab with a bag full of blood and threw it at Cohn's feet. "Here," he said simply, and stood on the bag to demonstrate its strength. Cohn, as usual, acted nonplussed.

The plastic bag and associated tubing revolutionized blood collection. It eliminated the risk of bacterial contamination and could be stored in much less space than a glass bottle. Combat medics loved it. Rather than hold a glass bottle high as they transfused under fire, making themselves a target, they could stay crouched, infusing the blood by squeezing the bag, or even by slipping it under the patient. In the worst of conditions, there was no danger of infusing air into the patient along with blood or plasma. Finally, the plastic surfaces repelled blood cells in a manner that protected them from damage.

The blood bag was more than a simple container: It was the heart of a flexible blood-processing *system* that made all subsequent advances in blood processing possible. Doctors could connect several bags into an airtight arrangement of bags and plastic tubes, separate the components of blood, and easily squeeze them from one chamber to the next—all with no exposure to the air. To produce the bags, Walter founded a company called Fenwal (a neighbor named T. Legar Fenn provided the financing). Located in a Boston suburb, the company was a model of progressive management, with liberal profit-sharing and hiring of the handicapped. Later Baxter Laboratories of Chicago bought Walter's company, cementing its position as a powerhouse in the world trade of blood derivatives and equipment.

Cohn incorporated some of Walter's technologies in a machine he was developing to separate the cellular elements. Blood, he had learned, is inherently self-destructive. Since it contains enzymes that attack its cellular components, the longer it remains outside the body the less usable those elements become. It thus became important to minimize the time between collection and processing. Cohn felt that, rather than having blood shipped to the laboratory, as had been done during the war, a fleet of mobile laboratories could bring the process to the donors, and derive the components while the blood was vital and fresh.

This in itself required several new inventions. To avoid the cell-damaging contact with glass, Cohn's team coated all surfaces with silicone; later they used the plastics that Walter had adapted. To avoid using citrate, they borrowed another of Walter's inventions—an exchange resin that, incorporated into hard little beads like water softener, selectively removed calcium from the blood and stalled the clotting reaction. They also modified the fractionation process by using zinc instead of alcohol to precipitate the proteins, which freed them of the necessity to run the procedure at subzero temperatures. It was all completely mechanized and enclosed, so that no human hand, no air, no bacteria made any contact with the blood.

Funded by the Atomic Energy Commission, Cohn supervised the construction of several prototypes of the Mobile Blood Processing Laboratory. An early model rolled up to the General Electric laboratories in Schenectady, New York, for a demonstration at the annual meeting of the National Academy of Sciences. Filling a conventional trailer truck—a "veritable wonderland on wheels," as one reporter described it—the contraption hummed for several hours after donors primed it with blood, then produced a dozen blood-derived compounds. A year later, the machine had been shrunk to the size of a clothes washer. This was the device Cohn had demonstrated in Lisbon, with its tubes, heat exchangers, and sealed, whirling centrifuges. His team produced a dozen of these prototypes, which found homes in laboratories in America and in France.

Cohn's lab fairly hummed with innovation and industry. Every corner had someone developing new machinery, isolating proteins, preserving the blood cells. The population of lab workers swelled, as young scientists from the world over came to do research and study his techniques. Many of these visitors became leading researchers in their own countries after their stints at Harvard.

Cohn worked as hard as everyone around him. "It's a great life," he told a reporter. "I'm exhausted. But I love it." But his strength was not limitless. For years he had been ignoring his doctors' advice to try not to aggravate his chronically high blood pressure; twice, in fact, they had checked him into a hospital, but he had signed himself out.

He also suffered tension-related asthma attacks, usually before public appearances. At such moments he would summon one of the doctors in the lab—usually Sam Gibson—to inject a large dose of adrenaline to reduce the swelling in his windpipe. "He'd take off his fancy vest, roll up his fancy cuffs, and say, 'Go ahead, shoot it all in,' " Gibson would later recall. Adrenaline triggers a nervous reaction, and the doses he gave Cohn were enough to make the average man jump out of his skin. Yet Cohn went back to work without so much as an uptick in blood pressure. After Cohn's death, an autopsy revealed that a tumor on the adrenal gland had caused his insensitivity to the hormone, but at the time it must have seemed to his colleagues that Cohn was so naturally high-strung that nothing could crank him up any further.

Now, however, having turned sixty, Cohn was beginning to show strain. His behavior, always difficult, became occasionally irrational, and even more obsessive than before. One of his obsessions involved his idea that he was on the verge of discovering a new principle of protein structure, a kind of grand, unified theory of biology. That in itself was not unusual: Cohn could make intellectual leaps that seemed unlikely until others caught up with him. But his affect surrounding this obsession had changed. Previously he would rarely present new ideas to the press; when he did, he would speak in the most cautionary, conservative terms. Now he fairly touted his new theory as the next revolution in protein science. Previously he would invite colleagues to challenge his ideas, even prolonging the discussion for the sheer joy of defending them. Now he grew defensive, shutting out friends who dared to disagree.

Colleagues recall a colloquium he hosted in a villa on the shores of Lake Geneva in Switzerland. The setting was bucolic and the atmosphere convivial as one speaker after the next presented the status of his research. Yet Cohn's presentation, for the first time that anyone remembered, made no sense to anyone in the room. Another scientist, Frederick Sanger of Britain, presented a theory of protein structure that seemed eminently more logical. Later, when Sanger came to Harvard for a visit, Cohn barred his students from attending the lecture, because he felt Sanger's ideas would mislead them. (Sanger later won two Nobel Prizes.) Cohn made life so difficult for his colleagues that one of his most trusted, biochemist John T. Edsall, transferred to another lab.

Colleagues did not know it at the time, but medical problems accounted for his behavior. Apparently Cohn suffered a series of small strokes. The condition became more apparent in time, making his utterances increasingly arbitrary. Sometimes he would blank for a minute or two, unable to speak. Yet he still maintained enough foresight and control to establish another organization to carry on his work in investigating the fundamental properties of blood.

The end came on a fall morning in Boston. Cohn was at his office that day, working on a manuscript, when he paused for a phone call with his friend George Scatchard. Scatchard, a chemistry professor at the Massachusetts Institute of Technology, had been a comrade since college days, and performed chemical analyses for Cohn. Cohn was chatting with his friend when he suffered a final, massive stroke. As the telephone earpiece fell from his hand, he moaned, "Georgie, I can't hear you." He died in the hospital a few days later, on October 1, 1953, at the age of sixty-one.

They held a service for Cohn at Harvard's Memorial Church, a classical structure with squat, solid columns built to honor the war dead. Several of his colleagues spoke admiringly of Cohn. John Edsall spoke too, despite the pain over their ruptured relationship. "I cannot recall what I said," he would say later. "My emotions were very mixed." When the National Academy of Sciences asked Edsall to write an appreciation of Cohn, he had to wait years to gain sufficient perspective.

What can one say about such a towering and difficult personality? Certainly the man had his failings. The wounds he inflicted on some of his colleagues remained painful for years, and not all of his creations found permanent success. Yet the good he did is almost immeasurable. He was not the first to see the value in plasma or to prepare it for industrial processing—John Elliott and Charles Drew did that—nor did he pioneer freeze-drying the liquid into a form used by thousands of GIs, as Max Strumia and others did. But by fractionating plasma, he set the stage for all the blood work that was to follow. By finding ways to use all parts of the blood—including the plasma fractions and the cellular components—he laid the groundwork for blood-component therapy as practiced today. He taught a generation of scientists about blood, for the people he trained became leaders in the field throughout the world. His ability to forge alliances between university and industry researchers presaged the modern era of corporate-academic science, yet his insistence that the patents be assigned for the good of humanity reflected the values of a bygone age.

George Scatchard liked to say that his friend was not one man but two, "one of whom occupied the center stage while the other sat in the gallery and watched the first." Similarly, Cohn was a man of two centuries—the modern, driven scientist in the corporate mold, and the scientific "noble" of a previous era, elegantly cultured, highly principled, whose sense of duty would not allow him to profit from his inventions. He never won the Nobel Prize that he felt he deserved, yet he won international respect and even a small measure of fame. People remembered the images of GIs bearing bottles of plasma and albumin, and of the prevaccine days of measles and polio. They saw how blood had become an increasingly vital medical commodity, useful in new and multiple ways. Indeed, many considered plasma fractionation, along with antibiotics, the greatest medical advance of World War II. "Dr. Cohn's fractionation of blood is one of the glories of American science," said a commentary in *The New York Times*. "It was work done with no other end in view than the enrichment of human knowledge, yet work that has proved to be as important [as] and more difficult than that which gave us the antibiotics." The allegorical quality of his work was inescapable. " 'Blood will tell' is a well-worn phrase because blood has always been a mysterious fluid. Dr. Cohn stripped away the symbolism and made blood tell what it never told before."

# CHAPTER 11

# THE BLOOD BOOM

In 1955, a man named Francis H. Bass decided to make blood banking his career. Bass aspired to be a self-made man. Having had no formal education past grammar school, he had worked as a tenant farmer in Illinois, a banjo-and-mandolin teacher in Indiana, and a used-car salesman and photo-shop manager in Missouri and Oklahoma. He was living in Houston with his wife, Margaret, running two photography studios, when he decided that blood was the future. After all, he had the business skills, and Margaret was a "registered nurse," although she had received her credentials through mail-order diplomas. So, with nothing more than the will and a dream, they began to put their plan together.

Margaret visited a blood center in Houston to observe blood-drawing and storage procedures, and they placed newspaper ads to raise seed money. They decided to set up in Kansas City, Missouri, where Margaret had previously worked, and hired a medical director—Dr. James W. Graham, a seventy-eight-year-old retired physician who had no hematology experience. Leasing a storefront in a seedy part of downtown, they hung out a sign reading "Cash for Blood" and offered to pay up to $5 a pint. There, under the name of Jackson County Blood and Plasma Service, which they changed to Midwest Blood and Plasma Center and later to World Blood Bank, the Basses officially established their business.

Before the blood bank actually opened its doors, Dr. Graham telephoned one of the city's leading pathologists, Dr. Victor B. Buhler, to establish collegial relations. Graham, a rank novice in the field, knew nothing of Buhler's long-standing opposition to for-profit blood banks. He described the new enterprise to Buhler, explaining how it could benefit the entire community.

"Isn't that wonderful?" Graham finally said.

"No! That is terrible!" said Buhler.

"Oh, is that so?"

"Yes!" thundered Buhler. Thereupon, he lectured Graham on the evils of paying blood donors, on how buying blood was an affront to human dignity and exploited society's most desperate classes. He added that a blood center required "someone knowledgeable and expert" to serve as director, not a retired general practitioner.

That small but embarrassing exchange presaged more than a decade of conflict between the doctors of Kansas City and the for-profit blood bank that came to their community, as well as a series of conflicts that began to emerge throughout the nation. In the 1950s, a new class of blood dealers arose—not people who had dedicated themselves to science and a cause, like Cohn, Strumia, Diamond, and Hemphill, but untrained profiteers. Seeing a new market, they set up collection centers in town after town, earning the sobriquet "booze for ooze" because of their reliance on derelict donors. Though never constituting a large part of the blood business, they represented for a time the fastest-growing segment. Everywhere they went came protests from local doctors who were disapproving of the centers yet chagrined at having no other local source of blood. The situation was bound to boil over, and it did in Kansas City, where a local conflict snowballed into a federal case. The outcome would affect the nation's blood laws for the rest of the century.

The Basses could not have picked a better time for their little enterprise. All across America, blood was becoming a growth market. Medicine had advanced mightily after the war, with open-heart surgery, artificial kidneys, and improved trauma care, all of which consumed huge amounts of blood. (The heart-lung pump, for example, required up to twelve pints of blood for priming alone.) With patients surviving longer and requiring more sophisticated equipment, hospitals guzzled blood as voraciously as cars of that era consumed gasoline. Unlike gasoline, however, blood had become more, not less, difficult to obtain. The sense of community service had diminished in the aftermath of World War II and the Korean War, with fewer people willing to give blood. Larger social forces came into play as well. As medical care became more technological and costly, it concentrated in the nation's great medical complexes, most of which were located in the cities. Cities thus became sinkholes for blood. At the same time, middle-class donors began migrating to the suburbs. A "blood gap" developed that voluntary banks, for all their expansion, found impossible to fill.

"Blood gap" only begins to describe the problems that quickly evolved into a national mess. Unlike most Western European countries, which had enacted blood legislation following the war, the Americans had never established a national policy. They left blood collection to the marketplace, with no central coordination, inventory, or control. The American Red Cross and American Association of Blood Banks each attempted to monopolize the market, yet neither was able to prevail. Their inability to cooperate divided the nation into a patchwork of territories. Each system had its own set of rules—free blood in the Red Cross areas and "loaned" blood in territories of the AABB. Neither respected the rules of the other. If a patient from a Red Cross territory fell victim to an accident in a region managed by the AABB, he would have to purchase his blood, just like any pharmaceutical.

The blood supply thus developed erratically—organized in some places, chaotic in others. Individual hospitals tried to take care of their own needs by encouraging donations and setting up policies by which a patient could "replace" the blood he had consumed by recruiting donations from family and friends. Often as not, these practices failed. Hospitals found it impossible to maintain stable inventories of the blood types they needed, given that they were dealing with a perishable commodity that came from unpredictable volunteers. Some developed surpluses; others ran short. Surgeons postponed operations for lack of the appropriate blood.

Just how disorganized the situation had become was revealed in a study of the New York City blood system in 1956. New York had been a transfusion pioneer—the city of Landsteiner and Carrel; of the Blood Betterment Association in the 1920s, and the Plasma for Britain project in the late 1930s. But by the 1950s, the city's blood supply had become a world of "greed, waste, chaos and danger," according to Dr. August H. Groeschel, chairman of the New York Academy of Medicine panel that investigated the system. More than 150 blood banks operated in the city, with no uniform policy, coordination, or control. Hospitals billed patients anywhere from nothing to $60 for a unit of blood; those that permitted blood-for-blood replacement charged anywhere from one to five pints to replace each pint the patient had consumed. No one knew what anyone else was doing. One hospital might postpone an operation for lack of a certain blood type that another up the road had in great supply. Near midnight of one St. Patrick's Day, for example, one hospital had to broadcast an emergency television appeal for blood, even though a hospital five minutes away had a surplus of the same blood type. The situation proved almost criminally wasteful: Each year in New York, hospitals discarded about ten thousand units of outdated blood.

These gaps and disorder on the national scale gave rise to a new generation of entrepreneurs who dealt with blood strictly as a commodity, buying it from donors and selling it for profit. Setting up near medical facilities, military bases, and poorer urban neighborhoods, they would solicit a variety of donors, from medical students to laid-off longshoremen. But the people with whom they became most associated were the perpetually unemployed, the down-and-out. Blood centers came to form part of the weary landscape of America's skid rows, with winos and drug addicts lingering outside and shady practitioners working within.

These for-profit blood centers never constituted more than a fifth of the nation's collection capacity, but in some areas they became particularly strong. In New York, for example, at least 42 percent of the city's blood came from professionals during the mid-to-late 1950s. These were not the carefully monitored citizens who gave in the old days of the Blood Transfusion Betterment Association, when the health department held them to a strict lifestyle and quasi-moral code, but an underclass, desperate and down on their luck. Made sickly by their living conditions and tempted to lie by the promise of cash, these donors represented high risks for hepatitis and malaria. On one occasion a patient developed malaria after receiving blood during an open-heart operation. Attempting to find the source of the infection, the doctors traced several units to a center on the Bowery. But they never could pinpoint the source of the contagion, because all the donors had used fictitious names. Dr. Aaron Kellner, a leader in blood reform in the city, used to tell the story (apocryphal perhaps, for Kellner was a salty and enthusiastic *raconteur*) of how one day he was emerging from a meeting with the commissioner of health and New York Governor Nelson A. Rockefeller when they all passed a drunk lying on the curb. Kellner peered at the man and said, "That's one of our donors!"

Like many for-profits, the Basses started out with a reasonable business plan. Moving to Kansas City, a place with an inadequate blood supply, they would provide blood to anyone who needed it, purchasing blood from professional donors and selling it to local hospitals for a profit. That was one level of their operation. They also set up a blood-assurance plan to help clients conform to the blood-replacement rules of local hospitals. Under the plan, the client paid the Basses an annual fee and donation of blood. If at any time during that year he or his family used blood at a local hospital, the Basses would replace it. It was all a matter of collecting from a large group of clients to service a few, just as an insurance company can make occasional large settlements because most of its paying customers never have an accident.

The assurance plan was elegant, profitable, and simple. All it required was that local hospitals accept the blood the Basses had collected in lieu of their patients' blood or replacement fees. This would require a measure of goodwill, something that evaded the Basses from the start. After Dr. Graham's telephone fiasco with Dr. Buhler, the Basses invited Buhler to come for a personal inspection. The visit began cordially enough. But when he asked Franklin Bass how much he intended to pay donors, Bass lost his temper.

From then on, everything the Basses did alienated the local medical establishment. Unschooled in the niceties of medical decorum, they relied on sales techniques more suited to flimflam artists than to medical professionals. They deluged hospital administrators with promotional letters and advertised their blood bank with crass, misleading newspaper ads. (One claimed that the Midwest Blood and Plasma Center had been formed "to eliminate profiteering in blood.") Often they tried to foist their blood on unwitting hospital personnel. Sometimes they would show up at a hospital blood bank late at night, when only the junior staff were in attendance, and cajole a technician into accepting their blood as part of a patient's replacement fee. On one occasion, a messenger from Midwest showed up at a hospital with three bottles of blood, two visibly clotted, and the third testing positive for syphilis. When the staff physician told the messenger to take the bottles back, the man "threatened him in a filthy manner."

If the Basses' sales techniques eroded confidence, stories about their hygienic conditions did even more. Doctors driving past Midwest on their way to the hospitals would see winos and the homeless loitering outside. One time, students who were attempting to donate blood for a friend reported that the waiting room at Midwest was swarming with ants. Former employees reported that the staff drew blood in a careless and unhygienic manner, taking it from anyone who was not visibly drunk.

All this confirmed to local doctors that the people of Kansas City desperately needed a safe, nonprofit community blood bank. For years they had discussed such an enterprise, but differences of opinion over issues of structure and management had always prevented them from moving forward. Now they saw the hazards of inaction. Casting aside the issues that had divided them, they established the Community Blood Bank of Kansas City Area. The new blood bank depended on voluntary donors, and made no profits from distributing blood. Managed by a board of eminent people from medical and civic life, it operated as a community resource, providing specialized derivatives such as red cells and platelets as well as whole blood, and training medical students and technologists. It was exactly what the area needed. When Community opened its doors in 1958, six of the area's hospitals shut their own blood banks and signed on with the new blood bank, and over the next few years twenty-six more hospitals signed on. By the early 1960s, Community had gained a virtual monopoly over Kansas City's blood.

Doctors viewed this development with relief. After all, blood should be managed as a community-wide resource. Its types, perishability, and dependence on volunteer donors necessitate community-wide collection and distribution. But the Basses did not see it that way. They saw themselves as a small business in a trade war, an independent David versus a burgeoning Goliath. And so they fought back. They hired a new salesman—a resourceful individual named James W. Remer, who previously had worked as a bartender at a Kansas City establishment called the Red Garter. Remer did virtually everything at Midwest, from screening patients to withdrawing blood; but his real virtuosity was in working the phone. A born telemarketer before the term became popular, Remer sold hundreds of blood-insurance plans to unions, church groups, and government employees. He developed clever public-relations schemes, such as a "Benevolent Fund," promising "free" blood to rural hospitals, which, after various fees were added, actually received no discount at all.

Now the Basses started attacking the hospitals. Franklin Bass had written to the county medical society, complaining that his "institution has received the most severe, unfair and unwarranted persecution, motivated by those whom we believe to have special interests in mind." He had lodged a complaint with the Kansas City Better Business Bureau. Now he began sending "memory joggers" to his clients. Each time a client received blood in a local hospital, he would receive a letter from the Basses reminding him that if the doctors refused to accept Midwest blood as replacement the patient would be "fully justified" in filing a lawsuit.

The Basses had certainly become a local irritant. But, then, such people would always be lurking, trying to profit from a community's need.

The doctors assumed that, by diligently turning away the Basses' business, they could keep the situation in check. Little did they know that what they considered sound and ethical medical advice could be seen, in some circles, as an illegal conspiracy.

Their troubles arrived in the person of C. F. Snavely, an investigator from the Federal Trade Commission. Snavely had come to determine whether the doctors in the Kansas City Medical Society had mounted an illegal economic boycott against the Basses. The doctors did not worry at first, since they were convinced that once he heard their side of the story he would dismiss the Basses' charges out of hand. But the investigator did not. Whereas the doctors saw the dispute in medical and public-health terms, Snavely saw it purely as an issue of free trade. Time and again the doctors tried to explain the nature of their disagreement, only to have Snavely ignore them. The exchanges grew snappish as the two sides talked past one another, neither comprehending the other's priorities. Physicians took to calling him "Snakely." In one confrontation later recounted at a federal hearing, Snavely asked Dr. Ferdinand C. Helwig, a renowned pathologist at St. Luke's Hospital, under what circumstances the doctors in the city could imagine using commercially obtained blood. Helwig conceded that he might consider paid blood in certain emergencies. Snavely pressed further: Could Helwig imagine a situation in which he could feel satisfied taking blood from Skid Row? Helwig bristled. "Mr. Snavely, would *you* be satisfied to take Skid Row blood?"

"That is not what I am asking you," Snavely replied. "Would *you* be satisfied?"

"That is not what I am *telling* you," Helwig insisted. "Would *you* be satisfied?"

At that point the two men simply sat back and glared, separated by an unbreachable gulf in training, ethics, and worldview.

Sometime later, Snavely left town. Any relief the doctors may have felt soon disappeared, when the Federal Trade Commission charged them with illegally conspiring to interfere with free trade. If guilty, the defendants—including the Community Blood Bank of Kansas City, the Kansas City Area Hospital Association, and sixteen local pathologists—would be forced to do business with the Basses or face a penalty of $5,000 a day.

The proceedings took the form of a trial, with attorneys presenting nearly one hundred witnesses and a thousand exhibits to a Federal Trade Commission hearing examiner. The government's case was simple: The hospitals and pathologists served by Community Blood Bank conspired to freeze out a competitor, thereby giving their own nonprofit blood the power of a regional monopoly. The doctors had thus violated the Federal Trade Commission Act by mounting a commercial boycott against the Basses to put them out of business.

The defendants argued that they had acted not as economic competitors but as professionals with a sense of public responsibility. The doctors knew of the conditions at Midwest, and morally objected to paying donors for blood. Given the doctors' knowledge and beliefs, it would have been irresponsible *not* to eschew blood from Midwest. Besides, one could hardly call their actions a boycott: Nobody *schemed* to drive Midwest out of business; what appeared to be a conspiracy was merely the collective impact of individual medical decisions.

Beyond the particular arguments at hand, though, lay a deeper and more fundamental disagreement. In order to make its case, the government defined blood as an economic commodity—something that could be bought, sold, and processed like any other drug. As such it would fall subject to the normal trade laws, forbidding economic boycotts and restraint of trade. This carried implications far beyond the current case. If blood became a product—like soda pop, for example—then other product-related statutes would kick in. The most threatening among them was the Uniform Product Code, a federal regulation adopted by all the states mandating that anything sold as an article of commerce carries an implied warranty. This means that the seller guarantees the safety of the product for the purposes intended—the soda pop, for example, must be fit to drink. If the consumer is harmed by using the product—if he, say, swallows a nail or harmful chemical in the soda—he can sue the manufacturer for violating the implied warranty. He does not have to prove negligence, the manufacturer should have known.

The prospect of an implied warranty sounded an alarm for blood bankers. Blood contained hazards that could escape even the most diligent physician. There was no way a doctor could determine whether the blood he dispensed contained the hepatitis virus, since no tests existed to detect it at the time, and even a seemingly healthy person could carry the disease. If the Federal Trade Commission's definition took hold, transfusionists could look forward to spending a good portion of their future combating lawsuits. To counter the government's arguments, the doctors pushed their own definition of blood—that it was not a commodity, but a living tissue, a human organ. Similarly, transfusion was not a commercial transaction, but a medical service akin to an operation. (Some went so far as to liken it to a transplant.) Granted, transfusion frequently involved money, as when a hospital paid service charges or when a patient paid a replacement fee. But no—the transfusion process was a medical *service* and should never be subject to commerce and trade laws. ("Service" is a key word here, for it excludes the doctor from the principle of implied warranty and demands a much higher legal standard for lawsuits, in which plaintiffs would have to prove actual negligence.) The very idea of blood as a product was unthinkable, a threat to good medicine and to public health.

It did not take long for blood bankers throughout the nation to perceive the threat in the Federal Trade Commission's case: If the FTC charges succeeded, no community would be safe from the likes of the Basses; no doctors would be free to make collective decisions. Meetings would become conspiracies; efforts to mobilize a community's blood, as the Red Cross had done during the war, could now be seen as regional monopolies, forbidden by federal law. The whole system, so painstakingly constructed in communities throughout the nation, would collapse under the weight of this legal attack. And so blood bankers from all over traveled to Kansas City to lend their support. Max Strumia flew in from Philadelphia to testify about the dangers of hepatitis. He eloquently explained how blood becomes a road map of every pathogen to which a donor has been exposed, and thus requires careful selection. Bernice Hemphill came from San Francisco to explain the workings of the National Blood Clearinghouse and how a well-coordinated exchange of credits could make blood a true community resource.

Despite the testimony of such authorities, a sense of unreality hung in the air. The doctors of Kansas City spent weeks talking about medicine and patient care, only to be told that the true issues centered on commodities and trade. The defendants felt increasingly desperate. In an open letter to colleagues in *Transfusion,* the journal of the AABB, Dr. Meyer L. Goldman, president of Community, pleaded for contributions for the blood bank's defense: "Is human blood a commodity to be bought and sold on the open market, and dealt with like beets, automobiles and clothing? . . . The position of the Federal Trade Commission on these basic issues is alarming and, in our opinion, is detrimental to the concept of blood procurement through voluntary donor efforts. . . . Success by the Federal Trade Commission would endanger not only the Community Blood Bank of the Kansas City Area, Inc. but could open to attack every similar organization in the United States."

Meanwhile, government lawyers, in making their case, deflated some of Community Blood Bank's righteousness. For example, when the Kansas City doctors criticized the Basses for using paid donors, the government revealed that Community often derived a portion of its blood from paid donors as well, sometimes as much as 40 percent. When the defendants criticized Midwest's use of prison blood, the government revealed that Community Blood Bank had set up a donors' club in Leavenworth Prison. These were not evil or cynical actions, but arose from the exigencies of blood banking at the time.

Ten months after the hearing concluded, the defendants received the verdict by mail. Just as they expected, the Basses had defeated them. Their blood bank, ruled hearing examiner William J. Bennet, had been a legal, fully licensed business entitled to the complete protection of the law. That the doctors thought the Basses immoral did not allow them to violate that law through "concerted refusal to deal." As to the more basic question of whether blood was a commodity, Bennet reasoned that freshly drawn blood may start out as a living tissue, but its nature soon changes through the addition of citrate anticoagulant. Adding this chemical makes the blood something other than virgin human tissue; it becomes a product—more specifically, a drug. As such, it is subject to all laws that govern commerce in the United States, including those that prohibit monopolies. Therefore, he ruled, the Community Blood Bank of Kansas City and its affiliated doctors and hospitals should cease and desist from boycotting the Basses.

The news of the decision triggered an outcry throughout the blood-banking community. "Monstrous!" gasped the chairman of the Red Cross blood center in Newport News, Virginia, in comments to a local newspaper. "It alarms me to think that operations of a voluntary, non-profit, lifesaving community service program could be curtailed because of an alleged competition with a commercial enterprise. I can't imagine such a thing happening. I am shocked." Dr. Byron A. Myhre, a pathologist and blood banker from Los Angeles, wrote that to classify blood as a "manufactured product" was nothing short of "sacrilegious." Dr. Charles M. Poser of the University of Missouri School of Medicine said no reasonable person would put "blood, eyes, bone for grafts, etc.," in the same category as "refrigerators or canned food." Dr. M. A. Meservey of the Iowa Association of Blood Banks warned that "blood banking as we have known it will cease to exist." In Geneva, Illinois, the medical staff at the Community Hospital passed a resolution affirming that blood is not a commodity to be bought and sold in trade.

Blood bankers were not the only people to worry about the consequences. Theologians were aghast that blood, rich in humanitarian and religious symbology, could be relegated to the status of a mere product, like cat food. While the decision was still pending, Thomas O'Donnell, a professor of moral theology at Woodstock College in Maryland, lamented, "This destructive and expensive litigation will solve nothing. I really believe that after all the passions and money have been spent, no one will know whether blood is to be considered exclusively a commodity or a tissue or whether these terms are mutually exclusive. . . . But whatever these painful litigations may prove in the future, they have already presented an indescribably valuable object lesson to all of us—they have shown what happens whenever human greed and jealousy invade an area of human activity which, of its nature, is only effective and meaningful when motivated by the love of a man for his fellow man."

O'Donnell was echoing the sentiments of Pope John XXIII, who, a few years before, had warned about the perils of commercializing blood. Addressing an international conference of blood bankers in Rome, the pope had asserted that blood donation "must be rooted in charity, which is love of God and love of brother. . . . Without charity acts of heroism would be like sounding brass and tinkling cymbals; with charity even a drop of blood acquires supernatural value before God."

One of the most eloquent protestations came from Dr. Tibor J. Greenwalt, director of the community blood bank in Milwaukee, a nationally respected organizer and editor of *Transfusion*. Greenwalt set forth a logical, impassioned argument why the FTC decision undermined the very principle of voluntary, community blood banking. He explained that, in order for a community to establish a blood bank, the local hospitals, medical societies, and civic groups must meet many times to coordinate their activities: That could now be seen as collusion. Hospitals would be forced to take blood from anyone in order to avoid a local monopoly, even the commercial firms that dealt in "ooze for booze." Such practices would increasingly exploit "the undernourished and unwashed segments of our population," he wrote. "Can it be considered proper to bleed and exploit the skid row inhabitants whose nutrition is so poor . . . ? In the case of manmade or natural disasters are we to rely upon the undernourished to supply the blood needs of the Nation? . . . If blood is to be treated like any other pharmaceutical product, all the efforts of those who have worked so hard to assure adequate supplies of blood from volunteer donors during the past 25 years will have been wasted."

To many observers, the Kansas City decision characterized a federal agency run amok, regulating activities it had no business touching. In Washington, Senator Eugene V. Long, whose district included Kansas City, argued that the Federal Trade Commission had not been established "to tamper with a lifesaving medical practice." He pointed out that organ-transplant technology was rapidly advancing, and that soon human organs would be "banked" just like blood. Would people like the Basses someday purchase eyes and kidneys from the poor, and would doctors collectively be powerless to avoid them? It was all too grotesque to consider. To forestall such scenarios, Long introduced legislation specifically to exempt blood banks from antitrust laws. The FTC decision could not be allowed to stand.

As the drama played out in Kansas City and Washington, events of near-seismic proportions were shaking up the blood bankers thousands of miles away. The postwar decades had been good for Japan, as the country rose from the war's ruination, and good for Dr. Naito, the country's premier for-profit blood banker. From the first collection-and-processing laboratory center in Osaka, Naito's Japan Blood Bank had grown to become the nation's largest purchaser of blood, with laboratories, branch offices, and collection centers in every region. Admittedly, he felt qualms about buying blood, but social conditions had left him no choice. With the national insurance system assuring free blood, the Japanese middle class did not have to think about the need to provide it. This attitude was reinforced by the lack of a Western-style philanthropic tradition. Even the Japanese Red Cross, which began as a nonremunerated service, found it impossible to attract volunteers and began paying its donors in 1955. Thus, the Japanese middle class thrived on blood that had been purchased from the poor.

Nevertheless, Naito could feel proud of his accomplishments, for he, as much as anyone in Japan, had brought blood banking and processing into the modern medical era. Indeed, he was so proud that he organized his colleagues to host the 1960 meeting of the International Society of Blood Transfusion (ISBT). This worldwide association of blood bankers and hematologists met in a different country every other year to discuss the latest in a variety of topics, from new laboratory procedures, to newly detected diseases of the blood, to methods of attracting donors. Eminent specialists addressed the society; the pope had made a speech at the conference when the delegates had convened in Rome. Hosting the meeting was more than a demonstration of hospitality; it was a sign that the host nation had arrived.

As delegates arrived at the ISBT's eighth annual meeting in Tokyo, the Japanese greeted them with pride and excitement. The event was so auspicious that Her Imperial Majesty Empress Nagako composed a poem for the occasion. Naito delighted in escorting visitors to tourist sites, his business, and even his home, especially those who had hosted him in their countries. Then, for the final day, the organizers moved the proceedings to Hakone National Park, in the shadow of Mount Fuji. Then, during a seminar entitled "Problems in Blood Transfusion," something unsettling occurred. A colleague of Naito's, Dr. Yoichi Azuma of the Kyushu Welfare Hospital in Fukuoka, presented a paper in which he frankly discussed Japan's blood-for-money system. He described the conditions that fostered it, how the overdrawing of blood caused anemia in the donors, and how not even that system could keep pace with the demand. Azuma's paper was not news to the delegates—they had known about Japan's paid-donor system and wanted to see it reformed—but it gave them a chance to voice their concerns, which, in the course of the discussion, became unusually sharp. The South African delegate remarked, "If kidney transplantation becomes popular will you allow people to buy and sell their own kidneys?" The Dutch representative wondered if the Japanese underclass would someday be selling their eyes. Naito told the audience that he shared Azuma's sentiments and hoped to promote a system of "blood depositing" based on the model of the AABB. But his comments could not take the edge off the rebuke. "My heart sank to the bottom," wrote Naito, "and was filled with anxiety for the company's future."

The press got wind of the delegates' remarks. Soon grim portrayals began to appear in newspapers about the ghettos of Tokyo and Osaka, where the blood centers did most of their recruiting. There legions of ghostly people tramped onto the buses sent by the blood companies to be transported like cattle to the collection centers. Reporters described donors as poor and addicted, their arms so scarred that their blood had to be withdrawn from their hands and feet. Many donors, known as *tako,* or octopus, drank home-brewed concoctions to boost their blood-iron levels, consuming chicken blood, spinach, or iron filings mixed with water. Organized crime members, or yakuza, would gamble in front of the blood banks to be near a ready source of cash when they ran into debt.

The press reports found an audience among university students who saw blood selling as yet another example of class inequalities in Japan. Led by Masanori Kimura of Waseda University in Tokyo, the students agitated for an all-volunteer system, organizing blood drives and demonstrations and recruiting celebrities to set an example. Sometimes they took a tougher edge, conducting undercover investigations and confronting civic officials with their evidence. Slowly they began to change Japanese society. Voluntary donations climbed from less than 1 percent of the total in 1960 to 2.5 percent in 1963. The following year, the Red Cross director, Dr. Shozo Murakami, announced that the organization would no longer pay donors. He admitted that the decision might bring some hardship, since removing money from the system could decrease the staff salaries, but the organization no longer had a choice. "If any of you do not agree with me," he told his staff, "write a resignation letter now. You can leave."

Murakami's decision may have sounded momentous, but it had little effect on the Japanese blood supply, since nearly all was provided by commercial firms, such as Naito's. Changing the situation would require a genuine jolt to public opinion, something that would illustrate, once and for all, the hazards of commercial blood trafficking. This came with the attempted assassination of the most admired American living in Japan.

Edwin O. Reischauer, the popular American ambassador to Japan, was born in Tokyo to American missionaries and educated at universities in America. He had earned a rare position of respect in Japan thanks to his fluency in the language, his respect for traditional values, and his Japanese wife. This popularity served him well under Presidents Kennedy and Johnson, especially when he was called on to defend, contrary to his personal beliefs, America's military involvement in Vietnam. He was leaving the embassy at noon on Tuesday, March 24, 1964, when he bumped into a small, slight Japanese man wearing an old raincoat. "Where is this man going?" Reischauer asked the people around him. Before anyone could react, the young man looked up, recognized the ambassador, and plunged a butcher knife deep into his thigh. As the embassy staff subdued the assailant, Reischauer called for a tourniquet, tied it around his hip, and allowed himself to be carried to the embassy car. "It was a gory scene, with the three men inside the car soaked with blood," he wrote in his memoirs. "Luckily the Toranomon Hospital, one of Tokyo's best, was only a couple of blocks away."

The assassination attempt mortified the Japanese, who, proud of their civility, had not seen a foreign diplomat injured since 1891, when an assassin tried to kill the visiting tsarevich of Russia, Nicholas II (the father of the hemophiliac Alexis, who died with his family at the hands of the Bolsheviks). An outpouring of apologies followed, and the public-safety commissioner resigned in contrition. Reischauer had been scheduled to appear the next day with Prime Minister Hayato Ikeda on the first live television broadcast from Japan to the United States via satellite; Ikeda now used the occasion to broadcast an apology to the American public.

Surgeons labored over Reischauer for hours, repairing the damaged arteries and nerves, and replacing the blood he had lost from a lot they had purchased from a commercial firm called the Fuji Organ Pharmaceutical Company. Knowing the blood carried a risk of hepatitis, they injected him with large doses of disease-resisting gamma globulin. Thousands of origami cranes filled the ambassador's room—the traditional get-well gift, the symbol of long life. During his recovery, Reischauer released a statement of unusual perception and sensitivity. Knowing that the Japanese felt a terrible loss of face, he cast the incident in a positive light, saying that he felt closer to the nation than ever before because Japanese blood now flowed in his veins.

Reischauer remained in the hospital for a few weeks, then traveled to Hawaii to complete his rehabilitation. Several weeks later, he developed hepatitis. The blood he had received must have contained such a high concentration of the virus that it was able to infect him, despite the gamma globulin.

The news of his infection doubly humiliated the Japanese. Not only had their honored guest suffered an assault in their land, but the blood they had given him—the very blood that had bound them together—was infected, *unclean*. The press escalated its attacks on the paid-donor centers. Amid the public explosion of outrage, the national Diet, or parliament, made voluntary donation the official Japanese-government policy. They designated the Red Cross as the official blood-collecting agency, setting aside 85 million yen for the Red Cross to buy equipment and bloodmobiles. They did not make paid blood illegal; to do so would have thrown the nation into a shortage. Instead, they planned to phase out the entrepreneurs by strengthening the voluntary sector.

A week after the Cabinet decision, Ryoichi Naito met with the Japan Blood Bank's board of directors. They knew the proposed changes could have a disastrous impact on the Japan Blood Bank, eliminating more than 40 percent of the company's business. The question was, should they fight the changes or find a way to accommodate them? The company had already begun to set up a system of American-style blood banks, where people could replace the blood they had used or pay a fee, rather than simply purchasing blood from the poor. Maybe by expanding this system they could abandon their exploitive centers and yet still retain a toehold in the lucrative whole-blood business. Then Naito rose and addressed the board. "This is the company that I founded," he began, "so, please, everyone, listen to me." Naito had always had a strong sense of history, and now as he spoke he reached back into Japan's ancient past. He compared his company to the *bakufu*, the feudal warrior class of Japan, who, centuries before, had brought order from chaos and, in resisting the Mongols, made it possible for Japan to develop as a distinctive society. It was the *bakufu* who gave rise to the noble samurai and the military Bushido code; it was during periods of *bakufu* dominance that the greatest advances in art and culture had taken place. Like these warriors, the men of the Japan Blood Bank had created an orderly and disciplined system of blood centers where only chaos had existed before. But the *bakufu*'s time had passed. In the late nineteenth century, opposing forces overthrew the old order and imposed a new system of parliamentary law under a divine emperor. Similarly, said Naito, it was time for the Japan Blood Bank to yield to the reformers.

Naito's soliloquy, with its rich evocation of Japan's heroic and tragic legacy, won over the board of directors, and they agreed to scrap their commercial enterprise in whole blood. Henceforth they would focus on collecting and fractionating plasma. They changed the company name to Green Cross (Midori Juji)—green being the color of peacefulness and growth, and "cross" indicating a beneficial purpose. Naito's decision was politically correct and economically fortuitous. Within two years, the company had phased out the troublesome commerce in whole blood and had moved toward becoming one of the world's leading fractionators. Other collectors followed Naito's example. By the end of the decade, whole blood was no longer for sale in Japan.

On August 18, 1964, Senator Eugene V. Long convened hearings to discuss his proposed legislation to exempt blood banks from federal antitrust laws. These were the hearings for which blood bankers had been waiting, which would excuse them, they hoped, from being forced to do business with people like the Basses. Although the subject involved specific legislation, the testimony would range over the spectrum of moral and economic issues surrounding blood, in order to establish a framework for managing it. As Long put it, "Is human blood properly an item of commerce to be peddled like maple syrup?" The answer, he said, should be "a firm and appropriate 'no. . . .' "

Long opened the hearings with a summary of the issues. Reviewing the facts of the Kansas City fiasco, he characterized the Federal Trade Commission decision as a "serious threat" to American blood banking. It worried him that an agency whose job involved economic regulation was meddling in the practice of medicine. The result had been a "threat of disaster," with blood banks everywhere "in danger of being crippled or put out of business. . . ."

Long planned to make his case with the help of several witnesses, including Bernice Hemphill and Dr. Rosser L. Mainwaring, both respected blood bankers and officers with the American Association of Blood Banks. But if the proceedings began predictably, they quickly veered away. Rather than being received deferentially, Hemphill and

Mainwaring came under attack. Their chief antagonist was a tough young attorney named S. Jerry Cohen, chief counsel to the subcommittee's chairman, Michigan Senator Philip A. Hart. In contrast to Long, who championed the exemption, Hart felt antitrust laws provided an important public service, and had never supported an exemption in his life. In fact, Hart did not even show up for the proceedings, trusting his chief counsel to make sure the issue went nowhere. "Hell, we were the *anti*trust subcommittee," Cohen would explain. "There was no way in the world this exemption would get past us."

Cohen attacked Long's witnesses by challenging their knowledge of the law and the logic by which they had come to oppose it. Hemphill, for example, had asserted that blood banking, because of its community-wide nature, could not survive under strict antitrust laws. Cohen forced her to re-examine her assumption. "Is it your feeling," he asked her, "that community blood banks cannot grow unless the law allows them to band together and boycott other blood banks?"

Hemphill replied that the law would make it impossible for doctors to organize on a community-wide basis, since under it those actions would be seen as collusion.

"Of course, the law has nothing to do with forming blood banks," Cohen responded. "It only goes to whether or not you can meet together and plan together and talk together to boycott another blood bank."

"Yes. I understand that," said Hemphill, "but how much does it take to be a boycott? There can be two doctors. One doctor has a bad experience and he tells the other doctor about his experience. . . . From there on someone else hears about it. . . . You cannot say it is a formal meeting but there is a rumor, and so forth, and that still is effective."

"So what bothers you really is not the law so much as the uncertainty" as to how people might interpret it, said Cohen.

Hemphill replied, "I guess that is right"—thus admitting that the law might not be quite as destructive as she had thought.

Cohen was equally deft with Mainwaring, who testified that the profit motive in commercial blood banks influenced them to use substandard donors and processing techniques, making their blood less hygienic. Such practices would proliferate in a federal trade policy that equated blood with a product. Under such a system, he argued, blood banking would be reduced to a strictly commercial enterprise, in which "nonprofit hospital and community blood banks and voluntary blood programs would be things of the past."

Mainwaring's words may have been passionate, but did little to impress his interrogator. Cohen pointed out that Mainwaring himself directed a blood bank in Detroit that, although nonprofit, used mostly paid donors. Cohen wondered how the doctor could condemn "ooze-for-booze" centers while his own nonprofit center also paid for blood.

Mainwaring grew testy. "I happen to be on the Board of Directors of the Detroit Blood Service because I felt that as long as we had to pay donors, I wanted to make sure that the quality was the best I could get . . . ," he said. "I do not like buying from paid donors but I am in this position." The blood service was opened, he said, only because the Red Cross could not fulfill the city's needs.

Cohen used the answer as a wedge. "My only point is this," he said. "The sole fact that they are paid donors doesn't in your mind make their blood any more dangerous than other blood, or of course you wouldn't be selling it."

"I am not happy about it," insisted Mainwaring. "We do the best we can with paid donors. I prefer voluntary donors."

"I understand that," Cohen answered patiently. "I just wondered whether you felt your blood was less safe because you do use paid donors."

Mainwaring evaded him. "I would say that I am not happy about it, and I would prefer voluntary donors because I do not always trust the paid donor." Then he paused and said, "The answer is 'yes.' "

"You don't trust the paid donors whose blood you sell?"

"I don't trust the paid donors who provide the blood I sell either; no sir. I do the best I can."

The lawyer bore in, using the wedge to crack open the blood banker's contentions. "As I understand it," he said, "you pay a donor, then you sell the blood, making sufficient profit to cover your expenses and salaries. . . . Doesn't this indicate to you that you are dealing with a commodity that is being used in commerce?"

"No," insisted Mainwaring. "There are various ways of looking at this; I will have to admit that. I have always considered this as a service rather than a commodity."

Mainwaring was stumbling over the two most important words in antitrust legislation—"service" and "commodity"—and Senator Long felt he must interrupt to save him. Long complained that Cohen was confusing the issue of products and commodities, comparing blood to a resource like iron or coal. "You are not touching on the moral question that blood is not a commodity and should be dealt with differently." Blood, after all, was live human tissue.

Cohen was not easily derailed by interruption, nor was he one to ignore it blithely. He listened to Long, and then adopted the senator's point to harry the witness. "One last question, Doctor," he said, turning to Mainwaring. "We have had the morality of this question brought up. Would you be in the business of buying blood if you thought it was immoral?"

"I think it is immoral," Mainwaring answered stubbornly. "I also think it is immoral to allow patients to die if they don't have blood."

"In balancing the two immoralities, you think the lesser immorality is selling the blood, is that right?"

"That is exactly right. . . . I would close up Detroit Blood Service tomorrow if voluntary donations could supply our blood needs."

"And inasmuch as they can't," replied Cohen, "the commercial banks perform a useful service to the citizens of this country, is that correct?"

"I don't consider myself a commercial blood bank," protested Mainwaring.

Cohen: "Would you say that blood banks that sell blood are performing a useful service to the country?"

"Oh, yes; they are definitely performing a service, no question about it."

Mainwaring had just been maneuvered into abandoning his position. Cohen concluded: "I have no further questions."

Cohen's machinations were completely unexpected. As one witness after another crumbled, Long tried to salvage the hearings, asking Cohen and the other staff attorneys to stop behaving like inquisitors. "I want the evidence presented on both sides fairly and properly," he said.

Yet the subject went deeper than the antitrust beliefs of the two opposing senators. Larger issues were hobbling the witnesses. By the time of the hearings in the mid-1960s, blood bankers in America were slipping from public favor. Once seen as the symbol of selfless patriotism, all were becoming tainted by the misdeeds of a few, whose scurrilous exploits made for titillating copy. The public became further jaded by the squabbling of the American Red Cross and American Association of Blood Banks, two otherwise admirable organizations with a congenital inability to cooperate. Blood bankers could no longer assume that the nation's legislators would support them.

Another confounder involved the increasing complexity of blood banking itself. Unlike the Europeans, with their national blood systems, the American blood-services complex, as it came to be called, was maturing under free-market conditions. Driven by freewheeling competitive forces, it was expanding rapidly, wildly. Regulations had developed in a patchwork manner as well, with only seven states licensing blood banks and only five states inspecting them. The result was that, though America led the world in blood-use innovation, the nation faced periodic crises in quality and supply and the recurring perception that the system verged on chaos. That was why even highly moral people like Mainwaring occasionally had to resort to buying blood, and why less scrupulous blood bankers could sell blood from Skid Row.

The members of the subcommittee may not have perceived all these complications. What they could see was that the blood-banking mess was too tangled to solve with an exemption to the federal antitrust laws. The bill never made it to the full Senate floor. Three years later, when Long reintroduced the legislation, the second bill died as quickly as the first.

Back in Kansas City, the doctors did not wait for salvation from Washington. They took their case to the next level of appeal—a full panel of five Federal Trade Commission hearing examiners. One commissioner wrote that he found the case "atypical to the point of freakishness. . . . Regulating the professional conduct of doctors is not our business." But by a vote of three to two the Federal Trade Commission upheld the original finding. Astonished and outraged, the Kansas City doctors filed an appeal in Federal District Court.

Meanwhile, more scandal tarnished the blood industry. In Texas, the proprietor of the for-profit Dallas Blood Bank and his wife were convicted of changing the blood-type labels on bottles and falsifying expiration dates. In New Jersey, a commercial bank had its license suspended for buying blood from known hepatitis carriers.

With the increasingly negative publicity, it came as an anticlimax when the Kansas City doctors finally won. In 1969, a Federal Appeals Court ruled that the Federal Trade Commission had overstepped its authority in imposing its rules on the Kansas City physicians. But even that victory was muted. In the original hearings, the defendants had argued that, as nonprofit institutions, the hospitals and Community Blood Bank should be exempted from federal trade law, just as they were excused from paying taxes. The hearing examiner had tossed out that argument. This time, however, the federal judge ruled that their nonprofit status placed them outside the FTC's jurisdiction. Thus, fourteen years after the Basses had come to town, and seven years after the start of litigation, the Kansas City doctors, hospitals, and nonprofit blood banks won the right to continue operating as before, even if it meant boycotting a local "ooze-for-booze" center—whose proprietors, by the way, had by now left the blood business.

Yet, if the doctors in Kansas City felt a sense of relief, there was no such sentiment in the business as a whole. The judge's decision had not really solved anything. In deciding the matter strictly on jurisdictional grounds, the court left unanswered the larger questions of whether human blood was a service or a commodity and whether doctors who dispensed it should be held to an implied warranty. To protect themselves against strict liability, medical associations persuaded almost all states in the nation to enact blood "shield" laws, specifying that blood was a service and not a commodity. The fruits of that labor ripened decades later, when thousands of victims of blood-borne AIDS or hepatitis found it all but impossible to sue.

Meanwhile, the national news continued to worsen. Newspapers ran exposés about winos and drug addicts selling their blood, and the concomitant rise of blood-borne hepatitis. The last twenty years had seen a change in the American blood business, from a heroic wartime activity to an ordinary part of daily life. Now, as the industry approached its maturity, a sense of scandal was hanging in the air. It became clear to the blood bankers in America that, notwithstanding the victory in Kansas City, their real troubles were about to begin.

# CHAPTER 12

# BAD BLOOD

The blood business boomed in the 1960s and '70s. The enterprise had become so decentralized by now that no one knew how much was collected, although most estimates put it well above six million pints a year in the U.S. alone, easily surpassing peak collections during the war. The liquid's uses multiplied as well, as Cohn's dream of component therapy approached realization. Rather than using whole-blood transfusions, doctors increasingly administered individual components such as red cells, white cells, platelets, and plasma. Plasma itself was giving way to an increasing number of fractionation products, including albumin, gamma globulins, blood-typing sera, and clotting factors for people with hemophilia. Doctors used more blood in more ways than ever before.

At this point the blood business divided. Hospitals and blood banks continued collecting whole blood, but plasma became an industrial affair. A new process called plasmapheresis propelled the separation. The system involved removing blood from the donor, centrifuging it to separate the plasma, and then re-infusing the red cells. The procedure was uncomfortable and could take a couple of hours (at least until it was automated in the mid-1980s), which made it necessary to pay the donors.

Plasmapheresis proved invaluable to the drug industry, allowing manufacturers to harvest greater volumes of the raw plasma they desired. The process was safer than harvesting whole blood, since removing only plasma did not lead to anemia. Furthermore, while it takes weeks for the body to regenerate red cells, plasma can be replenished in a couple of days. All this meant that drug firms could collect more often than before: Previously they had had to wait a couple of months between purchases of blood from a given donor; now they could buy from him twice a week—104 times a year instead of 6.

What happened next can best be envisioned by imagining that someone invented a very fast and cheap way of drilling for oil at the same time that the industry discovered petrochemicals. Almost overnight, the collection business boomed. Hundreds of new plasma centers sprang up to meet the demands of the burgeoning "biologics" industry, as it came to be called. Some belonged to the drug firms that had pioneered fractionation under Cohn; others belonged to small independents, specializing in collecting and selling the raw material. Like drilling rigs at an oil field, they sprouted wherever the resource seemed promising—around army bases and college campuses, in downtrodden neighborhoods, and along the Mexican-American border. From there the "source plasma" was sent to the nation's biologics manufacturers, who, in order to process it economically, pooled it in vats containing thousands of pints.

New classes of people became involved—shadier buyers, more desperate sellers. Experts had warned about the potential for abuse. During a 1966 conference at Cohn's Protein Foundation, Dr. Tibor Greenwalt, a leader in nonprofit blood banking, cautioned against "exploiting for its proteins a population which is least able to donate them"—yet that gave little pause to commercial entrepreneurs. Tom Asher, a fifty-year veteran of the plasma industry who worked as a manager for the Hyland division of Baxter Laboratories, ruefully recalled that his company set up its first center at Fourth and Town streets in Los Angeles—"absolute dead center, Skid Row. We'd immunize donors with tetanus to increase their antibodies for tetanus gamma globulin. When hurried, our doctor, who was also the bouncer, would occasionally give them shots of tetanus antigen right through their trousers." Later the company took to "bankrolling all sorts of characters" to meet the booming demand for source plasma, many with questionable ethics. Another Los Angeles center, called Doctors Blood Bank and run by two local pathologists, paid donors in chits redeemable at a local liquor store.

Stuart Bauer, a writer for *New York* magazine, investigated the world of down-and-out plasma sellers by becoming one himself. After a loved one died of transfusion-related hepatitis, Bauer went undercover, donning old clothes and selling his plasma thirteen times over a period of seven weeks. His tale was a bleak one of hardened collectors and avaricious doctors, and of the winos, addicts, malnourished, and destitute whose plasma they "farmed" at the center in Times Square. Among the chilling scenes in his article is one about the experience of donation:

> The pain of insertion comes in three overlapping waves. The first two waves—the puncturing of the skin and the piercing of the radial vein—are dicey enough, but stubbing a toe or biting the tongue are really worse. It is the third wave, the least painful part, that carries the freight. For when the body of the catheter is fitted inside the vein, distending it, it catches you—of all places—in the heart, which registers the intrusion with a chilly *ping*. When the next beat comes the heart's resumption has a choked rhythm . . . and you resolve from here on in to cater to your heart. But the only favor it occurs to you you can do is not to breathe too deeply. So you take in air in miserly little sniffs. And root for your heart as you would for a long-distance runner who had stumbled. . . .
>
> "Ever wonder what it's like?" [he asks the nurse].
>
> ". . . Wonder what what's like?"
>
> "Being on the other end of a hollow needle the size of a swizzle stick? . . . It's like being impaled on the antenna of a car radio, that's what it's like."

Later he describes a scene in which the doctor at the center finds an elderly donor lying quite still with his mouth and eyes open. "How are we today, Sydney?" he asks the old man. But Sydney is dead. After the body is removed, the doctor remarks that during his years of association with the center the man had donated almost half a million cubic centimeters of blood. " 'One always hates to lose a veteran donor with a gamma globulin like his. . . .' "

Not all those who sold their plasma were exploited. Some donors, with rare blood types or immunity factors, could sell their plasma at a premium. This was especially true of women who had developed a sensitivity to the Rh factor, the condition in which a baby with Rh-positive blood triggers an immune reaction in its Rh-negative mother. Two disciples of Karl Landsteiner, Drs. Philip Levine and Alexander Weiner, had shown that an Rh-negative woman could be immunized against the disease by injecting her with Rh antibodies immediately after the birth of her first Rh-positive child, and by the late 1960s this injection became commercially available. The source of this rare antibody—called "Big D" in blood-banking circles—was other Rh-negative mothers who had given birth to an Rh-positive child. The women most prized for plasma donation were "high-titre" mothers whose antibody concentrations were unusually high. A woman with such a rare combination of biology and circumstance could become wealthy from selling her plasma several times a month. One such woman, Dorothy Garber of Miami, Florida, had such a high concentration of the Big D antibody that she was able to earn more than $80,000 a year.

For every Dorothy Garber, however, there were thousands of less fortunate sellers—the unemployed, indigent, and substance-addicted—who would line up outside the centers in ragged neighborhoods to sell their plasma for $10 a pint. A "high percentage of our donors are either illiterate or functionally illiterate," the director of a South Carolina plasma center run by Cutter Laboratories wrote in an undated memo. "They have great difficulties reading words with more than two syllables and even more trouble understanding the meaning [of] those words. I am fairly sure most of the other Plasmacenters have the same problems."

The most disenfranchised group of donors was prisoners, who became an important source of plasma-derived products, mainly gamma globulin. Gamma globulins can be fractionated from anybody's plasma, but the best way to gather them is to find someone who has been exposed to a disease and has produced a high concentration of the antibodies in question. One way to collect gamma globulins would be to comb the population for survivors of diseases such as rabies or tetanus. A far more practical method is to inject a donor with a light dose of the pathogen and wait a few days for his immune system to gear up. This "hyperimmune" plasma can be fractionated to produce a highly concentrated and specific gamma globulin.

Prisoners proved ideal for this procedure. They were desperate enough to need the money (or furlough time, the reward in some prisons) but not likely to disappear, as were the transients from Skid Row. Soon prisons became an important source of gamma globulins for pharmaceutical firms such as Cutter and Hyland and the subcontractors who served them. Unfortunately, they operated in a regulatory vacuum. Under so-called short-supply provisions governing vital resources, drug companies could buy certain materials from unlicensed, uninspected vendors. Plasma was one such vital material. So, although federal health and safety rules covered the drug companies that *processed* the plasma, they exempted the smaller firms that merely *collected* it. A dangerous situation developed in which drug companies maintained reasonably safe and hygienic prison centers but the subcontractors who supplied them often did not.

The most notorious of these cases involved a chain of prison facilities owned by an Oklahoma physician named Austin R. Stough. Stough was a prison doctor for the Oklahoma State Penitentiary when he became aware of the emerging market for plasma. He opened a plasma center in the penitentiary, then expanded to institutions in Arkansas and Alabama. There he injected volunteer prisoners with the antigens for several diseases, collected their hyperimmune plasma, and sold it as raw material to the major biologics firms. By the mid-1960s, Stough had set up centers in five prisons in the South and was supplying the raw material for a quarter of the nation's hyperimmune gamma globulin.

Soon the prison donors started getting sick. One man nearly died when a technician reinfused him with someone else's red cells; another expired after a series of injections designed to boost his antibodies to whooping cough. Hepatitis rates jumped at several of the prisons. Five months after Stough's center opened in Kilby Prison in Alabama, the hepatitis rate among inmates soared from zero or one case a month to fifteen, then entered a sustained rate of twenty to thirty cases a month, including four deaths. Then forty-two men became sick at two other prisons in Alabama. "They're dropping like flies out here," said a penciled note from an inmate at Kilby. By the time the epidemic had run its course, the National Communicable Disease Center in Atlanta (the forerunner to the Centers for Disease Control) reported that 544 cases could firmly be linked to Stough's operation and that the real number probably approached a thousand. They could not make an exact estimate, because many health records had been lost or destroyed.

There was no doubt as to the cause of the infections. Stough ran a sloppy operation, with poorly trained technicians and unsanitary equipment. Even his customers knew it—an inspector for Cutter Laboratories reported that he was "appalled" by the conditions. Yet, right until the time that Stough was forced to abandon his plasma business, the major drug firms remained a loyal clientele. To them it was a question of supply. Having cultivated corrections officials with generous retainers, Stough had gained unparalleled access to the resource. Besides, reasoned the companies and federal officials, gamma globulins did not transmit hepatitis—as far as they could tell, the products derived from prison plasma were safe. Coldly and legally speaking, what happened to the prisoners was not really their concern.

Such indifference could not last for long. Dark stories were emerging about commercial blood and plasma in America, about a system that was poorly regulated and out of control. It also became evident that American blood products were not entirely safe. They had become tainted by a virus that, spread through transfusions and contaminated plasma, was killing hundreds, perhaps thousands, per year.

Hepatitis (the word derives from the Greek *hepatikos,* "liver") has a history touching the highs and lows of medical practice. Often known as "jaundice," because in advanced cases the patient turns yellow, the disease had been described since Babylonian times as a cause of fever, malaise, lassitude, stomach problems, and sometimes death. Hepatitis caused frequent pandemics in Europe, ranking only behind cholera and plague. Like the other medieval diseases, it seemed especially likely to break out in crowded, dirty conditions. A disease known as "campaign jaundice" plagued armies and civilians during medieval wars, and remained a scourge of the military for centuries. The French called the disease *jaunisse des camps;* to the Germans it was *Soldatengelb-schut;* scientists called it *icterus,* in reference to a yellow bird from Greek mythology, the sight of which was said to cure the disease. Jaundice decimated Napoleon's army during its Egyptian campaign, struck tens of thousands of soldiers during the American Civil War, and broke out among millions of soldiers and civilians during the Franco-Prussian War and World Wars I and II.

Despite the obvious patterns of the disease, it was centuries before doctors realized it was contagious. Autopsies of jaundice victims had revealed a swelling of the bile duct. Confusing the symptom with the cause, doctors decided that the jaundice arose from a blockage in the duct emerging from the liver. They called it "catarrhal jaundice," referring to the mucous lining that they found to be swollen.

The earliest evidence that jaundice could be spread by injections came in the late nineteenth century, when nearly two hundred workers at a shipbuilding company in Bremen, Germany, came down with the disease. A public-health officer named Dr. Lürman went to the factory to determine what caused the outbreak. In his introduction to a subsequent study, he wrote: "The etiology of these epidemics is obscure. Some believe the causes to be from noxious vapors. Others think the disease is a form of gastrointestinal catarrh. In only one case . . . does [a physician] state that the epidemic had the appearance of an infectious disease."

Like any good public-health official, Lürman began examining one possible source after another. The outbreak was not connected to the patients' socioeconomic status, since all classes of workers were affected, from laborers to supervisors to office personnel. Fumes could not have caused the disease—the factory, perched on a bluff overlooking the Weser River, was well ventilated—nor could the water have spread the disease, since many of the victims did not drink from the company wells. He ruled out nutrition: The patients came from a large assortment of families with a broad variety of food; even "the schnapps drunk by most of the workers came from various sources." Indeed, "none of the etiological events leading to an icterus epidemic thus far described fit this picture."

One possibility, however, intrigued him. A few months preceding the outbreak, in August 1883, nearly thirteen hundred workers at the factory had been inoculated against smallpox. The inoculant was prepared in the manner of the time: Doctors would prick the blisters of patients who had contracted cowpox, a relatively mild disease, drain off the discharge, mix it in pools, and add glycerine as a stabilizer. They would then vaccinate new patients by scraping the skin and applying the inoculant with a quill. Lürman wondered whether the doctors had transmitted other unknown factors as well. Working through the records of the victims at the factory, he found that, regardless of which building they had worked in or which among six doctors had given the vaccination, they all had been inoculated with vaccine from the same pharmacy. In contrast, none of the workers who were hired subsequent to the vaccination became jaundiced, nor did any who had been vaccinated elsewhere. Eliminating every conceivable possibility, he found himself left with exactly one. "Considering the distribution of cases," he concluded, ". . . one must take into account the [vaccination] . . . as the etiological source of the icterus epidemic."

Outbreaks of injection-induced hepatitis continued to appear well into the next century. Sometimes they took place at the newly emerging clinics for syphilis, diabetes, and arthritis, where doctors would inject their medications with insufficiently sterilized needles, or in which they prepared the medications in pools of human plasma. It was not until after World War I that researchers had gathered enough evidence to show that an infectious agent caused jaundice, and only during World War II that they identified that agent as some kind of virus. In fact, the worst single-source outbreak in history occurred during World War II. The U.S. Army had contracted the Rockefeller Institute to produce millions of doses of yellow-fever vaccine, which the institute produced in a solution of pooled human serum (that is, plasma from which the clotting factors had been removed). Evidence had already surfaced by then that plasma might carry a risk of hepatitis, so the institute took precautions. They bought the plasma from two unimpeachable sources—the Blood Transfusion Betterment Association in New York and the Johns Hopkins Medical School in Maryland—and heated it for an hour at 132 degrees Fahrenheit. But that was not enough. Shortly after the vaccinations began, hepatitis broke out at several army bases, in such widely separated locations as California, Hawaii, Iceland, and England. The infection rate was so high in the Third Armored Division at Camp Polk, Louisiana, that the entire unit was unable to go abroad. By the time the epidemic had run its course, 28,585 soldiers had been stricken and sixty-two had died. Doctors traced the infection to nine lots of serum pooled from the medical students, nurses, and interns at Johns Hopkins. No one had thought to ask at the time whether the blood donors had ever had jaundice.

It was difficult to link hepatitis and transfusion during the chaos of war. Battlefield conditions made record-keeping difficult—especially in the front lines, where plasma was most used. One field surgeon, attempting to track the rising rates of hepatitis, was able to divide his patients into groups no more precise than "Probably Transfused," "Possibly Transfused," and "Probably not Transfused." It was unusual when a physician could clearly document a direct correlation. One such case, of a blood donor directly transferring a severe case of hepatitis to a recipient, was reported by a physician in the Mediterranean Theater late in the war:

> The sergeant who made the donation was a 235-lb., strong, well-muscled member of a general hospital medical detachment, with an entirely negative previous history. On 8 May he played a game of baseball and knocked a home run. On the next day, he acted as a donor, and, on 10 May, his blood was given to a 19-year-old rifleman who had sustained a gunshot wound of the right lower abdomen. . . .
>
> On the next day, 11 May, the donor reported to sick call, and he died on 14 May, of fulminating infectious hepatitis, confirmed by the clinical course, the laboratory findings and the necropsy findings. There was no doubt of the diagnosis. . . .
>
> As soon as it became known that the donor had hepatitis, the recipient was transferred to a special ward, where he was kept under close observation. He remained perfectly well . . . until 21 May, when he began to complain of lower abdominal pain and generalized discomfort. The temperature was 99.4' F. On 23 May, a blood smear showed a few of the abnormal toxic lymphocytes ordinarily seen in early infectious hepatitis. Thereafter, the clinical course, as well as

the laboratory findings, were entirely typical of infectious hepatitis, except that jaundice did not appear until 1 June. The patient was critically ill for the next several days, but, after 8 June, his condition gradually improved and he went on to an apparent normal recovery.

Few case reports were so dramatic or direct. More typically, doctors would get a sense after a time that hepatitis was rising in proportion to transfusions. In an effort to get a handle on the problem, the army conducted a single-day survey of all the hepatitis patients in its hospitals in the United States on June 1, 1945. This one-day snapshot of the disease revealed that, of the 1,762 patients in hospitals with hepatitis that day, five hundred reported that they had recently been transfused with blood or plasma. It also became obvious that soldiers who had received plasma were more likely to get sick than those who had received whole-blood transfusions. The difference was due to the methods of preparation: Whole blood was prepared one unit at a time, whereas plasma was collected in pools of fifty units or more in order to make freeze-drying economical. A soldier who received one unit of blood found himself exposed to one donor; he who received one unit of plasma found himself exposed to the equivalent of fifty.

The Red Cross attempted to limit the problem by rejecting any donors who had a history of hepatitis during the previous six months. It was a laudable but inadequate precaution. Carriers might not exhibit clear symptoms and would not realize they had the disease. The British took another approach, limiting their plasma pools to ten units each in order to minimize the risk of contamination. But the Americans, driven by the need to provide massive quantities of medical supplies, found they could not conform to such limits, and used more and more freeze-dried plasma made from ever-larger pools. As Douglas Kendrick, by now a general and chief of the army's blood service, wrote, "Saving the patient's life was obviously more important than protecting him against the remote possibility of his contracting hepatitis."

After the war, when the army gave the surplus plasma to the American Red Cross, they knew the material carried a risk of hepatitis, but the army surgeon general, Norman T. Kirk, downplayed the concern. "I am writing you concerning the questions recently raised as to the possible risk of jaundice and hepatitis following the administration of the Army and Navy plasma declared surplus and accepted by the Red Cross for distribution for civilian use," he wrote to Red Cross President Basil O'Connor on February 13, 1946. "Some apprehension has arisen as a result, and it is feared that it may spread and thus jeopardize the whole program which the Red Cross has formulated for the proper distribution . . . of this surplus plasma." Kirk asserted that, though the "causative agent" of hepatitis "can be transmitted by plasma," it could also be carried by blood or any other biologic product. Besides, the benefits of pooled plasma outweighed the threat.

Kirk, who had earlier miscalculated when he opposed the shipment of whole blood to Normandy, did so again in pushing for the civilian use of the war-surplus plasma. Locked in the paradigm of battlefield medicine, he saw the issue as survival with contaminated plasma versus certain death. But civilian patients had other options available. Furthermore, death by hepatitis in a civilian hospital attracted more attention than a similar death in the chaos of the battlefield—in some cases it was followed by lawsuits. And so, immediately after the first cases appeared, the Red Cross recalled the thousands of cans of yellow freeze-dried powder.

During the Korean War, doctors looked at a variety of techniques to make plasma safe. One method involved treating the plasma with ultraviolet light to destroy a still-unidentified pathogen that was causing the disease. Assumed to be effective, this treatment failed miserably. Nearly 22 percent of the soldiers who received plasma transfusions during the Korean War contracted hepatitis, almost triple the rate in World War II. Part of the increase was attributable to a new, more sensitive way of detecting the disease; but part arose from the fact that plasma pools had now grown to four hundred units. In 1952, the National Institutes of Health recommended that the pools be reduced to World War II levels of no more than fifty units. The following year, because of the continuing high rates of hepatitis, the Department of the Army directed military doctors that, unless other solutions were unavailable, plasma should not be used "to support blood volume" in wounded soldiers. Not that all plasma products carried hepatitis; indeed, albumin, because it was heated, and gamma globulin, because of its antibody content, remained essentially hepatitis-free. It was only the whole plasma—pooled, freeze-dried, and reconstituted—that seemed to carry the disease.

By the early 1950s, doctors had learned that two strains of virus were causing hepatitis. The first strain, hepatitis A, causes a specific disease called "infectious hepatitis," a relatively mild form of the disease. It corresponds to the "campaign jaundice" of old, and is spread through contaminated water and food such as meat, salad, and shellfish. The symptoms appear quickly, generally do not progress to the jaundice stage, and linger for a few weeks or months. Another strain, hepatitis B, causes a more serious form of the disease. Known as "serum hepatitis" or "homologous serum jaundice," it spreads through bodily fluids—sexual contact, contaminated needles, and blood and plasma transfusion. The virus can lie dormant for months, and then afflicts its victim with a terrible lassitude, fever, appetite loss, vomiting, and a striking revulsion for alcohol or tobacco. A small percentage of those who contract hepatitis B develop long-term liver damage, and of those, anywhere from 1 to 3 percent die. The disease strikes tens of thousands in America and kills hundreds each year. Yet even after doctors identified the virus—they could take its mug shot with an electron microscope and its fingerprint by testing for a protein in its shell—they found themselves virtually powerless to prevent it.

By all accounts, Dr. J. Garrott Allen was an agreeable individual. Tall and genial, with sandy hair and a mild disposition, Allen, a surgeon at Stanford University Medical School, generally was admired by his colleagues, except for one tiny character flaw: He tended to be obsessive. This was not just in the manner of your everyday scientist, who enjoys pursuing an idea. No—once Allen got hold of a notion, he found it impossible to let go. He would gnaw on it, worry about it, and inject it into every social interaction. Family members who hadn't heard from him in a while would find it the topic of their phone conversation: A dinner date that started conventionally enough inevitably arrived at the problem at hand. For upward of three decades, Allen's obsessiveness—low-key but constant—made him a key player in the blood-banking drama. The subject that obsessed him was hepatitis in blood.

Allen did not set out to mount a jeremiad on the subject of hepatitis—the subject, as with most crusaders, seems to have found him. Born in West Virginia and trained in medicine at Harvard, Allen had been performing blood-related research since the war, when he was a surgeon at the University of Chicago. There, as part of the Manhattan Project, he performed army-sponsored research on how radiation affects blood. (His findings, critical in the use of postwar blood supplies, showed that people were much more susceptible to whole-body exposure to radiation than previously thought.) After the war, in addition to his surgical duties he managed the blood bank serving the University of Chicago Clinics. Cognizant of the problems with the war-surplus plasma, he experimented with new ways to make plasma hepatitis-free, and accidentally came upon a method. He drained the clear liquid from outdated bottles of whole blood, mixed it into pools of up to thirty units, and stored the bottles on a shelf. The shelf was rather warm, being near the ceiling, and the bottles would sit for months at a time.

Allen found that none of his patients who received this aged plasma contracted hepatitis. Scientists knew that viruses were highly resistant to cold and desiccation—that was why freeze-drying plasma preserved the viruses instead of killing them, and why the war-surplus plasma remained contaminated. They also knew that heat could destroy viruses, which was why Cohn was able to make albumin safe. What prevented them from treating whole plasma this way is that heat denatures complex proteins in plasma. Allen, however, found that long-term gentle heating—just above room temperature—eliminated the viruses without wrecking the proteins. The process destroyed the clotting factors, he noted, but aside from that left the plasma intact.

Allen used the technique for several years, and found that no one who received this liquid alone came down with hepatitis. Yet his method was never widely adopted. Even if effective, storing plasma for months would be an inconvenience, and probably an expensive one. Furthermore, other studies failed to replicate his findings. One, by the U.S. Division of Biologics Standards, found only a 50 percent hepatitis reduction using Allen's method. Another, by Dr. Allan G. Redeker at the University of Southern California, found almost no reduction at all. Allen contended that both studies were inadequate. In the government study, researchers used plasma they had stored not at the warm conditions Allen had specified, but only at room temperature—"a full 10° C. lower than our published reports," he wrote. "Unfortunately . . . they did not appear to appreciate the importance of duplicating the temperature at which our plasma had been held." In Redeker's study, the researcher used plasma that two pharmaceutical firms, Hyland and Courtland Laboratories, had collected from highly infectious Skid Row populations, under storage conditions that no one had carefully examined. Allen made a surprise visit to the companies. "I found numerous cases of plasma sitting outside in parking lots at a time when they were scheduled to be under incubation," he wrote. Allen complained, "It does no good to carry out a plasma study in which the conditions of storage of the plasma cannot be certified. . . ." He argued that the government should re-examine his methods under only the most scrupulous conditions. If they wanted to reject them at that point, so be it.

But no such examination was to take place. By the time Allen published his critique, the National Research Council had concluded that Redeker's results raised "serious doubt . . . on the safety of all pooled human plasma preparations." That, as well as some shortcomings with plasma—the destruction of clotting factors using Allen's method, and the ready availability of an alternative in albumin—made a persuasive case against using whole, pooled plasma. In 1968, the government revoked all licenses for the sale to consumers of whole plasma prepared from multidonor pools.

The experience frustrated Allen. It was not the rejection of his method that most bothered him, but his impression that people had not listened to the facts. "He just felt that, if he did his scientific work and kept at it, the truth would finally come out," said Dr. Edward Stemmer, a longtime colleague and friend. It would take him decades to vindicate this belief.

When Allen conducted his research on plasma storage, no method existed to directly detect the hepatitis virus. All scientists could do to understand the efficacy of their methods was to conduct what are known as retrospective studies. In these studies, researchers would gather statistics on the incidence of hepatitis among various groups of patients. Then, working backward through medical records, they would look for hints as to the source of the disease. In this way they might find certain correlations—that patients who had received prison plasma, for example, seemed to have high rates of hepatitis. Conversely, they might find that patients who had received disinfected plasma remained free of the disease. Such studies could take years, and ideally suited a personality like Allen's. During his nearly twenty years of research, he patiently conducted a series of surveys comparing the recipients of his plasma with those who had received whole blood and freeze-dried plasma from his blood bank. Each survey was larger than the one that preceded it, as more and more patients came through his blood bank. By the end, he had surveyed more than twelve thousand people who had received whole blood or plasma.

Like many blood bankers, Allen had found that his job had gotten more difficult after the war. By the late 1940s, in order to maintain supplies he had to buy blood and plasma from a local prison. (By the mid-1950s, in fact, prison blood constituted 69 percent of his product.) In the course of his surveys over the years, Allen found that the overall rate of hepatitis had grown almost in lock-step with the proportion of prison blood. In other words, there were two critical factors determining whether plasma transmitted hepatitis: the source of the plasma and the size of the pools. (The same would have been true with blood, except that, because red cells were not pooled, the second factor never became an issue.) This was no surprise—doctors had long worried about blood that came from prisoners or paid donors—but no one had conducted so large a survey or seen such a dramatic correlation.

Allen first detected the prison effect in a survey he reported in 1958, and it immediately prompted him to conduct several more. Enlisting the help of a statistician and of credit bureaus to track former patients (privacy laws were less restrictive than today), he studied a representative sample of 12,598 patients who had received a total of 42,407 units of blood in Chicago over a period of ten years, his tables filling more than 250 pages of a book he wrote about his work with hepatitis. He had moved to Stanford University Medical School, and in 1966 published his findings in *California Medicine*. The results were unnerving. Doctors had long known that professional blood carried somewhat higher risks of hepatitis, but "somewhat" turned out to be hardly the right word. Allen found *ten times* more hepatitis in recipients of professional blood than in those who had received blood from volunteers. And that involved only single transfusions. Recipients of multiple pints of professional blood would risk correspondingly higher levels of the disease. Extrapolating his results from prison donors to all paid donors of blood, Allen warned that plasma and blood centers would have to change their modes of operation rapidly. Doctors should limit their use of blood products "by giving one transfusion instead of two, two instead of three," and by avoiding all prison and Skid Row donors. In cases where no other source of blood products could be found, the bottles should be labeled as containing blood products collected from high-risk populations. After all, "when the source of blood is not known, the patient cannot be informed of the magnitude of the potential risk of hepatitis from the blood he is to receive."

Allen's recommendations did not sit well with certain members of the blood-banking establishment. In fact, a few months after he published his study, the California and Los Angeles Medical Associations denounced his suggestions as "impractical, unworkable, and cause for concern." Blood bankers recognized the danger of hepatitis, but worried more about their problems of supply, which, if the paid donor was eliminated, could escalate into a national crisis. Nor did his suggestion about labeling comfort them, for along with rising rates of hepatitis had come a rising level of damage litigation. Can you imagine, they asked, what would happen to the physician who used a bottle labeled as coming from a Skid Row or prison donor? The case would be a litigator's dream. As one blood banker told a reporter from *Science,* the journal of the American Association for the Advancement of Science, "If you label it paid you may as well pour it down the drain."

Yet Allen was not alone in making the link between commercial blood products and hepatitis. Other researchers had shown that commercially obtained blood carried at least three times the risk of volunteer blood. Eliminating paid donors, according to some estimates, could reduce the hepatitis rates by 85 percent.

Allen, meanwhile, continued his surveys. He wrote about the residents of Skid Row, whose use of alcohol, drugs, and unsterilized needles made them prime hepatitis carriers. In a letter to a colleague who had also linked paid blood to hepatitis, Allen described some time he spent in San Francisco interviewing young people who sold their blood and plasma for drug money:

> In 1967 and 1968, I spent a few afternoons in the Haight-Ashbury with the then "flower children." The pattern of events was fairly consistent for each one that I interviewed. Most had come from middle class or upper middle class homes, and were runaways. By about the fourth month, they had lost contact with their families, or vice versa. Money had become an acute problem, not only to support their drug habits but also for food. They freely admitted selling their blood in the Bay area and, when turned down by one blood bank, were generally accepted by another. . . . Putting this together with your data suggests that as their drug habits became more important to them, this group has been readily preyed upon by those dealing in paid blood and plasma. Again, according to your data, it would appear that nearly 80 percent of posttransfusion hepatitis . . . comes from carrier donors with an active self-injection addiction, or at least a drug habit in the past.

It became clear to Allen that the question of paid blood had moved beyond the realm of science inquiry and into the arena of action. And so, while maintaining his surgical duties at Stanford (and serving as editor of the medical journal *Archives of Surgery*), he began to churn out a volume of correspondence, writing to everyone he thought could influence public policy. He wrote to the U.S. Food and Drug Administration; to the nation's largest labor union, the AFL-CIO; to the American Red Cross; to the American Medical Association; and to congressmen in whose districts hepatitis had been publicized by the media. In clear, patient, yet passionate prose, he explained why the nation must convert to an all-volunteer blood supply, and institute the labeling of blood in the meantime.

Soon the media caught on to the problem. In 1970, *The New York Times* asserted that the blood-and-plasma industry was engaging in a game of "transfusion roulette" with blood products that might transmit hepatitis. In another investigation, a young Chicago *Tribune* reporter named Philip Caputo (who later became known for his Vietnam memoir, *A Rumor of War*) disguised himself as a vagrant and peddled his blood. Defying all conventions of good medical practices, several centers in a row bought his blood "even though the scab on [my] arm was still fresh from the day before." In another exposé, a popular NBC television-news program, "Chronolog," portrayed the "procurement of blood plasma on an assembly-line basis." Millions of viewers saw images of what had been disturbing people like Allen for years: derelicts lining up to sell plasma in the Skid Row neighborhoods of Los Angeles. They also saw the faces of the victims—not only the destitute who felt compelled to sell their blood, but the transfusion patients who unwittingly and innocently had contracted hepatitis.

One group of patients who merited special concern was the nation's twenty-six thousand hemophiliacs. Their lot had improved since the famous case of the Tsarevich Alexis. In the course of a mere generation, their treatment had progressed from ice packs and blood transfusions to home-based infusions of fresh-frozen plasma, and their life expectancy had jumped from twenty or so years to well into the fifties. Their lives, however, could not be called easy. Huge volumes of plasma were required to quell an episode of bleeding, sometimes more than the patient's circulatory system could take. (One doctor, in order to get enough clotting factor circulating through the body, resorted to draining blood from one arm while infusing plasma into the other.) Even in the best cases, treating the hemophiliac could mean hours of infusing the plasma as he writhed in pain from internal bleeding.

Robert K. Massie, who told the story of the Tsarevich Alexis's hemophilia in *Nicholas and Alexandra,* later wrote a memoir with his wife, Suzanne, about raising Bobby, their hemophiliac son. (It was his experiences with Bobby that prompted Massie to research the story of Alexis.) In this book, *Journey,* the Massies recounted the experience of being a family with hemophilia. They described the shock of learning their baby had hemophilia (there had been no trace of the disease in either family) and life in a perpetual state of emergency. They told about minor accidents that escalated into crises, shattering the routine of everyday life. Once, when Bobby was two and a half, he began bleeding spontaneously in the brain. With a police escort, his parents raced him to the hospital. He stayed there for a week and a half while doctors gave him constant transfusions.

Later, during his growth years, he suffered crippling joint bleeds, like most hemophiliacs. When blood flows into the confined space of a hip, knee, wrist, or other joint, it causes stiffening and twisting as the limb contorts to provide a larger space for the fluid. The liquid presses on the nerves, causing almost unendurable pain. In time the blood corrodes the cartilage and bones, causing premature arthritis and crippling. (This deterioration, common in the hip, causes a characteristic symptom known as "hemophilia limp.") Joint bleeding kept Bobby in and out of a wheelchair for years, and in some ways was the worst of the ordeal. Suzanne Massie wrote of their endless nights sitting up with the child as he thrashed and screamed, "No more pain! No more pain! . . . Through those sleepless nights, I sat by Bobby's bed. I soothed his forehead. I held his hand while he moaned, asking him to tighten his hold on my hand so as to forget the pain. . . . He would do this and, although he was a child, his grip would nearly crack the bones in my hand. . . . Impotence, helplessness, choked my throat. I sat there numbly, hour after hour, as if the very act of witness would somehow help; but the pain continued implacably." The Massie family's experience typified those of tens of thousands of others touched by hemophilia.

The circulatory system uses several overlapping systems to seal itself after an injury. Immediately after the wound is inflicted, muscle fibers in the blood-vessel wall contract to limit the tear. Then platelets move into the opening to clog it temporarily. Some platelets rupture, releasing a chemical that combines with proteins and enzymes in the plasma to form a tough, fibrous, long-lasting clot. All this happens in a series of steps known to hematologists as the clotting cascade—once it begins, it is virtually unstoppable. (That was why Richard Lewisohn's work with anticoagulants in the early part of the century represented such a breakthrough.) Yet in other ways blood clotting is a fragile process: If only one element of the sequence in missing, the rest of the process cannot take place.

Bobby, like most other severe hemophiliacs, lacked a single protein in the cascade known as Factor VIII, or Antihemophilic Factor (AHF). (A much smaller percentage lack Factor IX, which causes an identical form of the disease.) Scientists had spent decades trying to find a replacement for these components. The men in Cohn's lab had produced a concentrate of fibrinogen, a substance rich in clotting factors that did not prove effective enough to justify its expense. In 1965, Judith Graham Pool of Stanford University made an important advance. She found that by freezing plasma and then slowly thawing it she could collect a white residue rich in AHF. The sediment—cryoprecipitate, or "cryo" for short—had ten times the clotting power of plasma, delivering a high concentration of coagulation proteins. Blood centers could make it cheaply and easily, by freezing and centrifuging one bag or a few bags of plasma at a time.

Cryo became popular among families like the Massies, who kept a stash of it in the deep freeze in their basement. Rather than rushing to the hospital at each bleed, they would thaw out a bagful and wait for the doctor to infuse it at home. It had definite drawbacks, however. The liquid could take a long time to thaw—maddening when their son sat screaming from the pain of a joint bleed—and impurities sometimes caused allergic reactions. And in fact it carried a risk of hepatitis, since hemophiliacs injected it often. Furthermore, because cryo had to be kept frozen, it severely restricted the family's mobility. "One could not stray far from the deep freeze; we could only travel in sudden dashes, with the cryo packed in dry ice . . . ," wrote Robert Massie. "And so, grateful as we were for cryo, we still continued to dream of a stable, dried concentrate . . . one that could be stored in the refrigerator among the milk and vegetables, so to speak."

In the late 1960s, a new and more concentrated form of Factor VIII was developed. Drs. Kenneth M. Brinkhous of the University of North Carolina and Edward Shanbrom of Hyland laboratories produced the Factor VIII concentrate by pooling hundreds—later thousands—of units of plasma. From these pools they made large quantities of cryo. Then they redissolved the cryo, treated it with chemicals, filtered it, and centrifuged it—all to produce a white crystalline powder of pure, highly concentrated Factor VIII.

Hemophiliacs and their families greeted the new product with jubilation. Carried in a vial the size of a salt shaker, the concentrate had one hundred times the clotting power of raw plasma—so concentrated that the patient could inject it with a syringe if he wanted, instead of with a blood bag. No longer would patients remain psychologically tethered to the hospital or even the ice box; they could store the vial in any refrigerator, and later a pocket. At the first sign of a bleed, they could dissolve the white crystals and quickly inject, in a highly concentrated form, massive quantities of the blood-clotting factor. By injecting "early and often," as manufacturers suggested, hemophilia patients could quickly control their bleeds, avoid the devastating episodes of joint pain, and gain enough freedom to go on vacations.

"Today Bobby handles all of his own transfusions," wrote Robert Massie a few years later. "He travels alone with his medical supplies in his suitcase and his AHF concentrate in a small insulated ice bag. He is independent and makes all medical decisions for himself. . . ."

The new product, though, was horrendously expensive, costing severe hemophiliacs tens of thousands of dollars per year. But the hepatitis risk was worse. The raw material for the concentrate came from the industry's plasmapheresis centers. Many were in the nation's "hot zones" for hepatitis, where plasma was bought from destitute populations. In order to process the plasma economically, the companies mixed it into pools—not of fifty or a few hundred units, as the army had done, but thousands of units—a procedure guaranteeing that plasma from one infected donor would contaminate the entire pool. (Ironically, this dramatic enlargement of the pools occurred virtually at the same time that the federal government prohibited any further use of pooled *whole* plasma, on the grounds that it posed a threat of hepatitis and that safer alternatives such as albumin existed. Pooled Factor VIII did not come up for prohibition, because no comparable alternatives were available.) The companies could not sterilize the concentrate, for they assumed heating would destroy the fragile clotting factor. And so, even though hemophiliacs found their lives happily transformed by the new concentrates, the vast majority came down with hepatitis. Still later, most would become infected with HIV.

In 1971, Allen initiated a correspondence with the chief health-enforcement officer in the United States, Elliot Richardson, head of the U.S. Department of Health, Education, and Welfare. (Richardson, a public servant of unquestioned integrity, later became one of the few heroes of the Watergate scandal, resigning his post as U.S. attorney general rather than fire Special Prosecutor Archibald Cox, as directed by President Nixon.) As head of HEW, Richardson presided over the Division of Biologics Standards (DBS), which regulated the blood trade, albeit feebly. In an effort to get Richardson to stiffen his standards, Allen sent statistics illustrating the link between paid blood and hepatitis. Each time he sent a new analysis, he would attach a provocative letter. "Is there any good reason why commercial blood should not be labeled as such . . . ?" he wrote when he mailed his first batch of statistics. Later he wrote: "All we need to do is improve our national volunteer blood program and our rates of hepatitis will drop." Still later: "The facts in my table elaborate and speak for themselves." When the Chicago *Tribune* ran its series about paid blood donors, Allen sent a copy to the secretary. In the accompanying letter he warned:

> . . . Chicago is by no means unique among the large cities when it comes to commercial blood. . . . Similar conditions are easily found in Boston, New York, Philadelphia, Baltimore, Miami, and the majority of our large cities.
>
> The situation persists and appears to be deteriorating. . . . Perhaps DBS is not aware of how serious the problem is, for if they were, they

would have implemented socio-economic screening methods many
years ago. . . .

Richardson was well aware of the problem, but he was in no position at the moment to act. The DBS had come under congressional
attack for allowing drug manufacturers to market watered-down
influenza vaccine, an issue that was becoming a national scandal. It was
hardly time to ignite a new controversy. And so Richardson's under-
lings replied with the same bland counterclaims Allen had been hearing
for years—that labeling presented "a number of legal problems," and
that prohibiting the use of paid donors would create a supply crisis.

Allen, by all accounts, rarely showed anger. Although he was deeply
caring and passionate by nature, his mode of operation was always
methodical, wearing down problems through the relentless application
of intellectual force. Whether collecting data for his surveys or trying
to influence public debate, Allen came across as grindingly deliberate.
(One of his sons, who had published a book of marine photography,
likened his father to a starfish, which, by gentle but relentless tugging
with its suction cups, eventually pries open the clam.) A journal editor
recalled that when you received a letter from Allen you could either
publish it or refute it. If you published it, you would face months of
controversy as others wrote counterclaims and Allen doggedly disputed every one. On the other hand, if you refuted it, you had better
prepare for a lifetime of correspondence.

In December 1971, Allen mailed Secretary Richardson a new book
by British sociologist Richard M. Titmuss, which compared the blood-
banking systems of Great Britain and the United States. Titmuss, a
respected professor at the London School of Economics, had compared social policies of the two nations before. A large part of his previous book, *Commitment to Welfare,* had contrasted the health-care
establishments of the two nations and painted the American system as
one that neglected the very people it was supposed to assist. Now, in
his new book, *The Gift Relationship: From Human Blood to Social Policy,* Titmuss turned his critical eye to the blood-banking policies of the
two nations. America came up short.

The British system, as we have seen, grew out of the flood of social
and health reforms legislated after World War II. It followed the social-
welfare mode, with blood as a free community resource, collected and
distributed by the state. The American system developed as a mixed-
blood economy, from the community-resource model of the American
Red Cross to the commodity model of the for-profit blood bankers and

plasma industry. Titmuss portrayed the differences between the two systems almost starkly as good versus evil.

Basing his findings on extensive reviews of the scientific literature—including many papers that Allen had sent him—and donor surveys that he conducted himself, Titmuss methodically contrasted the two systems, in their medical and social effects. The British system, with its emphasis on community spirit and altruism, effectively drew people in, and the percentage of the population who donated blood steadily increased over the decades. The American system, with its emphasis on profits, seemed to repel people, and the percentage of donors steadily declined, slipping ever further behind the rising demand. The British system attracted people from society's mainstream—donors who "broadly resemble the population [as a whole] in respect of age, sex and marital status." The American system attracted marginal populations—"a strikingly high proportion . . . from the ranks of the unemployed." The British system wasted less than 2 percent of the blood, which became outdated, as compared with a 28 percent wastage rate in America. Finally, as a result of the kinds of donors they attracted, the systems produced noticeably different qualities of blood: Hepatitis rates in America were steadily rising, but those in Britain remained consistently low, with the transfusion-related rate generally below .1 percent.

Titmuss found so many problems with the American blood system—from shortages to wastage to a rise in litigation—that he saw it as the symbol of everything wrong with American-style capitalism. His well-documented and stolidly written arguments rose along a disapproving crescendo. He concluded with an extended tirade against the American way of doing business with blood:

> . . . the commercialization of blood and donor relationships represses the expression of altruism, erodes the sense of community, lowers scientific standards, limits both personal and professional freedoms, sanctions the making of profits in hospitals and clinical laboratories, legalizes hostility between doctors and patient, subjects critical areas of medicine to the laws of the marketplace, places immense social costs on those least able to bear them—the poor, the sick and the inept—increases the danger of unethical behaviour in various sectors of medical science and practice, and results in situations in which proportionally more and more blood is supplied by the poor, the unskilled, the unemployed, Negroes and other low income groups and categories of exploited human populations of high blood yielders. Redistribution . . . of blood and blood products

from the poor to the rich appears to be one of the dominant effects of the American blood banking systems.

In retrospect, Titmuss's critique was unfair. His thesis largely ignored the American Red Cross, which accounted for about 40 percent of the blood collected in America. He also dismissed the blood banks belonging to the AABB, asserting that the replacement fee tainted their collections as nonvolunteer. Instead he focused on the booming plasma industry and the rising number of for-profit blood banks. What he criticized was not the complex reality of America's blood resource, but a caricature—albeit an affecting one. At a time when Americans were questioning their blood system (and questioning much about America in general, for this was at the height of the Vietnam War), Titmuss's book hit a public nerve. It generated scores of reviews in the news media and scholarly journals. The publicity created a ripple effect. Soon after the Titmuss book came out in early 1971, waves of exposés appeared. Gone were the images of "Women at War" who managed the blood banks while their men fought overseas; or of the stalwart young medic holding aloft a bottle of plasma under enemy fire; or of the elderly veteran lining up to return the gift of blood he had received. The public now saw the derelict and the prisoner: people like "No-Surf Murph" of Miami's Skid Row, whose plasma sales kept him stocked with Swiss Colony wine; or Robert Irby, an unemployed truck driver who, revealed as a hepatitis carrier at one Chicago hospital, left and sold his blood to another.

Even the nonprofit blood bankers were losing stature, as their two principal organizations continued to snipe publicly at each other. Operating under competing philosophies—individual versus community responsibility—the American Association of Blood Banks and the American Red Cross had Balkanized America into a jumble of territories with incompatible rules. (The geography fractured even more when a half-dozen large hospital blood banks, though retaining their membership in the AABB, formed their own organization, the Council of Community Blood Centers.) The major groups had tried to coordinate their activities with treaties and commissions over the years, but each effort had collapsed under their mutual antagonism.

Several things now happened in quick succession. In early 1972, Richardson, having read the Titmuss book during his Christmas vacation, directed his staff to form a task force to look at new ways of managing the American blood supply. A couple of months later, President Nixon, declaring blood "a unique national resource," ordered the Department of Health, Education, and Welfare to make an intensive study of better ways to manage it. Several congressmen introduced bills to reform the nation's blood-services complex. In May, Richardson, testifying in Congress, made a seat-of-the-pants declaration that completely upended the regulation of blood. He announced that he would remove the job of regulating blood banks from the toothless Division of Biologics Standards and give it to the Food and Drug Administration. The FDA had much broader powers than the DBS; now, instead of regulating and inspecting the few hundred facilities engaged in interstate commerce, the government would oversee every blood and plasma center in the land—all seven thousand of them.

Meanwhile, several studies were confirming the public's doubts about the blood system. In the first comprehensive survey of blood use in the United States, the National Institutes of Health found that blood banking had become undisciplined and wasteful: Of the 9.3 million pints of blood collected every year, 29 percent went bad before it could be used. Then Richardson's task force produced its own evaluation, criticizing blood bankers along several fronts. Blood was so poorly distributed, they found, that many people had little access to the resource. They criticized the anarchy in blood pricing, whereby a pint of blood could cost a hospital $7.50 in Cleveland and $20 in San Jose. (The price to the patient, or "service fee," was always much higher.) They reiterated Allen's early contention, now widely held, that blood products were causing an epidemic of hepatitis—seventeen thousand cases every year, according to the task force, including 850 deaths. (Their estimate was low. The Centers for Disease Control put transfusion-related hepatitis deaths at thirty-five hundred a year; many doctors put it at ten times that number.) The disease cost the nation $86 million annually in sickness and lost productivity. In terms of human suffering, the costs were immeasurable.

Here was a case in which capitalism had failed. In mobilizing blood, a community resource, the free market had failed to provide products that were adequate, accessible, or, most important, safe. Yet, in stressing the need for reform, the government—more specifically, the Republican Nixon administration—put itself in a philosophical bind. Clearly the industry needed reorganization, probably around some centralized authority, perhaps even around vaguely socialistic principles, as in England, Holland, and France. Yet such a step was anathema to the Republicans, wedded as they were to principles of the free market. The last thing they wanted was to establish a new federal bureaucracy or to usurp local blood banks' control. So, rather than reorganize the blood-and-plasma business as other nations had done, government officials turned to the blood bankers themselves. They asked the industry to come up with a set of operating principles, including an all-volunteer donor system and a regionally coordinated use of the resource. Once these new guidelines were in place, they hoped, the industry would unite in a common, coordinated way of doing business. It all seemed sensible enough, idealistic and yet practical in an American kind of way. No one who knew anything about the blood trade in America gave the plan more than a remote chance of success.

# CHAPTER 13

# WILDCAT DAYS

Managua, Nicaragua, was one of the world's wretched places in the mid-1970s. Downtown was a ruin of rubble and weeds, having been flattened by an earthquake in 1972. That the effects of the disaster lingered so visibly bore witness to the corruption of the nation's dictator, Anastasio Somoza Debayle. It was Somoza who, as relief money poured in, channeled it into his family-owned businesses rather than redevelop the town. He left the nation's capital a blasted urban wilderness, a city with no heart.

His actions were hardly unprecedented, at least as far as his family was concerned. In the decades since his father had seized power, the Somoza family's legacy had blossomed, from a ramshackle coffee plantation to holdings valued at $500 million, including 60 percent of the nation's arable land and a controlling interest in almost every major Nicaraguan business. But now, after two generations of oppression, corruption, and egregious mismanagement, Nicaragua teetered on the brink of revolution, the residents of Managua for the most part afflicted with poverty, illiteracy, disease, and malnutrition.

In the dilapidated landscape, a few buildings stood relatively unscathed. One was the Hotel Intercontinental, a massive stepped pyramid of concrete and glass designed to invoke the grandeur of the Mayans. Around the corner stood another set of buildings, an inconspicuous trio of squat stucco structures painted a starched and hygienic white. Surrounding the compound was a high wall that could only be

231

entered through a heavily guarded gate. Inside usually stood a sad and ragged queue of the desperately poor who had come to sell their plasma.

The facility, Compañía Centroamericana de Plasmaféresis, was one of the few lucrative businesses in town. Owned by an exiled Cuban named Dr. Pedro Ramos Quiroz and built on property leased from the Somozas, it stood for a time as the world's largest plasmapheresis center. Nearly two hundred beds filled the facility, which engaged two dozen doctors and hundreds of other employees. At its peak, the center took plasma from up to a thousand people a day. The facility was modern and clean, and provided free meals and income to the Nicaraguans who came there. Yet many local people called it *casa de vampiros*— house of the vampires. To them it embodied the unfathomable greed of the Somoza regime: Having taken all the resources his country could offer, the dictator had now resorted to profiting from his citizens' blood.

The Nicaraguan center was one of many plasma facilities that sprang up in the Third World during the industry's wildcat period, when the boom in plasma products that had swept across America spread over most of the industrialized world. Driven by a skyrocketing appetite for albumin and the new clotting factors, demand climbed by double-digit percentages every year. Most of the products to meet this requirement came from a handful of American firms, including Armour, Cutter, and the Hyland Division of Baxter Laboratories, which obtained their raw material from American donors, whether students, prisoners, or the denizens of Skid Row. Yet even this supply had its limits, they found. Plasmapheresis, although harmless and lucrative, was the kind of activity the middle class shunned. Moreover, competition was increasing for the limited donor pool. "Good pickings in the United States had become difficult," recalled Fred Marquart, retired president of Hyland. As the market for plasma products grew internationally, the need for source plasma outstripped what American donors provided. And so, just as the oil industry had long scouted the world for new sources of petroleum, the drug firms now looked overseas as well.

There were differences, of course. Unlike the oil companies, the pharmaceutical firms never imported the majority of their material; most still came from American donors. Nor did the drug manufacturers buy from sheikhs in desert kingdoms. Instead, they turned to where the resource was most plentiful—the crowded urban slums of some of the world's poorest nations. This factor, however, eventually undermined them. No matter how much they might have paid for their plasma or how well they might have treated their donors, the fractionators never escaped the stigma of reaping profits by bleeding the poor.

It is difficult to say when the first shipments of plasma arrived from the Third World, although FDA records show that by the beginning of the 1970s a lively trade was already under way. The first center to receive public attention was a facility in Haiti called Hemo Caribbean. Located in the most impoverished capital in the Western Hemisphere, Port-au-Prince, it was established by Joseph B. Gorinstein, a stockbroker from Miami, in association with Luckner Cambronne, Haiti's feared minister of the interior and national defense. Technicians at the center would collect plasma from hundreds of donors a day, freeze it, and ship it out via Air Haiti, of which Cambronne owned a share. They exported up to six thousand liters a month to drug firms in America, Germany, and Sweden.

The collectors paid well for donations—$3 a liter, or about three times the average daily wage—but the condition of the donors was deplorable. A pitiable assemblage, they would line up starting at six-thirty in the morning, many in rags and wearing no shoes. In a nation of stark poverty and primitive health care, many "had medical problems of their own," observed Richard Severo, the *New York Times* reporter who exposed the operation in January 1972. "The prevailing diseases include tuberculosis, tetanus, gastrointestinal diseases and malnutrition. The caloric intake of Haitians is one of the lowest in America but [according to the company's technical director] only 1 to 2 percent are rejected because they are too weak."

Gorinstein energetically defended the facility as an important source of income to donors and employees. Besides, he asserted, his plasma was "a hell of a lot cleaner than that which comes from the slums of some American cities," which, in some extreme cases, may have been true. But the exposure stung Haiti's dictator, Jean-Claude "Baby Doc" Duvalier, who had been trying to cultivate a more moderate image than that of his father. After only twenty-two months of the center's ten-year contract, he summarily closed Hemo Caribbean.

Gorinstein's failure did nothing to discourage others from collecting plasma in Latin America. The region, in fact, had become a favorite harvesting ground, with its proximity, low cost of living, and large numbers of poor and willing donors. Moreover, despite the region's rampant malnutrition, donors reportedly gave high-quality plasma because of their protein-rich bean diet—or so said the collectors. Over the years, American firms bought plasma from more than half a dozen Latin American countries including Mexico, Belize, the Dominican Republic, Costa Rica, El Salvador, Colombia, and Nicaragua. "We were peddling so much plasma out of Costa Rica," joked Marquart, "that it must have been their second-biggest export after bananas." Meanwhile, American drug companies set up their own plasmapheresis centers along the United States' southern border so Mexicans could walk across and save them the trouble of importing the liquid.

A particular feature of this wildcat period was the wily and flamboyant entrepreneurs it attracted. One of the more successful and colorful of these people was a businessman named Delfino de la Garza, who came from a well-connected family in Mexico and lived on a beautiful estate in Costa Rica. In the late 1960s, de la Garza approached Hyland with a plan to set up a chain of plasma facilities in Central America. He flew his managers up to Los Angeles, where Hyland technicians trained them in plasma collection and hygiene, then opened a collection center in Costa Rica. "He had a donor center with eighty-five to a hundred beds," recalled Marquart. "The place was spotless, everything first-class." Soon afterward, de la Garza opened two other centers, in Guatemala and El Salvador, exporting to Hyland an estimated six thousand liters of plasma a month.

What made de la Garza memorable to his American associates was not so much his medical expertise (he had none) as his way of doing business, Central American style. De la Garza established his enterprises with a combination of connections, payoffs, and personal charm. "It was amazing, he was so smooth," said Tom Asher, a former Hyland manager and retired president of the Hemacare company in southern California. Asher recalled a time in the early 1970s when, as an independent plasma collector, he was scouting several Central American countries, conferring with local officials and businessmen, trying to make deals to establish facilities. (The arrangements, he recalled, generally involved requests for bribes.) In each case, he found that de la Garza had procured exclusive rights beforehand. Finally, Asher struck a deal in El Salvador, even though de la Garza had established a competing center there. Asher's business had been going for just under a year when "all of a sudden the minister of justice, the minister of health, and about fifteen armed soldiers marched in and said, 'Everybody out!' " Apparently, de la Garza wielded enough influence to have Asher shut down. It took Asher months to retrieve his equipment—and then only by having a crew smuggle it out in the dead of night.

Latin America was not the only place where one could buy industrial quantities of cheap plasma. Some came from Lesotho, the impoverished black homeland in South Africa, from a company called Scimitar, owned by a former blood banker named Dr. Ben G. Grobbelaar.

Grobbelaar had been one of South Africa's most prominent blood bankers, as medical director of the Natal Blood Transfusion Service, a member of the Executive Council of the International Society of Blood Transfusion, and vice-chairman of the World Federation of Hemophilia. He also established the first fractionation plant in the region. During his work in the voluntary sector, he had a change of heart. He came to believe that unpaid donation, as noble as it seemed, could never supply the world with enough plasma. His answer was to set up a for-profit plasmapheresis center in Lesotho. There he bought plasma for $5 a donation and sold it for more than five times that amount to reputable drug companies in Germany, Italy, and Spain. (Grobbelaar said he made a profit of $2 to $4 a pint after processing and shipping expenses.)

Grobbelaar's business made him a pariah in international blood-banking circles. Many former colleagues wondered how such a respected practitioner could engage in exploiting the poor. Yet to him that attitude was naïve. Volunteerism might work in Europe, because people there were relatively wealthy; they could afford to give blood and even plasma. But people in the Third World faced daily issues of basic survival, and didn't have the luxury of volunteering. Centers like his, he argued, offered a solution, harvesting plasma and providing income in areas of high unemployment. Nor did he accept the widely held allegation that businesses like his preyed on the impoverished. The $5 a donation he offered could easily double the average person's daily income. This meant that he could attract "average" citizens, as opposed to the down-and-outers who sold their plasma for relatively less money in the United States (an assumption that was not necessarily true, as we shall later see). In a letter defending his deeds and philosophy he wrote, "I find fault with the self-righteous Western Europeans. For forty years they have been selling tanks and machine guns to Africa," whereas he was exporting a lifesaving substance while paying his donors a decent living wage.

One company found it possible to import huge quantities of cheap plasma without exploiting anybody. The Institut Mérieux, a French pharmaceutical firm, imported tons of blood-rich placentas from maternity wards throughout the world. The idea had originated with Charles Mérieux, the *grand seigneur* of the company, who had set up some of the first clandestine transfusion centers in wartime France. After the war, he traveled to Edwin Cohn's lab in Boston and came away thrilled about the "industrial" use of plasma. While still in the United States, he learned of a plasma source that was both "ingenious and obvious" in the millions of placentas disposed of every year. As he wrote in his memoirs: "From this pocket of blood that one discards after each birth, we could extract all the albumin and gamma globulin we would ever need."

Over the decades, Mérieux set up a network of contacts to ship him placentas, to be squeezed out in modified wine presses for their plasma. At the height of production, placentas at the Institut Mérieux provided four-fifths of the gamma globulin in France and 8 percent of the world's albumin supply. With material flowing in from places as distant as Russia and China, Mérieux became the world's largest importer of placentas, processing up to fifteen tons of material per day—5 percent of the placentas in the world.

Of all the plasma centers of the time, the largest and most notorious was Compañía Centroamericana de Plasmaféresis, which thrived, as we have seen, in the ruined city of Managua. Twenty-four hours a day, the poor and indigent would traipse in, selling their plasma after submitting to urine and blood tests. Technicians froze the plasma and shipped it to way stations near airports in Mexico and Miami, where it was stored in a deep freeze pending shipment to pharmaceutical companies in Europe and America.

Ramos paid the sellers $5 to 7 dollars a liter, a handsome living wage in Nicaragua, then sold it to his customers for nearly five times that amount. The markups continued down the line. Tom Hecht, retired president of Continental Pharma Cryosan in Montreal and formerly the world's largest independent plasma broker, negotiated the original arrangements between Plasmaféresis and its customers. "We shipped huge quantities of source plasma to Travenol in Belgium [a sister firm of California-based Hyland]," he recalled. "Travenol took off the cryoprecipitate—we called it 'skimming rights.' Then we sold the cryo-poor plasma to Kabi Pharmacia [a Swedish biologicals company] for further fractionation." Hecht bought the plasma for $34 a liter, selling the skimming rights to Travenol for $12 a liter and the cryo-poor remainder for $38 in Sweden. The transaction netted his company about $240,000 dollars a month. "It was the biggest deal I'd ever negotiated," he said.

Hecht's business provides a window on the movement of plasma from the Third World, almost all of which went north to American and European drug companies. Some suppliers signed long-term contracts directly with the fractionators; others sold to the free-ranging brokers like Hecht, who, traveling the world, secured plasma on the spot market and sold it to the highest bidder. Huge quantities passed through Montreal and Zurich, which, because of lax bonding and transit laws, earned the two cities the sleazy reputation of being way stations for the international traffic in plasma.

Once the source plasma reached its destination, the drug companies mixed it into huge pools, fractionated the mixture, and sold the derivatives. Sometimes companies shipped finished products, such as gamma globulin or Factor VIII; sometimes they sold other companies the partially fractionated paste. Whatever the specifics, the effects were the same: The industry collected plasma from areas rife with poverty, malnutrition, and hepatitis, concentrated and processed the material, and sold it throughout the world.

As the developing world's largest plasma collector, the center in Managua offered benefits to many—material for the drug companies, profits for Ramos and Somoza, commissions for the middlemen, and some money for its donors. But there was also a downside, involving rumors of mistreatment and injury. In the fall of 1977, a woman from one of the barrios complained to the police that her son had disappeared. The young man, an alcoholic and a habitual donor named Mario Salazar Marques, had told her he was going to sell some of his plasma, just as he had the preceding month. When she went to the center to ask about her son, other donors told her they had seen him the day before, although no one could say what had become of him. Some suggested that Mario might have fainted—news that must have made the poor woman shudder. Mario had told her that donors who fainted or died were covered with a green sheet, were taken to the basement, and "disappeared."

Such occurrences, of course, were not uncommon in Nicaragua, but this one caught the attention of reporters at *La Prensa,* the opposition newspaper in Managua. Its publisher and editor, an earthy intellectual named Pedro Joaquín Chamorro Cardenal, had been jailed by the dictator at least half a dozen times for exposing the violence and excesses of the regime. The disappearance of Salazar and its connection to the center provided Chamorro yet another example of Somoza's villainy.

The newspaper sent a reporter to climb a nearby building to peer over the center's high and heavily guarded gates. There he observed long ragged lines of obviously unhealthy men. "Ninety percent of the donors come from the very poor classes," he wrote. "They usually are unemployed and in bad physical condition." More coverage followed. Day after day, under headlines such as "Why Are We Involved In This Bloody Business," "Information About the Dead Donors," and "PLASMAFÉRESIS MAKES THE COUNTRY SICK," the newspaper exposed a pattern of malpractice and abuse. The stories revealed how Ramos's employees bought plasma from known alcoholics, often ignoring screening tests their American clients assumed they performed. Reporters described how guards treated the donors like inmates, cursing and hurrying them roughly along. One donor claimed he was beaten. A committee of impoverished women from the barrio wrote to *La Prensa* that selling plasma was their only alternative to prostitution. "But when the needle enters our bodies we also feel we are losing part of our lives. The people who work in this center treat us in a very inhuman way. . . . We implore the people in charge . . . to investigate this awful situation."

The donors were not the only ones to complain. Doctors at a local hospital reported that they frequently had to treat habitual plasma sellers for "profound anemia and malnutrition," thereby subsidizing Ramos's profits with publicly financed medical care. Later, a committee of medical instructors who examined former donors in area hospitals found that several of the men tested positive for syphilis.

In fairness, one should question at least some of *La Prensa*'s reports. The stories tend to be vague in their medical details. One could question whether excessive plasma removal caused the anemia that local doctors observed, since a more likely result would be protein depletion and kidney damage. Moreover, Chamorro, a long and determined foe of Somoza, never claimed to be an impartial observer. Finally, people in the industry remember the center as one they admired—well equipped and professionally run. Industry insiders invoke the fact that the U.S. Food and Drug Administration licensed the center, thus giving the facility a stamp of approval.

One must distinguish, however, between what the FDA permits and how a facility performs. To hear veterans of the industry tell it, by the mid-1970s the Skid Row–type plasma center was becoming a mere memory, as government rules tightened and the industry became increasingly fastidious. That may have been so on paper, but daily operations often told a different tale. This held true even in America, as journalists learned when they disguised themselves as indigent donors. In late 1975, for example, a British television-documentary team, tracing a hepatitis outbreak among British hemophiliacs back to the American source of the product, visited several plasmapheresis centers owned by Hyland. In every facility they observed a discouraging collection of derelicts and alcoholics, along with careless and cynical screening procedures. They brought along a British hepatitis expert named Dr. Arie Zuckerman, who, after visiting the Hyland facility in Los Angeles, described it as "an offense to human dignity," with donors whom any British physician would have "rejected straightaway." Yet, when the producers interviewed Dr. Richard Wilbur, Hyland's vice-president for medical operations, he claimed to know nothing about those conditions, having made only one visit to any of his company's collection centers.

Perhaps Wilbur lied, hiding the fact that he and other leaders of the industry knew they were running substandard collection centers. The other possibility, equally unsettling, is that he really did not know, which would mean that industry executives did not understand the difference between what they saw in their paperwork and what took place on the ground.

If that was the case at major drug companies' facilities in America, imagine the situation in Nicaragua. No one could doubt that Plasmaféresis's employees took shortcuts, ignoring the well-being of the donors and skipping certain screening tests. Even Hecht, who made a fortune from the enterprise, remembers it with a degree of distaste. "The physical plant impressed me, but what bothered me most was the way the local management treated the people. They treated them with disdain, not like human beings." So one is inclined to believe *La Prensa*—at least in implication, if not in detail. For it is clear that this kind of center in this kind of location, with its oppressed and impoverished population, could never be anything but exploitive. That much was reflected in the treatment of the donors—the screaming, the shoving, the abuse. Chamorro labeled it a "shame on the nation," a crushing burden on his people's morale. An opposition senator likened the business to Dracula rising from his tomb "not only at nighttime, but during the day, weekends and Sundays too."

No one ever learned the fate of Mario Salazar Marques. But the stories continued, and the drama became ever more volatile and troubling, racing toward a violent and tragic conclusion.

In examining this period of the global plasma trade, roughly from the late 1960s to the late 1970s, it would be tempting to cast the Americans as the villains. After all, it was they who seemed to profit from an international commerce that exploited the poor. Indeed, that was what European blood bankers said at international meetings and congresses. But the realities of the plasma trade were far more complex.

In the wake of World War II, as we have seen, most industrialized nations had set up their blood systems along nonprofit lines for humanitarian and medical reasons. Indeed, in some nations, such as France, nonremuneration had become almost a religion. Later, when plasmapheresis was developed, several of these nations prohibited the practice, because it went hand in hand with the payment of donors.

Instead, the countries turned to a variety of methods to get the plasma they needed. Switzerland, for example, collected an excess of whole blood from volunteers, spun off the plasma, and threw out the red cells. Britain used plasma from outdated blood and plasma spun off from whole-blood collections. Others, like France and Germany, permitted plasmapheresis on a limited basis, allowing donors to give barely more than a quarter of the allowable American levels.

For a while, these arrangements worked well for the Europeans, allowing them to obtain the plasma they needed mainly from volunteers (except in Germany, where drug companies paid for plasmapheresis) and have the moral pleasure of condemning the Americans. But when the demand for plasma products exploded—first for albumin, later for Factor VIII—it became impossible to obtain enough volunteer plasma, and most European nations bought increasing amounts of plasma and its derivatives from the United States.

Britain, for example, found it necessary to import huge quantities of Factor VIII. Sad to say, the national volunteer-donor system that Titmuss had glorified had gradually become a "disorganized shambles," according to a prominent British critic—a collection of semi-autonomous regions that did little to help each other or share. Barely able to furnish enough whole blood, the system proved incapable of marshaling an adequate plasma supply. Beyond that, the nation's government-funded fractionation centers at Oxford and Elstree did not expand their capacities in time. Hemophilia specialists had begged the government to increase production, and the Ministry of Health had promised to take action by allocating money to update the facilities and setting goals for British plasma self-sufficiency. But they failed to meet every deadline they set. Thus, with neither the means to collect enough plasma nor the capacity to process it, Britain's Factor VIII stocks continually lagged behind demand, and the country imported more than half its supply from the United States. Similarly, France, though publicly boasting about its all-volunteer plasmapheresis program, quietly imported as much as 26 percent of its Factor VIII from America annually. Holland and Switzerland imported plasma and its derivatives. Outside Europe, Japan imported an astonishing 98 percent of its plasma products, notwithstanding the growth of Dr. Naito's Green Cross.

Thus, by the middle of the decade, America had become the OPEC of plasma. As Tom Drees, then president of the Alpha Therapeutic Corporation, later explained to a fractionators' conference, "As the U.S. feeds the world, so does the U.S. bleed for the world. Or, more correctly, the U.S. plasmaphereses itself for the rest of the world." Yet, even as America's production capacity grew, its ability to provide the source plasma plateaued, and the drug companies increasingly purchased Third World material.

So, if one is to condemn the American companies of that period as profiteers, one must also point out the hypocrisy of others. The same American drug companies that imported most of the material from the Third World also provided—from a variety of sources, including the Third World—most of Europe's plasma. Indeed, the Americans exported enough raw plasma and processed material to account for well over half of Europe's needs. As one seller put it: "How they [the Europeans] can consider the paying of donors as being immoral, and yet import plasma from the U.S.A. knowing that [it comes] from paid donors, has always dumbfounded me."

If plasma and its derivatives flowed from the United States to Europe, money flowed in the opposite direction. This particularly applied to the profits from Factor VIII concentrate, which sold for about 11 cents a unit in America (a year's supply cost the average hemophiliac thousands of dollars) and sold for at least triple that amount in Europe, making a dazzling target for pharmaceutical firms (and keeping down prices for American hemophiliacs). American companies descended on the continent, seeking a piece of the lucrative market. They offered special premiums to high-volume customers, including equipment, lab technicians' salaries, conferences, and perks such as cruises for executives. Some companies offered direct cash rebates to hemophilia treatment centers and societies.

Nowhere was the treatment more lavish than in West Germany. The nation's Factor VIII use was truly astonishing, with the average hemophiliac using up to four times the quantities of his American counterparts. As a nation, West Germany consumed more Factor VIII than all the other European countries combined; one hospital alone spent more on Factor VIII than did the entire United States.

The man most associated with the rise in German consumption was Dr. Hans Egli, the studious, mild-mannered director of the world's largest hemophilia treatment center, the Institute for Experimental Hematology and Blood Transfusion at the University of Bonn (Institut fur Experimentell Hämatologie und Bluttransfusionwesen der Universität Bonn), or the Bonn Hemophilia Center, as it was commonly called. Egli had visited one of the originators of home hemophilia care, Dr. Shelby Dietrich, in Los Angeles, and became an eager convert. Adapting her methods, he added what he thought were a couple of improvements. Under Dietrich's direction, patients gave themselves moderate doses of Factor VIII whenever it was determined that a bleed was beginning. She insisted they live in the Los Angeles area so she and her staff could monitor their progress. Egli, on the other hand, gave massive doses of concentrate and supervised patients who lived far from his clinic. Dietrich employed treatment on demand, in which patients would inject themselves in the event of a bleed; Egli instructed his patients to inject prophylactically—on a regular schedule, whether or not they suffered an episode. Such treatment, he felt, helped avoid joint bleeds, especially in young patients, whose joints were still growing. After starting his patients off with a two-week training session of rigorous physical therapy and pharmaceutical instruction, he would send them home to report back only a few times a year. They kept in touch by submitting detailed questionnaires and via a twenty-four-hour phone line.

He used even more radical therapy with "inhibitor" patients—those who develop antibodies in response to the injections. These people have always confounded their physicians: Their bodies destroy the one material that can help them. Dr. Hans Hermann Brackmann, Egli's nephew and the center's medical director, developed a method whereby he administered massive doses of Factor VIII in conjunction with other medications, in an effort to overwhelm the immune response. Brackmann reported that, of twenty "high responders" to Factor VIII after years of treatment, fifteen had lost the antibody response.

Egli's therapy gave hemophiliacs an unprecedented sense of independence, even more so than conventional home treatments. "At the age of sixteen I was confined to a wheelchair because of bleeding in my hip," recalled Dr. Werner Kalnins, one of Egli and Brackmann's early patients. "After two weeks of treatment, I was walking on crutches. Soon after that, I was absolutely well." Patients flocked to the Bonn clinic from distant parts of Germany, and from other countries as well. Frank Schnabel, founder of the World Federation of Hemophilia in Montreal, spent two weeks at the clinic in 1977. He marveled at the center's methods and technology, which had "brought hemophilia into the age of the computer."

> It is rather an interesting experience to be riding in Dr. Brackmann's car equipped with a sophisticated telephone system and hear a hemophiliac from Stuttgart, some 150 miles away, call regarding his treatment and the doctor, in turn, prescribes the course of treatment. The patient may only be reporting the daily improvements in his knee, in centimeters, as the swelling diminishes. Infusions continue until the joint is back to normal. Or, it may be a serious hemorrhage and the hemophiliac may find himself with insufficient AHF [Antihemophilic

Factor] concentrates in his home care inventory. Since it may take 12 hours for a special shipment to reach the hemophiliac, the Bonn Centre checks the computer and determines the amount of Factor VIII in the homes of hemophiliacs in the adjacent region. These hemophiliacs rush . . . sufficient material to control the crisis until the shipment arrives. The visitor to the Bonn Centre will see large maps graphically presenting the logistics of this operation.

"I have seen the epicenter," Schnabel concluded, "and hemophiliacs throughout the world will certainly benefit from the waves radiating from Bonn."

Despite such enthusiasm, Egli's methods raised questions. Other physicians, such as Professor Günter Landbeck at the University of Hamburg Children's Clinic, criticized Egli's long-distance therapy, stressing the need for a "competent doctor in the locality." At a conference in Heidelberg in 1977, Professor Klaus Schimpf, director of that city's hemophilia center, insisted that his patients thrived with a fraction of Egli's doses—one had even joined a table-tennis team. Another doctor questioned whether it was wise to inject young men with "hundreds of thousands of units of a foreign substance daily, weekly for the rest of their lives."

Physicians outside Germany worried as well. In October 1977, for example, an international delegation of home-therapy experts visited the center. Impressed by what they saw, they nonetheless worried about its scientific basis, especially since Egli and company had published so little. "We all regard precise documentation of your work as of the very greatest importance," they wrote to Egli. In 1979, the Council of Europe denounced the German hemophilia centers for overusing Factor VIII. Yet the Germans seem not to have considered any limits, in quantity, cost, or the risk of hepatitis.

Egli's group defended their practice as essential, but there was a less savory aspect to his colleagues' prescribing patterns: In handling massive amounts of Factor VIII, they enriched their own center in the process.

German hemophiliacs benefited from one of the world's most liberal systems for the handicapped. Under this system, in which handicapped people have the legal right to be fully mobile, hemophiliacs received without charge all the Factor VIII they needed, not only to survive but to be mobile and pain-free. Hemophilia centers ordered whatever quantities of Factor VIII they wished and billed the insurance companies (some of which are municipally owned) for reimbursement. The patient paid nothing.

As meticulously documented in the book *Böses Blut* (Bad Blood) by German journalist Egmont Koch, Egli's group manipulated this system into paying enormous amounts to the Bonn Hemophilia Center. It purchased almost all of its material from America, where it cost only a quarter to a third as much as the German-made product. Importers billed the group at high German prices—bills that the Bonn center passed on to the state insurance companies for reimbursement. Periodically, however, the importers would give Egli's group sizable rebates, in the form of "project-related" expenses such as technicians' salaries or research grants. Although these premiums effectively lowered the price, the center did nothing to notify the state. This behavior "had no recognizable purpose other than to keep those who bear the costs from full knowledge," according to a 1981 internal briefing document at Germany's largest municipal insurance agency, AOK (Allgemeine Ortskrankenkasse). Beyond that, Egli's group took an administrative fee of up to 15 percent of whatever it paid. (This charge alone netted the center nearly $9 million annually in state-paid reimbursements.) In other words, the more Factor VIII Egli's group used and the more money it paid for it, the more profits they received. According to a report issued by the German equivalent of the Federal Trade Commission (Bundeskartellamt), "The treatment centers have an outright interest in keeping factor VIII prices up."

The cozy relationship between industry and treatment providers did not confine itself to Germany. Indeed, it seemed the norm throughout most of the advanced world. Most of the World Federation of Hemophilia's budget was paid for by the fractionation companies, who picked up the tab for its lavish annual meetings—philanthropy or bribery, depending on how you look at it. In America, the National Hemophilia Foundation received anywhere from 15 to 25 percent of its operating budget from industry, as well as special grants for educational projects. Prominent hemophilia doctors who served as the foundation's medical directors received tens of thousands of dollars annually to run clinical studies of the companies' new clotting products and conduct industry-sponsored training groups and seminars. There was nothing illegal about this; drug companies often finance research. The doctors and the hemophilia organizations argued that the relationship was appropriate and collegial, not coercive. They saw it as a mutual exchange of medicine, money, and information to help their patients get as much clotting factor as they needed at the best prices. Yet patients would later claim that the financial links between the drug companies and the doctors influenced the treatment providers to be complacent about safety.

In Germany, however, this collegiality reached unprecedented levels, and crossed the boundaries of ethics and good judgment. In the late 1970s, the Bonn Hemophilia Center started doing business with a Swiss-based supplier called Lutz and Co. The importer met with Dr. Franz Etzel, a protégé of Egli's at the Bonn center whose job it was to order materials. As part of the deal, the Lutz representative offered Bonn direct cash rebates equivalent to 26 cents a unit—a handsome amount of money, considering the quantities. Beyond that, he offered Etzel a personal kickback of 4.5 cents a unit, discreetly deposited in a numbered Swiss bank account. When Etzel arrived at the bank to countersign, he found that $135,000 had already been deposited.

The profits from this enterprise grew to enormous proportions, netting the Bonn center $15 million in industry rebates since 1975. With Bonn as a respected and influential example, the profligate use spread throughout Germany. By the end of the decade, AOK was paying $133 million annually to sustain the nation's two thousand hemophiliacs. More than 40 percent of the total went to Egli's center. Reimbursements to his individual patients were shocking, often exceeding $800,000 annually. In the extreme cases of some patients with inhibitors, costs for one person exceeded $4 million a year. Overwhelmed by the complexity of these schemes and drained by their excesses, the insurance agencies began to leak stories to the press, publicizing a local hemophiliac as "the most expensive patient in town"— which, though doing no harm to the doctors, effectively served to humiliate the victims.

Still, the late 1970s were the glory days for Egli's treatment center in Bonn. His proudest moment may well have been in the fall of 1980, during the First International Hemophilia Conference in Bonn. No expense was deemed too lavish: The entire delegation from the World Hemophilia Federation's headquarters in Montreal had been flown to Rotterdam and taken on a cruise down the Rhine River to the conference, all at the organizers' expense. Egli delivered the keynote address, proclaiming the beginning of an era that "began with fog and is ending with sunshine." Frank Schnabel testified, "Fresh hope is expanding beyond the Rhineland."

It was almost exactly a year later, in October 1981, that the newspaper *Die Zeit* broke the first of several media scandals about "a mess of enormous dimensions" that allowed Egli's institute "to practically swim in gold." Later a criminal court would sentence Dr. Etzel to prison for tax evasion and "profiteering." Still later Schnabel would face his own struggle—a futile battle against AIDS, which he contracted from any of the thousands of donors to whom he was exposed through his use of the concentrate. For the time being, however, hope and imports both remained high.

Everyone knew that the transfer of plasma from the poor to the rich could not continue, as profitable and convenient as it may have been. Regardless of how much one paid to the donors, the practice, by its nature, was inherently exploitive—not to mention potentially unhealthy. Long before Chamorro's press crusade in Nicaragua, people in the international health field had been working to staunch the flow of "red gold," as they called it. Several years earlier, in fact, the American hepatitis activist Dr. J. Garrott Allen had raised the alarm, "not only because of my concern about the spread of hepatitis but also because of the further depletion of an already depleted population."

Reports kept surfacing about unscrupulous businessmen scouting new locations in Asia and Africa. Doctors in India said that plasma-company representatives approached them to seed businesses in several major cities. Similar stories emerged from Nigeria. In South Africa, Dr. Maurice Shapiro, the patriarch of South African blood bankers, reported that a company called Serocenter of America had approached him with an offer to establish a chain of commercial pheresis centers. The representative handed him a four-page proposal describing how each facility would provide income for the government and employment for the community, and would introduce "a new clinical pharmalogical [sic] industry" to South Africa. The company's real agenda appeared on the final page of the proposal: "the right to export the plasma and other such products as we shall derive without limitation."

The International Red Cross repeatedly condemned the spread of such industries among the poor. At its Inter-American Conference in Paraguay in 1974, the agency warned its members about "this new modality of exploitation of the most needy . . . a dangerous, scandalous, and unfitting traffic." That same year, the World Health Organization (WHO) sent a questionnaire about the practice to several of its poorer members. Of the twelve nations that responded, eleven said that commercial firms had approached them. Working stealthily to avoid industry interference, a committee of experts from the International Red Cross and a few other organizations assembled evidence about the trade in advance of the WHO's 1975 World Health Assembly meeting in Geneva. When they presented the information, along with an urgent resolution condemning the practice, the delegates unanimously voted to approve it.

Soon afterward the harvesters began shutting down, almost domino-style. Costa Rica had already nationalized Delfino de la Garza's plasmapheresis center, converting it to a nonprofit facility providing products for its citizens. (The wily de la Garza, say industry veterans, managed to sell it to an unsuspecting businessman just before Costa Ricans took it over.) Now the El Salvador and Colombia centers closed. Lesotho revoked Grobbelaar's permits; later he set up a center in Transkei, another black homeland, where he operated in a nonprofit capacity, plowing the profits into the homeland's growing transfusion infrastructure. He finally closed it and moved to Canada, with the ending of apartheid and abolition of the black homelands. Haiti, as we have seen, shut down Hemo Caribbean long before the declaration appeared.

No one has ever documented whether plasma from the Third World caused elevated hepatitis rates in the First. Too much intermingling of materials took place to determine clearly the source material's final destination. But there is no doubt about the corrosive effect that the "vampire" collectors had in poor countries, with their demoralizing and exploitive routines.

Nowhere was that clearer than in Nicaragua, where the dictator and his cronies continued to gather plasma, oblivious to the gathering political storm. The contrast between Nicaragua and the rest of the world pained editor Pedro Chamorro, and he hammered that point in a November 1977 editorial: "In Haiti plasmapheresis has been discontinued because the fat little dictator in Haiti has a bit of conscience and responsibility—that's why he listened to the WHO, Red Cross, and United Nations. Here we cannot hear those voices, because the ears that should be listening are interested in millions from Plasmaféresis. They take the money and try to silence those who defend human dignity."

Chamorro pursued the issue relentlessly. To him the center was not just a business, but a malignancy on the nation's morale. Interviewed by *The New York Times,* he denounced the business as "contributing to the disgrace of the country." Emboldened by his leadership, others took up the call. Doctors and intellectuals circulated petitions. Medical students at the Hospital San Vincente de León put up posters saying "Stop Plasmaféresis!" Public officials promised to launch probes. The inquiries, of course, proved nothing but a sham, since the business fell under Somoza's protection. Irked by the government's failure to investigate, opposition senators created their own commissions.

Seeing the whole city turning against him, Ramos clumsily attempted damage control, distributing circulars explaining the benefits of his facility and announcing a "special-help plan" of interest-free loans to his regular donors. Both tactics backfired. Doctors disparaged his letters as unprofessional, geared to mislead rather than inform, and the loan program blew up in his face when more than two hundred donors showed up for money; screaming and cursing, he drove them away, threatening to have them beaten by the guards.

The drumbeat continued. In November 1977, Chamorro revealed that, for all the center's profits, it paid no taxes. That was because the center had been classified in "Group A"—for industries so vital that they operated tax-free. Such a decision, of course, rested with Somoza, and he saved it for enterprises in which he held a share. Chamorro attacked the arrangement as further proof of the dictator's involvement.

Ramos sued Chamorro for slander. Reporting on the trial, *La Prensa* cruelly mocked Ramos as sweaty and obese, chewing tobacco and hiding his face behind fat, grubby fingers. His suit did not succeed.

On Tuesday, January 10, 1978, Chamorro was driving through downtown Managua on his way to *La Prensa*. He drove, as usual, at a leisurely pace. Suddenly a green Toyota darted from a side street and pulled alongside him. Before he could react, three men aimed shotguns out the window and fired point-blank.

Chamorro's assassination triggered the worst riots in more than a decade. Somoza expressed shock at the murder and denied any involvement, but no one believed him. As the funeral procession bore Chamorro's casket to the cemetery, tens of thousands of people converged. Shouting "Who killed Chamorro? Somoza!" they surged through the streets, setting fire to buildings and overturning cars. Police rushed in with tear gas and clubs. A contingent of rioters broke off and surrounded the plasma center. Shouting "*Casa de vampiros!,*" they stormed it, stoned it, and burned it to the ground.

Police arrested four men for the murder. One of them alleged that Ramos had hired them. But Ramos was gone, having fled to Miami before the killing. Interviewed there, he called the charges "stupid."

Some say Somoza was responsible for the crime; others say Ramos. Certainly, the assassins must have had Somoza's approval. Somoza himself, in a self-justifying memoir called *Nicaragua Betrayed,* offers his own assessment: "Pedro Joaquín Chamorro . . . thought he was the kingpin of Managua. He honestly believed that, through *La Prensa,* he could wipe the streets clean. . . . When he directed his venom at Ramos, he chose the wrong person. . . . Chamorro didn't understand that Ramos, being Cuban, had been brought up under a different code of ethics. . . . Obviously, he was more volatile and, apparently, concluded that 'personal satisfaction' was his only recourse."

Chamarro's murder and the riots that followed marked the beginning of the end for the Somoza regime. For the next eighteen months, the country exploded with strikes, uprisings, and rebel attacks, climaxing in Somoza's resignation and exile.

After the revolution, a jury tried Ramos *in absentia* and found him guilty of murder. Ramos had vowed to face his accusers, but he never returned. He surfaced for a while managing a plasma-for-export center in Belize, the only such center remaining in the Third World. He eventually died peacefully in Miami, where he had served the Cuban community for years. Associates described him as kind and beneficent.

The events in Nicaragua sent tremors through the plasma industry. After all, the center had provided a significant source of material to the manufacturers, and suddenly it was gone. They also realized that they had depended too heavily on outside providers. Even putting aside the risk of hepatitis, it was too much trouble to rely on volatile, impoverished nations, especially given the exploitive implications. As one dealer later put it, "You got tired of people calling you a vampire." It was all too difficult to manage and explain. And so the industry returned to using only American donors, regardless of the additional expense. For all practical purposes, the torching of Ramos's facility ended the industry's wildcat days.

Sometime later, at an office near Los Angeles, a plasma-industry market analyst presented a dispassionate assessment of the events in Managua. "A major point of reference on the Source Plasma Demand Cycle . . . occurred with the loss of the center in Nicaragua. While the quantities being imported only represented 8–10% of the total demand, a short-term shortage did occur in the market," wrote Jack Reasor, president of the Marketing Research Bureau. "Bidding for plasma was intensive. Prices were 12–15% higher than before the Nicaragua loss."

# CHAPTER 14

# THE BLOOD-SERVICES COMPLEX

Back in America, in the supposedly genteel world of nonprofit blood banking, antagonism had become the order of the day. It is hard to imagine that blood banking, which derived its material from citizen-volunteers rather than from paid donors or from the Third World, could be as cutthroat as the pharmaceutical sector, but the nonprofits competed every bit as fiercely. After all, whole blood represented a resource worth hundreds of millions of dollars—money that traveled from patients or their insurance companies to hospitals and to the blood banks, regardless of their "nonprofit" designation. That the blood banks obtained their resource for free only added an edge to their competition, for the industry's two dominant organizations—the Red Cross and the American Association of Blood Banks—fought not only over business and territory but over the moral high ground as well.

Their public bickering eroded their attempts to present an admirable public image and attract enough voluntary donors to maintain high levels of blood donations. Twice before, the two had unsuccessfully tried to reduce the competition and coordinate their activities. Now, in the 1970s, threatened by public resentment, blood-use inefficiency, and blood-borne hepatitis, they made a third, prolonged try. It would result in the most rancorous failure of all.

The effort began hopefully, as we have seen, when President Nixon, declaring blood "a unique national resource," ordered Elliott Richardson's Department of Health, Education, and Welfare to find a better way to manage it. In 1974, the HEW proposed a National Blood Policy, an attempt to reform and unify the blood system without resorting to federal control. The policy included a list of ten goals to increase the safety and efficiency of blood use—such as eliminating paid donors, setting up efficient blood-supply regions, finding better ways of sharing blood, and establishing uniform practices—and left it to the industry to devise a way to achieve them. Richardson's successor, Caspar W. Weinberger, a non-nonsense bureaucrat who later became secretary of defense, initiated the policy, pleading with the blood bankers to come up with a plan. They, in turn, founded the American Blood Commission, a nonprofit voluntary body, to serve as a forum for all blood-related debates, and to issue decisions that members would voluntarily follow. Modeled on the United Nations, the commission was designed to be broadly inclusive, comprising forty-three organizations, including blood bankers, plasma dealers, medical societies, consumer groups, labor unions—almost any group that had anything to say about blood. The hope was that the blood industry, meeting voluntarily, would formulate policies to eliminate the problems of waste, shortfalls, and blood-borne hepatitis.

The inaugural meeting took place at the Statler Hilton Hotel in Washington, D.C. Announced with great fanfare as "the most significant, non-scientific development in blood services in the United States since World War II," it was chaired by a management consultant named John J. Corson. Corson could see the job that lay before him, and, in an impassioned invocation, he exhorted the participants to rise above self-interest and "think for a moment" what their efforts could mean. "Think about what it can mean to the rank and file citizen who may face an urgent need for safe blood tomorrow, next week or next year. Think, please, what it can mean to the anemic, the hemophiliac, the leukemic, whose need for blood is relentless. . . . Surely, in 1975 you and I can recognize that even as no man is an island, no blood bank can or should operate in regal independence. We live in a compact society. . . ."

Everyone recognized the nobility of the sentiments, the call for a cause greater than their own—and immediately started arguing. As the years passed and one quarterly meeting gave way to the next, almost everyone in the American Blood Commission became furious with everyone else. Delegates recall walkouts, and red-faced representatives sputtering in frustration. "In a word, it was hostile," remembered Dr. Byron Myhre, chief of clinical pathology at the Harbor-UCLA Medical Center and a past president of the AABB. "I remember screaming matches. . . ." One major player, the American Blood Resources Association (ABRA), the trade group for the plasma industry, took an even more intransigent stance, and unsuccessfully sued Weinberger to dismantle the body.

One might wonder why the nonprofit blood bankers could not drop their hostilities and move forward. But this feud had been simmering for decades, ever since the beginning of mass blood banking in America, when the Red Cross announced its intention to monopolize blood collection and the AABB decided to resist. The Red Cross had grown mightily over the years. Operating from its headquarters in Washington, the elegant "Marble Palace" near the White House, the organization managed eighty-one blood centers in fifty-seven regions throughout the United States, not counting its mobile collection vehicles. The Red Cross collected more than half the blood used in America, or more than five million units a year. And despite the agency's nonprofit status, these collections translated into serious money. Even though the organization eschewed "profits" as such, it did not hold itself above collecting "processing fees." In 1972, the Red Cross's board of governors mandated its first rise in processing fees, in order to reduce the blood program's debt. The increases continued, and by 1977 the blood program's annual income—"excesses of revenue over expenditures," was how they put it—increased to more than $9 million. (That number would rise to $27 million by 1983.)

Such money and power frightened Bernice Hemphill, the founder and leader of the AABB. In truth, her group was not the underdog that she portrayed it to be. The AABB's more than two thousand members—hospitals and independent and community blood banks—collected more than 35 percent of all the blood used in the country. Acting as a trade organization, the group set standards, ran inspection programs, and held annual conferences attracting thousands of doctors. It also maintained the National Blood Clearinghouse—the national accounting center through which member institutions could exchange blood and blood credits. Yet she worried about the Red Cross's monolithic ambitions and its ability to wield power. In 1960, for example, the Red Cross reached a "statement of understanding" with the AFL-CIO, which lined up millions of union members to provide blood donors for the Red Cross and support for its policies. Later it would find an ally in the United Auto Workers. That kind of power, linked with the Red Cross's oft-stated goal to be the "total blood supplier to the nation," made the leaders of the AABB jump every time the Red Cross so much as twitched.

This struggle for dominance between the two nonprofit blood-banking giants turned what should have been a series of practical concerns about collecting adequate resources into a drawn-out and wasteful moral crusade. Nowhere was this clearer than in the imbroglio over the nonreplacement fee. This was the charge that AABB members would levy against those patients who did not arrange to replace the blood that they used. In practical terms, not much difference existed between those blood banks that employed the fee and those that did not—patients paid about the same for their blood. The two groups even used similar recruitment campaigns—give blood today so that you or a loved one can use it tomorrow. The blood bankers, however, acted as if the contrast in their philosophies made all the difference in the world.

Less than 5 percent of the qualified American public donated blood, which left blood banks continually begging. (In contrast, European nations, with their less transient populations, generally had donor rates of 10 percent or greater.) In order to address the donor-recruitment problem, the commission appointed a task force, which studied nine blood banks throughout the United States—six of which employed nonreplacement fees and three of which did not. Factually speaking, their study was inconclusive: All nine of the blood banks, whether they favored the AABB or the Red Cross philosophy, recruited donors with similar success. But the task-force members, most of whom favored the Red Cross, injected politics into the discussion. In the end, they presented an inflammatory document, concluding that charging nonreplacement fees was "tantamount to selling blood," and urging all blood banks to abandon it.

The findings struck Hemphill like a blow. "I had high hopes that the task force study would bring out the more salient facts about how to motivate donors," she told an American Blood Commission board-of-directors meeting in September 1977, but "such data had not been produced." In a minority report, she argued that the task force had found no factual basis for rejecting her philosophy of individual responsibility. She cited her own blood bank's thirty-six-year "heritage" of successfully using the nonreplacement fee as proof that it could work. At the same time, she supported the Red Cross's right to practice community responsibility in regions where it had proved effective. The whole point was that *either* doctrine could succeed, depending on the experience of blood banks in their region. "I am for pluralism in [donor] recruitment," she argued, ". . . freedom and the right of choice." She saw the task-force report as the Red Cross's attempt to crush the AABB and its allies, putting an end to the multiple blood systems in America, and uniting them all under the American Red Cross.

The tumultuous commission meetings reflected a much broader war for donors and territory. The first shot had been fired in 1976, when the Red Cross summarily withdrew from the AABB's clearinghouse, with which it had cooperated for sixteen years. The situation had seemed harmonious at first, with the two organizations exchanging blood and blood credits—"a big happy family," as Hemphill recalled it. But over the years the Red Cross ran up a debt, having borrowed more blood from the clearinghouse than it returned. Under the rules, it could repay what it owed in blood or money, but the Red Cross insisted on paying only in blood. The agency could not collect enough blood to meet its own needs, much less the demands of the clearing-house, and its deficit rose steadily. By the time of its withdrawal— strictly on principle, according to the press releases—the Red Cross owed the clearinghouse more than thirty-seven thousand units of blood, or the cash equivalent of about $300,000.

The Red Cross's sudden withdrawal confused and disturbed people. Over the years, many had given blood to the American Red Cross in order to build blood credits for themselves or their relatives living in areas served by the AABB. When the Red Cross withdrew, those donors' hard-earned credits were immediately wiped out. In one case, a retired army colonel named Melvin W. Ormes of Philadelphia had been giving to the Red Cross for twenty-five years, assuming his blood went into the national exchange. Sometime after the Red Cross pulled out, his wife underwent surgery, transfused with blood from a blood bank that belonged to the AABB clearinghouse. "We believed . . . we would be entitled to a cost reduction or an exchange," he told a con-gressional hearing. Instead he met with an unpleasant surprise—a blood charge of nearly $2,000.

The withdrawal caused a blot on the public image of blood banking, which Hemphill was quick to exploit. She organized a campaign to persuade the Red Cross to reverse its position, charging that the agency was "turning its back" on the thousands of donors who took part in the exchange. She publicized a letter she wrote Red Cross President George M. Elsey, saying that the withdrawal would lead to a "fragmen-tation of the nation's blood banking complex at a time when both organizations should be working cooperatively. . . ." Finally, in a move of unusual audacity, she urged donors living in areas served by the Red Cross to seek out AABB-affiliated blood banks instead.

The Red Cross was furious. Seeing her campaign as unwarranted and irresponsible, Robert G. Wick, the Red Cross's vice-president, wrote a letter of complaint to the American Blood Commission. The only real threat to the blood supply, he charged, was "the confusion and anxiety that Mrs. Hemphill is spreading in her attempt to bring public pressure on the Red Cross to stay in the Clearinghouse. . . ." The Red Cross dispatched one of its top managers to Los Angeles to revive the city's failing Red Cross center so it could compete with Hemphill's for dominance in California. "If we can prove it to the country that we can do it in Los Angeles," the administrator, Norman J. Kear, wrote in a memo, "a major battle for our entire program will have been won."

Others found more devious ways to strike at Hemphill and her blood bank in San Francisco, often called the AABB's "mother ship." Several Red Cross advocates in California goaded the state's young consumer-affairs commissioner into charging the Irwin Memorial Blood Bank and Hemphill with profiteering from donated blood. All the charges were eventually dismissed, but fighting them cost Irwin an estimated $700,000 and a temporary loss, through adverse publicity, of six thousand blood donors.

One cannot ignore the confusion such tactics must have imposed on the public. In explaining America's traditionally low donor rates, apologists for the industry often cite the nation's transient, multiethnic society, in which individuals feel little connection to the collective. That may be so. But one must also consider the lack of faith the blood bankers engendered in their shifting policies and unseemly competition. As Alvin Drake, a Massachusetts Institute of Technology systems-engineering professor who studied the blood system, told Congress in 1979, the "fractured nature of blood collection messages and efforts" made it difficult for people to acquire "consistent blood donation opportunities and habits." Such habits would require a steady, dependable approach to collections, in which blood donation became "a simple, routine, satisfying part of life." But blood bankers forced themselves into a crisis mentality, and cried wolf to maintain a sufficient supply—which would itself eventually prove self-defeating.

It was all such a mess. As the furor continued to roil in California, the American Blood Commission held several more meetings to discuss the task-force report, each more acrimonious than the preceding one. Finally, they voted to phase out the nonreplacement fee and unite all blood collection under the Red Cross's doctrine of community responsibility. Truth be told, the decision meant nothing, since the commission had been given no enforcement powers. Blood bankers simply ignored the decision and continued doing business as before. Aside from causing enmity, the proceedings had no other effect.

By the late 1970s, the American Blood Commission lay in ruins. Examining the failure, the General Accounting Office cited the "disagreement between the two largest blood suppliers" as a principal cause. In 1979, Senator Richard Schweiker made one more attempt to settle the dispute. He pushed for a law that would unify the blood system under a single nonprofit national blood exchange, open to all blood centers but supervised by the government. Like those who had come before in such efforts, he failed. Meanwhile, a chastened blood industry promised to reinvigorate the American Blood Commission and honor its pronouncements. But the commission died a long and insignificant death.

And so the old pattern continued: Two groups performing genuinely good work for the American public found it impossible to get along. Only later, when they faced a common enemy in AIDS, did they learn to cooperate—and then temporarily. Their lasting failure to join peacefully in recruiting donors has contributed to paralyzing shortages, even in some cases today.

It was outside the commission that the important advances in blood safety were taking place, especially in the control of hepatitis. In 1964, Dr. Baruch Blumberg, of the National Institutes of Health, had discovered a particle associated with the surface coating of the hepatitis B virus which, after a few weeks, triggered an antibody response. His work eventually led to a test for the antibodies that could screen donors who carried the disease. The test, mandated by the FDA in 1972, was only about 15 percent effective. A subsequent test, required by the government in 1975, screened with about 40 percent effectiveness. The majority of infected units were still getting through, but a start had been made in controlling the virus.

Meanwhile, a debate was taking place on other methods of protecting the nation's blood supply. For years, scientists like J. Garrott Allen had argued that one group in particular tended to carry a high risk for hepatitis—donors whose circumstances required them to sell their plasma or blood. Studies had shown that paid donors had hepatitis as much as three times more frequently than volunteers. Allen and others argued that removing paid donors would be the single most effective way to reduce the disease.

As straightforward as it sounded, a ban would be difficult. The plasma industry depended on remunerated donors, since few people volunteered for plasmapheresis. (It may not have been an attractive arrangement, but it was the only way to obtain sufficient plasma; witness the plasma deficits of Europe and Canada.) The AABB objected to a ban because some of its member centers relied on paid donors during shortages.

There was also some counterevidence to consider. Dr. Howard Taswell, who ran the Mayo Clinic blood bank, one of the nation's best-run and most hepatitis-free, argued that paying his donors had enabled him to demand the highest levels of professionalism and commitment. His staff knew donors personally and had been tracking their health histories for years. That was the secret to safe blood, he contended—not the moral aspect of voluntary giving, but the practical ones of close contact with donors and a thorough knowledge of their health. Others would argue that Taswell worked in an unrealistically ideal situation: Located in central Minnesota, his center drew from a homogenous, stable, and healthy group of donors. He could never do the same in Los Angeles or New York, with their huge and transient populations with generally higher disease rates. Blood bankers there felt that banning paid donors would help them eliminate the most likely carriers.

The debates over paid blood rumbled for years. Finally, in 1978, the FDA chose a surprisingly prescient compromise position. The agency required blood banks and plasma collectors simply to label their blood bags as "paid" or "volunteer," and let the marketplace accomplish the rest. Almost overnight, paid blood disappeared, since no hospital would buy blood that was implicitly inferior. (The fractionators continued using paid donors as before, screening the plasma with laboratory tests.)

Within just a few years, the hepatitis B rates plummeted. But the reduction of the B virus removed the mask from another lurking pathogen. Detection rates began climbing for another form of hepatitis—a mysterious strain called "non-A non-B" which, later renamed hepatitis C, was found to cause 90 percent of post-transfusion hepatitis cases. In 1984 alone the new virus infected an estimated 180,000 transfusion recipients, killing an estimated 1 percent of them. Among those who would die of the disease was the entertainer Danny Kaye. In retrospect, if there was a lesson learned from the hepatitis B story, it was not that a virus could be cleverly defeated, but that another would inevitably rise to take its place.

As the whole-blood collectors fell into disarray, the plasma industry consolidated as never before. Drug companies had learned from their experience in Nicaragua not to entrust collections to shady middlemen—especially in marginal, unstable nations—so they took direct control of the resource. By the late 1970s, major drug companies had bought out almost a third of the nearly four hundred collection centers in America. National plasma chains owned most of the remainder—organizations such as North American Biologicals, Inc. (NABI), the world's largest collector of source plasma. The wildcat collector had become a fixture of the past.

Yet, if the concentration of resources bought more professionalism, it proved unsettling in other ways. Four major companies had controlled most of the world's plasma. Based in the U.S., they included Cutter Laboratories of Berkeley, California; Alpha Therapeutic Corporation of Los Angeles; Armour Laboratories of Chicago; and Hyland in a suburb of Los Angeles. These firms represented a pharmaceutical tradition. Armour, as we have seen, had been around since the previous century, having spun off from the nationally known meatpacking business. For generations the firm had made animal-based pharmaceuticals, before helping Edwin Cohn kick-start the American biologics industry. Cutter, an old family business in northern California, boasted a colorful history of public involvement dating back to the earthquake of 1906, when the company provided tetanus and diphtheria antitoxins to the people of San Francisco. They too became one of the original fractionators under Cohn.

Now, with the Third World plasma mills shut down and rising fears of a worldwide "plasma crunch," foreign firms, eager to gain direct access to plasma and entry into the lucrative American drug market, began taking over the American producers. In 1978, Dr. Naito's Green Cross Company of Japan bought Alpha Therapeutic—a sensible move, since Japan used more fractionation products than any other nation yet collected less than half its needs domestically. Later Green Cross bought 50 percent of the Spanish fractionator Grifols, which American plasma also supplied. In 1977, Bayer AG, the German pharmaceutical giant, had taken over Cutter Laboratories. Armour passed from one owner to another until the French multinational Rhone-Poulenc held on to it. Of the major producers, only one remained in the hands of Americans—Hyland, which itself had been purchased by Baxter Travenol Laboratories, a multinational health-care conglomerate based in Chicago.

The Austrian fractionator Immuno bought an old Parke-Davis fractionation plant in upstate New York, along with a chain of more than a dozen collection centers. Tom Hecht, owner of Continental Cryosan, of Montreal, purchased NABI and its chain of plasma-collection centers in order to supply his international brokerage. Later he sold the chain to the Institut Mérieux of France, which was looking for new sources of material.

More than ever, plasma was becoming an integrated resource, mixed and distributed all over the world. A clue to the extent of that integration can be found in a diagram of the Cutter Laboratories network as it existed in 1978. Prepared by John "Newt" Ashworth, Cutter's vice-president of scientific affairs in the international division, the map is a maze of boxes and lines, as complex as an electrician's schematic and sprawling over a sheet of paper the size of a page from *The New York Times.* Most of the lines converge at a box labeled "Cutter Berkeley + Clayton." This was the hub of Cutter's global network—the company's main laboratories in California and North Carolina. Source plasma flowed to these laboratories from plasmapheresis centers throughout the United States, from sources as varied as prisons and college towns. After several months in the manufacturing pipeline, the material emerged as a variety of products, including albumin, two kinds of gamma globulin, and Factor VIII. These products traveled to Cutter distributors in America, Great Britain, and Japan. Cutter sent intermediate powders—unfinished forms of albumin and gamma globulin—to the Cutter subsidiary in Germany for further processing. It sent other plasma fractions to unaffiliated drug companies in Sweden and Canada.

Germany occupies the next-most-cluttered part of the diagram. There a Cutter subsidiary called Tropon received powders from America, source plasma from centers in the U.S. and Germany, and cryoprecipitate from Cutter in Mexico. Plastic plasma bags arrived from Cutter subsidiaries in the U.S., Puerto Rico, and Canada. Tropon in turn sent out albumin, Factor VIII, and gamma globulins.

Near the bottom of the diagram sits Cutter in Mexico, which shipped cryoprecipitate to Germany and Japan and finished products to South America. Farther below rests the Australian affiliate, which provided Cutter U.K. with plastic bags. North on the diagram sits Cutter of Canada, which sent bags to Germany and the Canadian Red Cross. Finally, hovering around the periphery of the diagram are various boxes with the names Behring, Biotest, Immuno, and Mérieux. They indicate Cutter's trade relationships with diverse companies in Germany, Austria, and France.

"The point is, this was not unique to Cutter," Ashworth explained many years later. "Everybody could have a map just like this." Complicating his diagram and that of other companies was the fact that prices and quantities of individual plasma components fluctuated, as supply and demand changed from one country to the next. So it would be better to see this map not as a static web but almost as a global circulatory system, its fluids pumping this way and that, to the zones of greatest profit and demand.

In contrast to plasma, blood tended not to travel internationally. Remember that red cells perish more easily than plasma, which makes them unsuitable for long-term storage and trade. Within American borders, however, a lively exchange in whole blood took place. The routes are too complex and fleeting to chart, because, in addition to working through their organization's clearinghouses, the directors of blood centers made countless *ad hoc* arrangements to get the quantities and blood types they needed.

Given the charitable nature of blood giving, it was surprising how businesslike these arrangements could become. Blood centers with an excess of blood found they could charge whatever they wanted to those facing shortages. Bidding wars could occur when certain blood types entered periods of particular demand. Some nonprofit blood banks found it so lucrative to sell to others that they collected more than their communities needed, staging "blood-crisis" campaigns to drive up donations without telling citizens about their true intent. Some tactics seemed to have nothing to do with charity. One, called "bundling" or "tying," involved linking two blood types in a single transaction, forcing a blood bank that needed certain blood types to buy others it did not need as a condition of the sale. Because "nonprofit" status precludes organizations from paying dividends to owners or shareholders, the money generally went into augmenting facilities, salaries, and the blood banks' savings accounts.

Many of these arrangements were later revealed by reporter Gilbert Gaul in a Pulitzer Prize–winning series in the Philadelphia *Inquirer* in 1989. Gaul's revelations and those of subsequent congressional hearings scandalized the public, a reaction that blood bankers could not understand. After all, *they* knew they were trading in a resource; didn't the public see it that way as well? The public, of course, had no such understanding, having been schooled by the nonprofit blood bankers in the ethos of strict noncommercialism.

One other small but important trade route deserves mention. Amid the rivers of plasma leaving America ran a countercurrent of blood from Europe to New York. Despite the dominance of commercial plasma-processing companies, several European nations maintained fractionation laboratories run by their governments or national Red Crosses. They obtained their source material the old-fashioned way—

bleeding volunteers, centrifuging the liquid, and saving the plasma. This gave them an excess of red cells, which they unthinkingly poured down the drain.

Dr. Aaron Kellner, the enterprising director of the New York Blood Center, witnessed this practice on a visit to colleagues in Amsterdam, when he noticed a tube running from their plasma-separation unit down to a drain. "What's going down there?" he asked. "Oh, that's the blood," his Dutch host replied. "We don't need it; we're discarding it."

Kellner was thunderstruck. And then, "like in the cartoons, an electric light went on right over my head," he later recalled. "I suddenly realized that here we were struggling in New York . . . and these fellows were throwing away these marvelous goodies."

Kellner offered to buy the wasted red cells, but after a year of negotiation the deal fell through. So he traveled to France, where he met with Dr. Jean Pierre Soulier, director of the Centre National de Transfusion Sanguine (CNTS) in Paris. Soulier was intrigued by Kellner's proposal, but, given the French ethos of benevolent blood giving, he could never consent to trade blood for money—although he might be able to trade it for blood bags and equipment. Kellner waited months for a response, only to be disappointed in the end. The decision had gone all the way to President de Gaulle, according to Kellner, "who said in a few very carefully chosen words that he would be damned if he would send good French blood to those blank-blank people in the United States."

Finally, Kellner traveled to Switzerland. There he sought out a blood banker as unfettered and resourceful as himself—Dr. Alfred Haessig, director of the Swiss Red Cross Central Laboratory. Haessig had established a fractionation center in Bern that, owned and operated by the Swiss Red Cross, obtained all its plasma from volunteers. Like his colleagues elsewhere in Europe, he poured thousand of gallons of red cells down the drain.

Kellner and Haessig shared a sense of entrepreneurship, a feeling that, despite their insistence on nonpayment of donors, blood should be used in the most intelligent, practical, and profitable way. They saw no conflict between nonprofit blood banking and the principles of good business. So it did not take them long to reach an agreement in which Haessig's employees, paid by the New York Blood Center and working under the Center's FDA license, would bleed volunteers, spin off the plasma, and ship the red cells to New York.

During the first year of the program, 1973, Kellner imported about twenty-two thousand units from Switzerland, or about 6 percent of his center's supply. Germany, Belgium, and Holland joined the program.

Some doctors predicted that their older Jewish patients would never accept German blood (a concern that never materialized); other New Yorkers distrusted the idea. "It's a stinking shame that we have to depend on Europeans to lend us blood," said New York City's outspoken mayor, Edward Koch. Yet Euroblood, as Kellner liked to call the material, was a notion whose time had come. Soon Kellner was importing more than a quarter of a million units per year, more than a third of New York City's supply. Kellner was receiving so much blood from Europe that he faced a glut rather than a deficit, and shared it with Los Angeles, New Orleans, and Chicago; although he had to stop sending to those centers when the Red Cross decided that as a matter of policy they should strive for self-sufficiency instead. Kellner took a more practical approach: "You do what you have to do in order to get the job done."

Haessig, for his part, was delighted with the results. The program brought money, and saved him the potential embarrassment of being seen to pour precious Swiss blood down the drain. The Swiss Red Cross began shipping red cells to Greece, where an endemic condition called "Mediterranean anemia" caused permanent shortages, and to needy nations in North Africa and the Middle East. Meanwhile, the managers of Central Laboratory found new ways to earn money. From plasma, they derived a form of gamma globulin that could be administered intravenously instead of by injection into the muscle—a more effective means of administration for which the Swiss Red Cross, for a short time, held a monopoly. To market the material, they formed a partnership with the Swiss drug company Sandoz. The two deals involving Sandoz and Euroblood proved so profitable that the Swiss Red Cross built a new central laboratory in Bern.

Kellner and Haessig were not alone. A new generation of blood bankers was arising who, though preaching the principles of nonremuneration, appreciated the value of blood as a commodity. In Germany, Dr. Waldemar Schneider, director of the Red Cross's Westphalian regional blood service, and Dr. Heinz Schmitt of the Lower Saxony region both marketed plasma products to compete with the imports. West German tax authorities felt that this commerce strayed beyond the bounds of the Red Cross's nonprofit status, and so revoked tax-exempt status for moneys received from the plasma operation.

In France, an entrepreneurial blood banker named Dr. Michel Garretta was rising through the ranks of CNTS to take over from the donnish Soulier. French imports of Factor VIII were hovering around 20 percent—a figure that, given the French passion for self-sufficiency, embarrassed the Ministry of Health. The government built a new fractionation plant outside Paris, and instructed Garretta to increase the production. At the same time, he was told to phase out the foreign blood products, something he had the power to do because, among the seven regional fractionation centers in France, Paris alone could authorize imports. Clearly Garretta was the man for the job, having studied management as well as medicine. Assuming the role of commercial-operations director, he vowed to make blood an "industrial" resource. He would rival the multinationals for European dominance.

Even as the Health Ministry was boosting its own blood centers, especially the one in Paris, it moved to eliminate competition. Since 1952, the privately owned Institut Mérieux in Lyon had been collecting plasma from its four plasmapheresis centers to produce albumin, gamma globulins, and vaccines for the French market. In the late 1970s, the Health Ministry, increasingly militant about nonremuneration, ordered the company to close the four centers and banned it from selling products within France. Henceforth it could import placentas and source plasma, but it would have to export all the products it made. From that point forward, the government-run system of Blood Transfusion Centers (CTS, or Centres de Transfusion Sanguine) had become a monopoly, with inordinate power resting with the "national" regional center in Paris, CNTS.

In America, several people within the Red Cross—including its president, George Elsey, and vice-president and blood-program director, Dr. Lewellys F. Barker—realized that, despite the agency's nonprofit orientation, they would have to manage blood in a businesslike way. We have seen how the Red Cross earned millions in the collection and distribution of red cells. Centrifuging the red cells of millions of volunteers left them with huge excesses of plasma, which they arranged to have fractionated under contracts with companies such as Hyland and Armour. The arrangement proved lucrative. In 1977, for example, the Red Cross paid an estimated $9.1 million to fractionate excess plasma that it later sold for more than three times that amount. Indeed, the plasma market proved so rich for the Red Cross that the agency made plans to build its own fractionation plant in conjunction with Hyland—a move that, given the organization's tax-exempt status, would have destroyed its competitors, commercial and nonprofit alike. "We're not after anyone's share of the market," Wick told *Business Week,* a remark that was sure to put no one at ease. The plan never materialized, but, with an estimated 15 to 30 percent of the nation's raw plasma, the nonprofit Red Cross maintained a powerful presence in the commercial marketplace.

In Canada, a strengthening current of entrepreneurship was eroding the old partnership between the Red Cross and Connaught Laboratories. Most Canadian hospitals at the time purchased clotting factors on the open market, since Connaught lacked the technology to produce it. But the Red Cross still relied on Connaught for albumin, gamma globulins, and other derivatives of the plasma of Canadian volunteers. The facility was aging badly, however: By 1974, it was losing 50 percent of the Red Cross's plasma to bacterial contamination.

In 1972, the University of Toronto sold the laboratory to private ownership. Later the Red Cross learned that, in order to bolster its financial condition, Connaught's managers had secretly been selling Red Cross material to plasma brokers for foreign distribution. One year was particularly galling: In 1974, when Canada was experiencing a shortage of albumin, Connaught sold $500,000 worth of products derived from the Red Cross's plasma. Disillusioned with the laboratory, the Red Cross applied for government funding to build its own fractionation lab and take charge of the nation's fractionation. Not to be frozen out of the business, Connaught applied for and received government approval to produce Factor VIII. The company also proposed to construct its own new fractionation plant, paid for with government funds.

Now that the old alliance was broken, the provincial governments got into the fray, vying to host a profitable and publicly funded business. After extensive negotiations, the provinces' health ministers agreed to build a network of three fractionation plants, in Ontario, Winnipeg, and Manitoba. The plants would be owned by private companies, including Connaught, but operated as nonprofits. Trumpeted as the key to Canadian self-sufficiency, the scheme had one astonishing deficiency: In focusing on the moneymaking *processing* of plasma, the ministers ignored the issue of supply. The Canadian Red Cross had never been able to collect enough plasma to feed even one plant, let alone three. Despite tens of millions of dollars expended, the plan was never brought to completion. Eventually Connaught updated its facility, but imported much of its source plasma from America. Indeed, throughout the 1970s and '80s, at least half the plasma products circulating in Canada originated from paid American donors.

This, then, was the blood-services complex on the eve of the AIDS epidemic. Whole blood, with the exception of Euroblood, generally remained within national boundaries, although in America that encompassed quite a bit of territory and regional exchange. Plasma and its derivatives traveled globally, with the United States serving as the most important source. Within the U.S., plasma came from a variety of sources, from colleges to prisons to indigent neighborhoods. Homosexuals also became an important source. Because gays had high rates of hepatitis, the drug companies valued their plasma for hepatitis antibodies. Before isolating the antibodies, the fractionators would "skim" off the cryoprecipitate and use it to produce clotting factors. This practice was not thought to spread hepatitis, since antibodies in the plasma neutralized the disease. What it did do, though, was selectively include plasma from high-risk populations—who were apt to carry other viruses as well—into the world's clotting-factor supply.

By the late 1970s, the blood-services complex had become an interlocking network that mingled the blood and plasma of millions of people who lived in regions thousands of miles apart. Generally this intermingling proved beneficial: More people received lifesaving blood products than ever before. But the distribution also bore peril. Integrating the world's blood products network did more than increase the efficiency of supply—it established optimal conditions for the spread of emerging viruses. America, as we have seen, was the OPEC of plasma. This meant that almost anyone in the Western world or Japan who received plasma-based medication made intimate contact with American donors, most of them professional donors living in the nation's hot zones for disease. For people who received large quantities of medicine—mainly hemophiliacs—the contact with tainted blood products was becoming inevitable.

# CHAPTER 15

# OUTBREAK

The U.S. Centers for Disease Control and Prevention (CDC) serves as America's early-warning system for disease. This nondescript cluster of brick buildings nestled into the campus of Emory University is home to thousands of scientists working in some of the world's most high-tech biological-containment facilities. Established to investigate malaria prevention during World War II (Atlanta was chosen as the heart of America's malaria belt), the agency has since expanded its mission to monitor all forms of microbial invaders, from the familiar diseases of flu and hepatitis to the enigmatic outbreaks of legionnaire's disease and toxic shock syndrome. The agency played a key role in several historic triumphs, including the global eradication of smallpox and the defeat of polio in the U.S. The centers' researchers see themselves as a strike force against epidemics wherever they may emerge, from Africa to India, Arizona to New York.

One of the centers' less well known functions over the years has been to distribute certain vaccines and rarely used drugs. One such drug was pentamidine, a medication used against a rare form of pneumonia. Doctors had treated most pneumonia cases with sulfa drugs, but one form, Pneumocystis carinii pneumonia (PCP), resisted everything except pentamidine. Until fifteen years ago, fewer than one hundred cases occurred annually, and the CDC kept track of them.

In January 1982, Dr. Bruce Evatt, the CDC's specialist in hemophilia, got a call from a Miami physician whose patient had died of

Pneumocystis pneumonia. The patient was a sixty-two-year-old married hemophiliac who lived in New York but spent winters in Florida.

The mention of Pneumocystis pneumonia and hemophilia together sounded an alarm bell for Evatt. He knew that PCP cases had recently inexplicably jumped, particularly among gay men. In June 1981, five cases had been detected among "previously healthy" male homosexuals; since then the caseload had risen to more than one hundred. The disease seemed to travel in concert with other ailments, including a rare cancer known as Kaposi's sarcoma (KS). These "opportunistic" infections, as they came to be called, attacked people whose immune systems had been compromised, perhaps by anal intercourse. So stunning was the sexual connection that the syndrome became known as GRID—gay-related immunodeficiency disease.

Evatt suspected that the Miami patient, although heterosexual, had also died of GRID. Indeed, the case might show that hemophiliacs were also at risk for the syndrome, probably through their exposure to blood products. Hemophiliacs were well known to suffer high rates of blood-borne diseases; witness their towering rates of hepatitis. Indeed, with their massive exposure to blood products, they had become unintentional canaries in the cave. But Evatt needed more cases to prove the connection between GRID and Factor VIII. "If it's real there will be more of them," the agency's director had told him. Evatt's one wedge into cracking the mystery lay in the fact that, as the nation's sole distributor of pentamidine, the CDC could track where every dose went. He notified the CDC's pentamidine office to alert him the next time a request from a hemophiliac's doctor came in.

Half a year later, the disease struck two more hemophiliacs. They, like the Miami case, seemed to be average Americans—a fifty-nine-year-old man from Denver and a twenty-seven-year-old man from Ohio. They did not fit the known risk groups, or have anything in common other than that they used Factor VIII. Evatt sent Dr. Dale Lawrence out to examine them. When he returned with a diagnosis of GRID, Evatt felt they had made the connection.

Evatt immediately called the U.S. Public Health Service, the four major fractionators, and representatives of the National Hemophilia Foundation. He laid out his findings in the CDC publication *Morbidity and Mortality Weekly Reports.* (Set up as a rapid-response scientific publication, the *MMWR,* as it is called, is read throughout the world.) Evatt felt the cases raised two possibilities: that the pathogen probably moved through the blood, and that the agent was most likely a virus, since the process of making Factor VIII filters out anything larger. On July 16, 1982, he articulated his concerns to an emergency working group of the U.S. Public Health Service, which included representatives of the CDC, the National Institutes of Health, the FDA, the National Hemophilia Foundation, and the major blood and plasma organizations.

By this time, GRID occurrences were forming the bell curve of a classic epidemic. The disease now afflicted more than 440 Americans overall. Doctors were astonished at the virulence of the syndrome—more than 50 percent of their patients had died. No one knew what caused the disease. Some disease experts thought that the pathogen had been transmitted by sperm; others suspected amyl nitrites, a stimulant then popular among gays. Some theorized that no single entity had caused the disease, but that the gay victims had been exposed to so many diseases—they had notoriously high rates of syphilis and hepatitis—that their immune systems had simply collapsed from the strain. With all these possibilities to choose from, the meeting participants were skeptical of Evatt's theory of blood-borne transmission, drawn from the data of just a few hemophiliacs. Most also did not want to believe the implications. Evatt's data would mean that the disease affecting gays in America could potentially infect the rest of the population. In a second meeting, later that month, the agencies set up a surveillance program to watch for new cases among blood recipients and hemophiliacs. They also agreed on a new name for the disease: acquired immune deficiency syndrome—AIDS.

Throughout the fall of 1982, with the number of hemophilia-AIDS cases rising, Evatt and his colleagues embarked on a campaign to warn the nation's medical establishment. The CDC was undergoing budgetary cutbacks, so Evatt used his own money for the travel. He worried that, given a lag time of several months before the symptoms appeared, a blood donor could have AIDS and not even know it. The safest course of action, he felt, would be to exclude the people most likely to carry AIDS—mainly, homosexuals—whether or not they exhibited the symptoms.

During his travels from one meeting to another, he received "mixed" responses to his unsettling news. Blood bankers, besieged with economic problems and continual shortfalls, did not want to add another crisis to their list—at least not until the evidence was definitive. Nor were they willing to exclude their gay donors. At a time of waning sense of community and declining blood donations, gays represented one of the most loyal donor groups. Generally well educated and civic-minded, they provided about a fifth of the Irwin Blood Bank's supply in San Francisco, for example. No one wanted to lose the best donors, or contribute in any way to antigay prejudice. "It was as though someone had wandered in from the desert and said, 'I've seen an extraterrestrial,' " Evatt recalled. "They listened, but they just didn't believe it."

The National Hemophilia Foundation, though alarmed at his findings, gave him a decidedly mixed response as well. Dr. Louis M. Aledort, codirector of the foundation's Medical and Scientific Advisory Board, had pioneered the concept of hemophilia therapy. Having seen his patients literally climb out of their wheelchairs, he was loath to abandon the therapy. Years before, he had developed the nation's first "comprehensive-care" center at New York's Mount Sinai Hospital; later, he lobbied Congress to fund a national network of hemophilia treatment centers. Still later, Aledort formed a consortium among treatment centers in the New York area to leverage the best prices from the drug companies and provide reliable quantities to patients. With such a large market at his command, he became a power in the medical-pharmaceutical world, attracting tens of thousands of dollars in grants every year. Aledort felt that he needed more data before recommending that patients change their treatment; after all, Evatt had found only three cases among the nation's twenty thousand hemophiliacs. He knew the good that the therapy could do, and worried about an overreaction. As Evatt understood it, "Here they were dealing with a strange disease that they really didn't want to believe existed."

Others in the National Hemophilia Foundation could not ignore Evatt's findings. After he spoke to the foundation in October, it passed a resolution urging the drug manufacturers to exclude from their plasma pools groups with a relatively high incidence of AIDS, including gays, IV drug users, and Haitians. At the same time, it presented a reassuring face, issuing a series of advisories that the risk of AIDS in Factor VIII was minimal and that hemophiliacs should continue their infusions as before.

Ironically, the most cooperative early response came from the profit-oriented drug companies (although some later would try to stall, as we shall see). Tom Drees, president of Alpha at the time, said he was "knocked off [his] chair" when Evatt addressed a group of fractionators. He immediately made plans to exclude high-risk donor groups, despite potential charges of discrimination. His and other companies also accelerated their research to find a way to decontaminate the product.

The companies did not react solely out of altruism. Living in the worlds of medicine, regulation, public relations, and law, they recognized certain hard, even cynical, realities. They, like the blood bankers, prized their gay donors—not for their civic-mindedness, but as prime sources of hepatitis B antibodies. During an early meeting with industry leaders, Dr. Dennis Donohue, director of the FDA's Division of Blood and Blood Products, had asked them if they could turn away high-risk donors from certain "hot spots," such as San Francisco, Los Angeles, and New York. "He is not basing this request on scientific concerns that such plasma or coagulation by-products transmits AIDS but believes the action is a political necessity to prevent national adverse publicity and . . . undue concerns in the hemophilic population," wrote Cutter official John Hink in an internal company memo. After informally surveying his competitors and finding that most intended to go along with the request, Hink concluded that Cutter should agree, for "political, moral and liability reasons." Meanwhile, in patient advisories and brochures, Cutter (and other drug companies) continued to describe the AIDS risk as minimal.

By late in the winter of 1982, the number of hemophilia-AIDS cases had risen to eight—more than a doubling in less than six months. Now came the first report of a blood-related AIDS case. A baby in San Francisco had died of the disease more than two years after receiving multiple transfusions of blood and blood products. Looking back through their records, officials at the Irwin Blood Bank discovered that one of the donors was a gay man who, seemingly healthy at the time, later died of AIDS. This information closed the case as far as Evatt was concerned. He and other public-health officials called for a summit meeting to discuss what to do about AIDS in the blood supply.

More than two hundred representatives of the blood industry, doctors, gay groups, and patient and hemophiliac groups came to the day-long session at CDC headquarters in Atlanta on January 4, 1983. After two preliminary presentations, Evatt laid out his most recent data. He told the assemblage that, in addition to the eight known hemophilia AIDS cases, he strongly suspected two more. Furthermore, a survey he had taken of more than one hundred hemophilia centers turned up thirty-seven more cases he planned to investigate. Thus, he said, the epidemic curve for AIDS in hemophiliacs was looking more and more as it had among gays, gathering momentum, and about to rise sharply. Furthermore, he confirmed details of the blood-related AIDS case of the San Francisco baby and said that two more transfusion cases were under study. Now that the problem was known, he said, the question was how to protect the blood system. It was especially crucial to act as soon as possible: With a known lag time of more than a year, someone who passed all the medical exams could still carry the disease and pass it on through blood or plasma.

The next speaker, Dr. Tom Spira of the CDC, presented some options. One thing that struck epidemiologists, he said, was the correlation between AIDS and hepatitis B—in fact, nearly 90 percent of the known AIDS sufferers in America had been exposed to hepatitis B. With such a tight correlation between hepatitis and AIDS, hepatitis could serve as a "surrogate marker" for AIDS—an indication of those who might harbor the disease. Blood banks routinely tested for hepatitis by screening for an antibody to the virus's surface coating, a test that only detected recent exposures. Spira found that another antibody—one that reacted to the virus's core—remained in the body for years. Thus, though no AIDS test existed, blood banks could provide an interim measure of safety by using the hepatitis B core test.

Evatt believed that he and his colleagues had presented a complete package—the disease, the risk groups, and the methods to exclude them. "We went into that meeting expecting it to be a snap," he recalled. "How could anybody doubt the data we'd accumulated, the *trends?* We thought it was a no-brainer."

He could not have been more mistaken. Gay representatives immediately objected that labeling them as unacceptable donors would trample their human rights. Roger Enlow of the National Gay Task Force said that the community had been educating its members and could be counted on to behave responsibly. To exclude them legally from donating, however, would put the stamp of approval on homophobia. Dr. Bruce Voeller, a member of the gay group Physicians for Human Rights, argued that, since AIDS seemed limited to promiscuous "fast-track" gays, excluding gays in general would "stigmatize . . . a whole group, only a tiny fraction of whom qualify as the problem. . . ." For that reason, he favored the use of the hepatitis B core test, since it relied on a laboratory procedure rather than invasive questioning. Enlow agreed: "We think screening blood, not people, is the way to go."

"I don't think anyone should be screened for donating blood on the basis of sexual preference," said Dr. Donald Armstrong of the Memorial Sloan-Kettering Cancer Center in New York. "I think that is wrong."

The drug companies disagreed. With their paid donors and enormous processing pools, they had to act quickly to check the contamination. A representative from Alpha said the company already had begun excluding gays, Haitians, and drug users, "because frankly, we don't have anything else to offer at this time." Seizing the public-relations advantage, he added, "I would hope everyone in the industry would follow suit." Other drug manufacturers agreed to the principle, which the National Hemophilia Foundation had been pushing for months. Commenting later to a medical publication, Aledort said, "I disagree vehemently with the National Gay Task Force. They may want to protect their rights, but what about the hemophiliacs' right to life?"

There was plenty of reason for everyone to feel defensive at the CDC meeting. For one thing, the setting put everyone on edge. The visitors had expected a scientific exchange, a sober and considered policy discussion. But when they entered the conference room with the horseshoe-shaped tables, they found themselves blinded by television klieg lights and battered with questions from aggressive reporters. This was hardly the stetting for a rational discussion of the delicate topics of blood, blood products, and sexual orientation, hardly the place for secrets to be revealed. When a couple of people raised the question of prison plasma, the Pharmaceutical Manufacturers' Association "stonewalled" it as "immaterial to the discussion," according to a memo by a drug-industry representative. In a postmeeting memo, one Cutter official wrote, "To exclude such plasma from the manufacture of our coagulation product . . . would presage further pressure to exclude plasma collected from the Mexican border and the paid donor."

Dr. Oscar Ratnoff, a renowned hemophilia physician from Cleveland, suggested that hemophiliacs sidestep the problem by suspending their use of Factor VIII. They could resume using safer cryoprecipitate, made from pools of ten donors or less. "Sure it'll cost more, but not as much as a funeral or the lawsuits we're going to get after more hemophiliac deaths," he said. The drug companies and other hemophilia doctors opposed him, arguing that after a large number of exposures to cryoprecipitate hemophiliacs would probably contract the disease anyway. (They turned out to be wrong, as we shall see.)

Many harbored doubts about the hepatitis core tests. Dr. Aaron Kellner, director of the New York Blood Center and engineer of the Euroblood program, complained that the test would cost his center $5 million a year and force his and other blood banks to turn away 5 percent of their donors. "This is a very serious problem," he stated, "but we ought not to do things that would jeopardize the community's blood supply."

Kellner and others felt that Spira's core test was not specific enough. It certainly *correlated* with most cases of AIDS, but it did not specifically *detect* the disease. What about those people who falsely tested positive, who had previously been exposed to hepatitis but were at no risk for AIDS? Imagine a donor's horror after he has been rejected by a surrogate AIDS test. As Dr. Joseph Bove, director of the Yale University

Blood Bank and then president of the AABB, later testified: "This was a major worry, that in a time of this AIDS concern, hysteria, whatever you want to call it, one out of twenty individuals walking into the blood bank . . . would be told 'You can no longer donate blood because your blood is positive by the core antibody test. Yes, that is the test we're using to screen out people that might—but don't worry, you don't have AIDS.' " The point, Kellner argued, was that the data were not strong enough. "What do we have in the way of evidence?" he asked. "Three cases at most and the evidence in two of these cases is very soft. . . . Don't overstate the facts." Added Bove: "We are contemplating all these wide-ranging measures because one baby got AIDS . . . and there may be a few other cases."

Evatt tried to explain that when a disease appears suddenly and spreads rapidly it meant that they were witnessing the birth of an epidemic.

Yet how could they take actions against a syndrome, some argued, when they had not even identified a cause? These immune deficiencies could be triggered by many things. "I'm concerned about the concept that we are convinced it is an agent . . . ," said Aledort. "We could be doing something through transfusion that causes it . . . or something in the patients' immune complex. Now in another six months we may . . ."

At that point Dr. Donald Francis erupted. Francis, assistant director of the CDC's hepatitis lab in Phoenix, Arizona, could not believe what he was hearing. He had chased epidemics from India to Zaire, but had never seen bureaucratic resistance like this. Pounding his fist on the table, he shouted, "How many people have to die? Is three enough? Is six? Is ten? Is a hundred enough? Just give us the number so we can set the threshold!"

Years later, still seething about the incident, Francis said, "I just couldn't believe these guys. It was something like having a bend in the train track and sitting there and you hear the whistles and the signals are blinking and the tracks are beginning to shake, and they're saying, 'There's no train coming.' "

"I think they were listening, but I just don't think they wanted to believe it," Evatt reflected. "The implications were so catastrophic for the whole industry they just wanted it to go away."

Certainly denial lay behind the resistance—the implications of a contaminated blood supply were virtually unthinkable—but blood bankers had reasons to doubt the CDC. For one thing, the recommendations they heard that day were by no means *official*. The suggestions about donor questionnaires and hepatitis core testing came from a few individuals *within* the CDC, whose superiors had not endorsed them at the meeting. (In the public-policy world, this small difference assumes galactic proportions.) Furthermore, the agency had cried wolf in the past. Many recalled how CDC experts had deeply embarrassed the Ford administration by sounding an alarm for the swine-flu epidemic that never came and pushing for a vaccine program that probably killed more people than it protected. They also knew that the agency faced budget cuts. As an American Red Cross official wrote in a post-meeting memo: "It has long been noted that CDC increasingly needs a major epidemic to justify its existence. This is especially true in light of Federal funding cuts and [the] fact that AIDS probably played some positive role in CDC's successful effort . . . to fund a new $15,000,000 virology lab. This CDC perspective is also obvious from the general 'marketing nature' of the January 4, 1983 Atlanta, [meeting, with the] abundant press. . . . In short, we can *not* depend on the CDC to provide scientific, objective, unbias[ed] leadership. . . ."

Beneath those reasons lay even deeper issues. Even though they were citizens of the same country, the adversaries in this discussion came from two different cultures, with contrasting sets of values and views. Francis and his colleagues were a fast-acting lot, sensitive to the slightest hint of a trend. These were the kind of people the CDC attracted—activist, Peace Corps types, ready to move, react, and respond. In contrast, the blood-banking culture resembled the business world. Its leaders, though scientists, concerned themselves with businesslike issues of inventory, quality control, and supply. Describing themselves as constitutionally conservative, they were loath to make quick decisions on scanty data.

This divergence of perspective gave the two groups radically different views of the AIDS-epidemic curve. To the CDC workers who had tracked it from the beginning, with the case rate doubling every six months, the response time was *now*. To the blood bankers who administered tens of millions of transfusions, these half-dozen or so cases were a blip, a troubling anomaly—certainly worth tracking but not enough to upset traditional collection methods. It is not surprising, then, that the meeting produced no decisions. Afterward, Francis wrote of his disappointment in a memo to his superiors. "I feel there is a strong possibility that some post-transfusion AIDS and much post–factor VIII receipt AIDS will occur in this country in the coming two years. . . . For hemophiliacs I fear it might be too late."

Two days after the meeting in Atlanta, the country's major blood-banking organizations (the American Red Cross, the AABB, and the

Council of Community Blood Banks) set aside decades of feuding about policy and territory and convened a Joint Task Force against AIDS. On January 13, the group issued its first Joint Statement on Acquired Immune Deficiency Syndrome Related to Transfusion. It was a conservative document, insisting that the case for blood-borne transmission was inconclusive, and offering several "reasonable" measures for blood banks and physicians to follow. These included educating donors about AIDS, allowing "autologous donations" in which patients could set aside their own blood for future use, and discouraging donations among groups "that may have a high incidence of AIDS." The statement did not recommend surrogate testing. The blood bankers added, "Direct or indirect questions about a donor's sexual preference are inappropriate."

Publicly the statement seemed reasonable, cautious, and reassuring. Privately, though, Bove harbored doubts. He hoped that the statement would "buy time" with the public while blood bankers figured out what they should do. "There is little doubt in my mind that additional transfusion related cases . . . will surface," he wrote in a memo to the AABB executive board. "Should this happen, we will be obliged to review our current stance and probably to move in the same direction as the commercial fractionators. By that I mean it will be essential for us to take some active steps to screen out donor populations who are at high risk for AIDS. For practical purposes this means gay males. . . ."

The fractionators had by now begun screening gay donors, but the National Hemophilia Foundation pushed them even harder. At a summit meeting with the industry in mid-January, the foundation stepped up pressure for tough donor screening—"lapel-grabbing, finger-pointing questions," as an industry veteran put it. They also asked companies to consider surrogate testing. The organization issued a dozen recommendations to hemophiliacs and their physicians designed to cut back the risk of overusing Factor VIII should it prove to carry an infectious agent. They urged hemophiliacs to postpone elective surgery, and physicians to use cryoprecipitates for newborns and other patients without previous Factor VIII exposure. Meanwhile, however, they continued to advise most hemophiliacs to keep using their factor as before.

The plasma companies agreed with the Hemophilia Foundation on the issue of screening, although some preferred a less confrontational, self-exclusion approach. They felt less sanguine about the foundation's demand for surrogate testing, which could cost $5 a test and eliminate 10 percent of the paid donors. Nor did they want to switch to small plasma pools, which would raise the price to prohibitive levels. ABRA, the plasma-industry trade group, urged all collectors to intensify their screening by requiring donors to read informational brochures about AIDS and to certify they did not belong to any risk groups. They did not recommend the hepatitis core test.

Through the early months of 1983, as the first halting steps were taken to protect the nation's blood supply, fear and indecision rose in tandem. Provocative articles appeared about the threat to the blood products. A story in *Rolling Stone* asked readers to "think about the unthinkable: Are our blood banks already contaminated? Is AIDS going to flow into your veins the next time you need a blood transfusion?" On Long Island, New York, the Roslyn Country Club Civic Association established its own members-only donor list. In San Diego, a group of lesbian donors formed the Blood Sisters Project; if gays and their partners harbored the infection, then lesbian blood would be unusually clean, since they never had sexual contact with men. (The group later received a national award.) Donations declined as some people worried they could catch AIDS by merely *giving* blood; by midsummer in New York, for example, donations had dropped by 25 percent.

Threatened by shortages and public hysteria, the leading blood banks gave contradictory messages. Kellner openly scoffed at the risk of blood-borne AIDS transmission. Yet, even as he did, his staff began an experimental program in which, after an interview, the donor could check a box on a form reading "my blood is only for studies" if he felt that he belonged to a high-risk group. In San Francisco, Dr. Herbert Perkins, medical director of the Irwin Blood Bank, rebuffed a public plea from a group of AIDS specialists at the University of California, San Francisco, to consider using the core antibody test. Perkins argued that there was no "rational evidence" it would effectively screen AIDS, and that it would eliminate enough blood to jeopardize the region's supply. Later, under increasing public pressure, Perkins agreed to try the test. Many centers added more searching inquiries about AIDS-related symptoms to their donor questionnaires. No one really knew what do to. No leadership was coming from the Reagan administration, which had not even officially acknowledged the epidemic.

Finally, on March 4, 1983, the U.S. Public Health Service issued its first official AIDS-related recommendations. By now more than twelve hundred cases had been detected, including eleven among hemophiliacs and about half a dozen possible transfusion cases. The Public Health Service statement urged citizens to avoid sex with multiple partners or with "persons known or suspected of having AIDS." It also asked members of high-risk groups—including "sexually active homosexual or bisexual men with multiple partners"—to refrain from donating plasma or blood. A few weeks later, the U.S. Food and Drug Administration issued specific guidelines for the blood industry with self-exclusion procedures to weed out members of high-risk groups. These generally took the form of informational materials, one-on-one interviews, and statements for recipients to sign saying they understood the risk of AIDS. Neither directive recommended surrogate lab tests or direct sexual questioning.

Those who had been tracking the epidemic believed that the government had done the absolute minimum, merely endorsing the existing consensus. (The Public Health Service had rejected an earlier draft prepared by the CDC including the surrogate blood test and the exclusion of gay donors, promiscuous or not.) But to Evatt the action marked a turning point—the beginning of the end of the period of denial. "I think it was gradual; but the fact that something actually came out [of the government] made a lot of people think, 'This really is happening.' Things began to change after that."

Evatt was right: Things had begun shifting, at least in terms of public policy. In terms of the physical resource, however, change came at the pace of making a U-turn with the *Titanic*. Millions of units of blood, plasma, and clotting factor collected the old way remained in use all over the country—in blood banks and drug companies; in hospitals and warehouses; in boxes and storage bins up and down the chain of distribution. Thousands upon thousands of infected bottles sat in the refrigerators of thousands of hemophiliacs, who would use them at the next sign of bleeding.

In the tiny rural town of Dolores, Colorado, forty-five-year-old Susie Quintana had just come home from some hiking and target practice in the woods. Hers was a bucolic existence, built around her husband, children, grandchildren, and community. Everyone in town knew Susie, and liked her. She had grown up in Dolores, met her husband at a dance, and gained some renown with her prize-winning crocheting. She was admired for her levelheadedness and cheerful personality. Returning from her hike on May 27, 1983, she was putting away her .22 rifle when it discharged, wounding her in the side. Later, at the hospital, her son Ron asked the doctors if the family could provide blood. Like millions of families throughout America, the Quintanas had heard news of the AIDS epidemic, and discussed it around the dinner table. They had all agreed that, if any of them ever needed blood, the others would provide it. Ron and his father had the same blood type as Susie. But the doctor dismissed him, telling him that the local blood was perfectly safe. "There are no gays or homosexuals in the county," he said.

Aside from the ignorance of the doctor's assumption, what he neglected to tell Ron was that the blood did not come from their county, or their state. The hospital purchased blood from the nation's second-largest blood-banking conglomerate—United Blood Services, based in Arizona. On April 18, the company had staged a blood drive at a school in Santa Fe, New Mexico. The collectors knew about the Public Health Service advisory and took it seriously, carefully examining and interviewing the donors. They distributed printed sheets, asking members of the high-risk groups, including "homosexually active males with numerous contacts," to refrain from donating. One donor, a teacher who happened to be gay, read the sheets and answered the questions. The phrase "numerous contacts" did not apply to him, so he gave blood with a clear conscience and the best of intentions. He had no way of knowing that he might be carrying AIDS. One month later, in a small hospital in rural Colorado, doctors infused his blood into Susie Quintana.

At a charity basketball game in Tennessee, Dana Kuhn, a forty-year-old seminary student and father of two, fell after jumping for a rebound, and broke one of the bones in his foot. Kuhn was a mild hemophiliac, and had never injected Factor VIII before. But doctors infused him, just to be safe.

In Los Angeles, Corey Dubin, a radio journalist and severe hemophiliac, was watching the evening news as he gave himself a Factor VIII injection. A hulking man with a Pancho Villa mustache and a savage intensity, Dubin had been infusing since he was a boy; he was among the first group of patients on whom Hyland had tested the product in the late 1960s. It had changed his life utterly, giving him the freedom he craved to hike the Muir Trail in the mountains of California and bushwhack through the jungles of Costa Rica. Now, as he infused another of the thousands of doses he had taken since childhood, he turned to his wife and said, "Shit. I just *know* I'm shooting myself up with AIDS."

Outside America, the issue of AIDS in the blood supply stirred up a complicated mixture of concern and denial. Many in other countries saw AIDS as an American disease; it had, after all, blossomed there first. "The initial reaction was that whatever happened in America won't happen here. After all, we don't have gay bathhouses," recalled David Watters, director of Britain's National Haemophilia Society. Yet, just as in the United States of a year or so before, cases were increasing among European gays, and had begun to appear among blood recipients and hemophiliacs. Deny it as they might, nations throughout the world would have to reconsider the safety of their blood systems. They would also face the uncomfortable reality that they imported most of their plasma products from a country with the world's highest AIDS numbers.

In Britain, concern about AIDS sparked yet another call for plasma self-sufficiency. The failure of England to reach Factor VIII independence had become an old and dreary news item by now. The government had set and missed numerous deadlines to modernize and expand the fractionation plant at Elstree. If they had succeeded, critics argued, it might have been possible to make safer clotting factors from small lots taken from well-screened British donors. What some critics found especially galling was that, just north of the border, the Scottish National Transfusion Service ran a clean, modern plant with excess capacity. The English could have collected plasma from their own donors and sent it for processing in Scotland. Doing so would have meant putting the facility on round-the-clock shifts and negotiating overtime pay with the unions—out of the question under the Conservative Thatcher government. And so the Scottish plant sat idle for a portion of each day while the English increased their imports from America. By the spring of 1983, when the first hemophilia AIDS cases began appearing in Britain, English hemophiliacs were getting about half their Factor VIII from commercial American firms.

Physicians watched the situation with dread, for they fully understood the hazards of the imports. In 1975 and 1978, for example, Dr. John Craske of the Public Health Laboratory in Manchester had published two studies implicating American clotting factor with hepatitis B outbreaks among British hemophiliacs. He and others knew that it was only a matter of time before AIDS, which followed the same routes as hepatitis, began to endanger hemophiliacs as well. But the AIDS risk was uncertain and the dangers of untreated hemophilia were clear. So, as they watched for the signs of a rising epidemic, doctors told their patients to keep using the clotting factors, imported or not. Someday, they hoped, safer products would become available. Until then, said Dr. Carl Rizza, chief of the Oxford Haemophilia Centre, the fate of hemophiliacs was "in the lap of the gods."

The Germans had plenty of early warning about AIDS. Not only had the federal health authorities been in close touch with the American CDC, but as early as May 1982 a patient at the Bonn clinic died of what appeared to be AIDS. The diagnosis, if accurate, would represent the world's second case of hemophilia-linked AIDS (after the January case reported by Bruce Evatt), and came at a sensitive time for the Bonn center. After all, Dr. Hans Egli and Hans Hermann Brackmann had built the center's reputation on the massive use of Factor VIII, which they were now being forced to defend against the insurance companies. This was also the time when German tax authorities were launching an investigation against the center's chief procurement officer, Dr. Etzel, in relation to his arrangements with the importers from Switzerland. It was not surprising, then, that Brackmann and his colleagues at the Bonn center hotly disputed the diagnosis, arguing that the patient had actually died of alcoholism and hepatitis. More cases inexorably followed. By the fall of 1983, the number in Germany had risen to six.

In Strasbourg, the Council of Ministers of the Council of Europe urged its member nations to react to the disease. Even though "no formal proof" linked AIDS and the plasma supply, the council urged its member nations to avoid blood products derived from commercial (i.e., American) donors, and try to eliminate imports altogether.

Germany could never hope to comply. It led the world in its prescription of the factor, the vast portion of which came from the United States. In November 1983, a special working group of federal health authorities, hemophilia treatment providers, and industry representatives debated the wisdom of continuing to prescribe high levels of clotting factors. Some doctors argued for an immediate cutback, but others, principally Brackmann, insisted that a virus did not cause the disease, that there must be some other "co-factor" at work. Brackmann and his colleagues prevailed in the end, and the Federal Health Authority (Bundesgesundheitsamt, or BGA) concluded that "a limitation of imports" was "out of the question."

During this time, the hundreds of Bonn patients received no indication of the ongoing debate. All they knew about the safety of their factor was what their doctors kept telling them in a reassuring series of "Dear Family" letters. The July 1983 greeting, for example, described hemophiliacs with AIDS as having been found "exclusively in America" and stated that "no patient who has been treated at our center has been affected," thus ignoring the death of at least one of their patients. One patient, Dr. Werner Kalnins, recalled that the center's attitude at the time was a strange mixture of blind faith and cynicism:

> As soon as I heard about AIDS, I said, "Doctor, wouldn't it be safer if I took European products, say from Immuno [the Austrian fractionator]?" He said, "It doesn't matter. They all get their plasma from America anyway."

If they had given me the chance to take cryoprecipitate, I must say I would have taken it. I would have said, "OK, I'll be careful and play no sports for one or two years . . . ," but I would have taken it.

Years later, Professor Hans Egli would testify in Parliament that three-quarters of his patients with severe hemophilia had become HIV-infected—a number he found "surprisingly large . . . alarmingly large." One was Werner Kalnins. Another was Frank Schnabel, the founder and president of the World Federation of Hemophilia, who, having lavishly praised the Bonn center for years, died of AIDS in 1987.

One of the pivotal events of the early AIDS years was the World Hemophilia Federation's annual meeting in June 1983. Convened at the Karolinska Institute in Stockholm and attended by delegates from all over the world, the conference embraced a wide-ranging program, from the "Psychosocial Effects of Hemophilia" to "Possibilities to Increase Yield of Factor VIII." To all who attended, though, the underlying agenda was clear. They had been hearing the rumors, and watching developments in the U.S.; this meeting would be their first chance to assemble as a group and compare notes about AIDS. Indeed, part of their purpose was to produce a resolution that they all could take back to their home countries for guidance.

Bruce Evatt had been invited to speak, but he felt himself set up in a way. Though Aledort was supposed to give a brief introduction, instead he swung into a lengthy discourse on how little scientists knew about the disease. Evatt, when his turn came, felt he had to defend how much they *did* know. By now he and his colleagues felt certain that the agent was a virus, transmitted though sexual contact and blood products. The only reason they had not observed more cases was that the disease displayed a mysterious lag time, but they had no doubt that more cases would come. It would be "prudent," he suggested, in the courtly language of scientific conferences, "to take measures to reduce the risk of acquiring and transmitting AIDS via blood and blood products. This may be especially pertinent to the Hemophilia patients."

A couple of days later, the group prepared to vote on a resolution. Dr. Shelby Dietrich, a member of the federation's medical-advisory board, had written the draft. Dietrich had a complicated experience with AIDS. As head of the hemophilia department of the Los Angeles Orthopedics Hospital, she was an enthusiastic Factor VIII proponent, having introduced home therapy to the West Coast. When AIDS came along, she reacted responsibly, suspending elective surgery for all the hemophiliacs under her care. After several months with no cases, however, she lifted the suspension and urged her patients to continue their infusions. At the same time, she continued to use cryo with "virgin" patients and small children. She had come to the meeting with a six-page summary of what scientists in America knew about AIDS. Given the uncertainty, she concluded, it really boiled down to a simple decision of whether patients would risk more by discontinuing their factor and suffering the known consequences of hemophilia, or by continuing their treatments and incurring the unknown risk of AIDS. She suggested, in part, the following resolution: "There is insufficient evidence to recommend, at this time, any changes in the treatment of hemophilia, therefore present treatment should continue with whatever blood products are available. . . ."

The wording outraged the Dutch representatives. Cees Smit, head of the Dutch Hemophilia Society, had extensively researched the international plasma trade, and what he found had scared him. Having helped a Dutch journalist named Piet Hagen research his 1982 book, *Blood: Gift or Merchandise,* Smit could see how a virus could spread through the global blood-products system, be it hepatitis or AIDS. Indeed, early in 1983 he had convinced Dutch medical authorities to curtail severely the use of imported Factor VIII. Now he and his countrymen tried to persuade the World Federation to take a more cautious approach to the use of the factor. Rather than give approval, for example, the federation could urge patients to use clotting factors only in cases of life-threatening emergencies, such as a brain bleed, or to revert, for a time, to cryoprecipitate. Granted, these measures would be inconvenient, but they might, in the end, save a few lives. "There was certainly enough proof to at least have warned the hemophilia community about what was going on," Smit later said. But almost no one supported the Dutch, and the resolution passed as originally worded.

The decision was tragic. Dietrich may have meant for the resolution to stand as an interim measure—to continue provisionally while doctors gathered more evidence—but many delegates took it as a *carte blanche* for the continued and unfettered use of Factor VIII. Indeed, some nations, including Britain, France, and Japan, escalated their use after the conference. As a result, thousands of hemophiliacs throughout the world freely infused themselves with a product that they should have been regarding with utmost suspicion.

Dr. Takeshi Abe, vice-president of Teikyo University, near Tokyo, was one of the experts at the Stockholm conference. Abe was a legendary figure among hemophiliacs in Japan. In a culture where disability meant disgrace, Abe treated his patients with dignity, extending to them the right to be rehabilitated, not scorned. When Factor VIII came on the market, he enthusiastically promoted it, becoming the nation's pioneer in home hemophilia care. He and a couple of colleagues had traveled the country, bringing the therapy to urban and rural populations alike. People still remember how Abe and a colleague, scouring the rural areas for hemophiliacs, came upon a little boy curled up in a barn, twisted from years of disfiguring joint bleeds, and brought him in for treatment.

It was no surprise then that Abe continued to promote Factor VIII, especially after the Stockholm conference. Others in Japan had come to value it too, and not only for therapeutic reasons. Its use had been climbing ever since the country's insurance commission began covering the cost of Factor VIII home infusion in February 1983. Indeed, as hemophilia doctors were debating the resolution in Stockholm, Factor VIII use in Japan had just begun to take off. But Abe had other things to consider as well. As leader of a newly formed government AIDS commission, Abe had to determine whether the epidemic would soon sweep into Japan, and if so how to act.

The Japanese had always prized their national purity, and considered AIDS an American disease. In July, a hemophiliac whom Abe had been treating died of multiple causes. Although suspicious that the man had died of AIDS, Abe hesitated to make a positive diagnosis, which would have meant acknowledging that the disease had arrived in Japan. That summer, when two American CDC specialists happened to be attending a conference in Japan, Abe met them and related the details. "We said that it seemed very similar to what we were seeing," recalled Dr. Tom Spira, one of the two. "We put it in the same context as the cases here." In other words, the patient had AIDS.

Meetings in Japan do not proceed under the same social rules as those in America. Fewer memos are exchanged; fewer formal agreements pass from hand to hand. With a cultural sense of shared understanding, corporate and policy decisions can move forward based on an assumed consensus, on a nod, a look, or a lack of objection. Therefore, the literal details of what happened next probably will remain somewhat obscure. According to witnesses in the Diet hearings that eventually followed, Abe initially agreed with the diagnosis. But when the media picked up on the story that AIDS had "landed" in Japan, Abe's commission voted to backtrack on the story and deny the patient had AIDS. Abe reportedly dissented, but for the sake of solidarity announced to the media that the patient had died of "quasi-AIDS" probably brought on by his use of steroids.

Thereafter the cover-up took on a life of its own. In August of 1984, after an experimental AIDS test had become available to scientists in America, Abe sent forty-eight blood samples from his patients to Dr. Robert Gallo of the U.S. National Institutes of Health. Twenty-three of the samples were HIV-positive. Abe informed the Health and Welfare Ministry, but for several months said nothing to his patients or to the public. Later that year, when another sampling showed that a high proportion of his patients had AIDS, Abe still kept the information quiet.

It is difficult to comprehend the extent of the medical establishment's denial in Japan. In 1983, when Japanese hemophiliacs asked the Ministry of Public Health about clotting factor, they were told, "Blood concentrates are safe, so the blood system does not have to be changed." In a "Proclamation of Safety" to Japanese hemophiliacs, Abe's commission pronounced: "The need to worry about AIDS is slight, so to worry about stopping the imports from the U.S. is probably too much. Imports will continue to be improved so hemophilia treatment will not be disrupted. . . . All of you are being saved because of blood products."

As far as the Japanese public was concerned, AIDS had not yet arrived in the motherland. And for all the hemophiliacs knew, their clotting factor was safe. Not until March 1985 did the Health Ministry's commission announce that they had found the first AIDS case, a man who had been living in New York but who after he got sick came home to Japan. They picked someone who was as atypical as possible. As sociologist Eric Feldman has written: "[He] was a homosexual, not a hemophiliac; an artist, not a salaried worker; a Japanese national living in the United States, not Japan; in short, a deviant, not an average Japanese, who was identified as the country's first AIDS patient."

Eventually Abe and the Japanese Health Ministry acknowledged that the first hemophiliac had indeed died of AIDS. By now, of course, hundreds had been infected and dozens diagnosed, most of whom had not been told of their condition. As late as 1988, in fact, Abe and at least some other hemophilia specialists did not tell their patients they had become HIV-positive. To a newspaper he explained: "Until we can have a procedure to conquer AIDS, we prefer to hide the real data from the HIV test. . . ." In one sense, this conformed to a Japanese paternalistic medical tradition of not directly telling patients that they have a terminal disease; the doctors generally tell close relatives instead. But in this case, the doctors told no one in the family. As a result, at least thirty HIV-positive hemophiliacs gave the infection to loved ones and spouses.

Thus, Japanese medical authorities kept secret the very existence of AIDS in Japan for two crucial years, from 1983 to 1985. During this period, their imports of Factor VIII and plasma continued to climb.

Jean Péron-Garvanoff was a living testament to the wonders of French hemophilia therapy. Born in Bulgaria, he emigrated to France after the war as a boy with his parents and two half-brothers. The move saved his life, for he had gone from one of the most backward nations in the treatment of hemophilia to one of the most advanced. In Bulgaria, there had been nothing for his bleeds except ice, improvised plaster wraps, and prayers. (Once, in order to stop a spreading hematoma, his parents had no choice but to throw him into the snow.) A new world opened up for him in France. There he received the most modern treatments. He remembered being treated by Dr. Arnault Tzanck, the father of French transfusion medicine, who, his kindly eyes twinkling, would sit next to the boy, turning the little crank on the arm-to-arm transfusion pump. After Tzanck died, Péron-Garvanoff became a patient of Dr. Jean Pierre Soulier, another luminary of French transfusion medicine. Soulier introduced him to the therapeutic marvels of plasma and cryoprecipitate.

Of all the doctors who helped him, however, he reserved his greatest affection for Dr. Jean-Pierre Allain. It was Allain who, as head of a school for hemophiliac boys, introduced hemophilia home therapy to France. Allain had impeccable credentials. In addition to doing pioneer work at the school, he had performed research at the University of North Carolina with Kenneth Brinkhous, the developer of Factor VIII. Allain later joined CNTS, where he became chief of anticoagulant research and development. Péron-Garvanoff found this therapy miraculous—not only for him and his two hemophiliac half-brothers, but also for their mother, who lived in anguish about the condition of her sons. It kept his joints supple enough to pursue his career as a boogie-woogie piano player. Furthermore, he *liked* Allain. With his rumpled appearance and boyish enthusiasms for tennis and jazz, Allain did not stand on the traditional pedestal; he even let his patients call him Jean-Pierre. Sometimes Péron-Garvanoff performed at the *soirées* Allain and his wife liked to host. The depth of Péron-Garvanoff's loyalty became apparent one night when an argument erupted between Allain and an unruly neighbor. The neighbor was becoming violent. Just as he reared back to lunge at the doctor, Péron-Garvanoff, risking a hemorrhage if anyone had struck him, threw himself between the doctor and his assailant, and shouted, "Don't touch this man! He's a saint!"

On June 19, 1983, Péron-Garvanoff heard a radio broadcast about a malady called AIDS among homosexuals and hemophiliacs in New York. "I remember that date because I wrote in my notebook," he later recalled. He called Allain and asked a simple question: Did he, as a hemophiliac, stand a risk of contracting the disease?

Allain too had begun to worry about AIDS, which had just begun appearing in France. In Paris, with its exuberant lifestyle and mixture of populations, chances were high that this "gay" syndrome might infect hemophiliacs as well. Allain had already seen the first hints of an invasion: In a survey that he and a colleague conducted of twenty-three hundred hemophiliacs, six exhibited AIDS-like symptoms of swollen lymph nodes and dramatic weight loss. The study also told him that no products were risk-free, even those made solely from French volunteer donors. Concerned by the results, Allain was forming an AIDS Hemophilia Study Group, a collection of more than four hundred hemophiliacs in which he would try to pin down which products seemed to be carrying the disease. Yet he did not feel ready to alarm his patient. And so, rather than reveal his fears, he dismissed all the AIDS talk as "journalistic gossip." Something, however, rang hollow to Péron-Garvanoff. He wrote in his notebook: *"Je ne suis pas rassuré du tout"*: I am not at all reassured.

Much has been written about France's notorious "contaminated-blood affair," in which four doctors were convicted for failing to protect the nation's hemophiliacs. But the scandal extends deeper and wider than the courts and the media have suggested. Far from affecting only hemophiliacs, negligence in the French medical establishment condemned thousands of French citizens to AIDS, hemophiliacs, and nonhemophiliacs alike. They did so in a slow-motion sequence of denial and deception that has never been publicly acknowledged.

The denial in France is easy to understand, given the country's transfusion history. Ever since the war, as we have seen, the country had collected its blood on the basis of a deeply held philosophy—*bénévolat, volontariat, anonymat*—blood freely, voluntarily, and anonymously given, with no profits earned anywhere along the way. The premise seemed sound enough on the face of it. After all, volunteer blood tended to be cleaner than paid blood, given the donors' social and medical backgrounds. Yet the French elevated that practical consideration to the level of dogma. They actually believed their blood was inherently safe, simply by dint of the *bénévolat* tradition. The assumption of purity extended to plasma products as well, since they too came from voluntary donors. Long after Paris became the AIDS capital of

Europe, bloodmobiles kept circulating through Beaubourg, the Latin Quarter, and Pigalle, collecting blood and plasma from their high-risk populations.

The first person publicly to question these practices was the most qualified man in the nation to do so—Jean Pierre Soulier, director general of CNTS. As he neared his retirement, Soulier could look back on a long and distinguished career. Trained by Tzanck in the early days of transfusion, he had become a professor, clinician, and internationally known researcher—discoverer of Factor IX, the missing protein in people with hemophilia B. Soulier had supported the Health Ministry's efforts to make the nation self-sufficient for Factor VIII. But now, with the specter of AIDS on the horizon, he began to worry about the pace of the program. He knew that benevolence did not ensure purity, even less so with pooled plasma products than with whole blood. Beyond that, he had always been a bit wary of clotting concentrates, or "comfort products," as the French had come to call them. "Every transfusion is a risk," he used to say, and hemophiliacs were infusing the equivalent of thousands. Now, as he saw the epidemic growing in America, he renewed his call for caution with Factor VIII, even to the point of using cryoprecipitate instead of the new concentrates. "It was a question of prudence," he would say later. He was not suggesting a return to the past; only a "retreat" for a couple of years until doctors could understand this new disease. In doing so, he joined other like-minded physicians, such as Oscar Ratnoff in Cleveland, Dr. Bernard Noël of the blood center in Chambéry, and most of the Belgian medical establishment. (The Belgian Red Cross, which distributed clotting factors, had cautiously stayed away from Factor VIII in favor of locally produced cryoprecipitate.)

Almost no one in France appreciated his suggestion. The Health Ministry frowned upon cryoprecipitates, which would draw plasma away from their showcase Factor VIII program. Hemophiliacs objected strenuously. They remembered the old days of feeling tethered to the hospital, waiting dreary hours for the liquid to infuse. Gone would be the ski weekends and soccer afternoons, the promise of a life without crippling and pain. In an angry rebuttal to Soulier's suggestion, André Leroux, president of the French Hemophilia Association, wrote in *L'Hémophile,* the association's journal, that patients should insist on receiving their products. If the transfusion service could not produce enough clotting factor, they should be forced to import it. If they still refused, then the patients should "protest to the point of threatening" the clinicians. Soulier argued that French hemophiliacs should "temper their enthusiasm," because "mercenary" products imported from America carried a greater risk of transmitting viral disease. (Little did he know that the French products would become every bit as tainted.)

Meanwhile, Soulier was challenging complacency about whole blood as well. French blood bankers always had relied on what they called the "serological shield"—a battery of lab tests to screen out diseases such as syphilis and hepatitis—but they avoided asking personal questions. "Donors were like gods," said Claudine Hossenlopp, longtime secretary of CNTS. "No one would have wanted to offend them." Since AIDS could not yet be detected in a lab test, Soulier favored the only technique that was available at the time—the kind of sociological screening that the Americans had introduced with their pamphlets and questionnaires. If blood banks could not yet identify the pathogen, they could exclude those groups most likely to carry it. That was why the British, Swedes, and others were beginning to use such techniques as well. In May 1983, Soulier proposed the nation's first donor questionnaires for CNTS in Paris, asking donors to acknowledge whether they used intravenous blood or had multiple homosexual partners.

The move provoked an immediate backlash from human-rights advocates and gays. "Faggots—an Undesirable Blood Group?" mocked a headline in the leftist newspaper *Libération*. Soulier tried to explain that homosexuals do not constitute a blood group, simply "a group of individuals at risk," but his argument did nothing to quell the storm. Bureaucrats complained that, though his intentions were laudable, the wording of his new questionnaire—with its references to tattoos, drugs, and multiple gay liaisons—was simply too "rough" for the general public.

A month later, the Health Ministry attempted a "smoother" approach. In a "circular" to the directors of the nation's nearly 170 regional and local blood centers, Director General of Health Jacques Roux gently suggested that they begin to consider the lifestyles of their donors. The notice did not go to the donors directly—only to the blood-center directors, whom he took great pains not to alarm, describing the AIDS risk through transfusion as "minimal." Roux did not require the exclusion of high-risk donors, or even the filling out of questionnaires; he simply left the matter to the directors' discretion.

Roux knew that he had little power to enforce such an edict, no matter how tactfully he worded it. Under French law, Roux, as director of the Health Ministry's policy office (the Direction Générale), could set prices and policies, but he lacked the authority to enforce them. Whatever policies he formulated were largely at the behest of the Consultative Commission, an advisory body consisting mainly of blood bankers.

(Indeed, the circular itself resulted from a decision by the Consultative Commission, which had viewed Soulier's experiment favorably.) As one government study later described it, the nation's blood system resembled less a centralized medical network than a "feudal system consisting of multiple baronies."

It was no surprise, then, that few paid attention to Roux's circular. Most blood-center directors saw AIDS as something foreign—an American disease, or perhaps a Parisian one. The mere suspicion of a "minimal" risk was not enough to force them to embarrass their donors, so they collected unscreened blood, just as before. So, at a time when other nations were taking the logical measures to contain the epidemic, lifestyle screening was "systematically forgotten," according to French sociologist Michel Setbon. A year and a half passed, during which only about half the nation's blood banks had instituted procedures. Finally, in January 1985, Roux issued another notice urging the "immediate and strict" application of screening, and warned the directors that they could be held legally responsible for the transmission of AIDS. Even though additional blood centers obeyed him, donor questionnaires never became rigorously enforced.

An excess of faith in the government's blood system and the purity of "benevolent" donors had set the stage for the spread of AIDS through the French blood system. Now an additional factor would come into play that would make a national public-health calamity inevitable.

For decades, French blood bankers had been staging collection drives in prisons. Born in the flash of postwar idealism, and expanded during the reforms of the 1960s, the program was thought to have "redemptive" significance; social scientists felt it humanized the prisoners by making them part of the larger social compact. Prisoners who gave blood enjoyed it, since it gave them a break from the monotony, with wine, sandwiches, and a brief change of scenery. Wardens felt it had a palliative effect. Blood bankers appreciated the custom as well, not so much for the quantities involved (prison blood never exceeded .5 percent of the national total) as for the fact that they could collect blood during traditional lean times of holidays and vacations.

As the first AIDS cases appeared in France, a few doctors began to question the collections. In March 1983, Dr. Luc Noël, director of the Versailles regional transfusion center, examined 212 of his volunteer prison donors. Subjecting them to several diagnostic procedures, including the hepatitis core antibody test, he found that 31.5 percent tested positive, making them prime candidates to transmit hepatitis and possibly AIDS. Astonished at the proportion, Noël wrote that to continue taking blood from these men was "ethically and economically inconceivable." He also alerted the Health Ministry. "They ignored me," he later recalled. Later that spring, Dr. Michel Garretta, of the Paris-based CNTS, ended his center's prison collections. Still later, the doctors at the regional transfusion centers in Strasbourg and Toulouse found high viral markers among prisoners they studied.

Despite these early warnings about prison blood, the nation kept using it as freely as before. All but a few of the blood centers kept visiting the prisons. Indeed, during the two years when AIDS found a foothold in France and matured into an epidemic, discussion of prison blood never became part of the national agenda. It remained conspicuously absent from ministry meetings and those of the Consultative Commission. Roux's famous "circulars" about high-risk donors in 1983 and 1985 made no mention of prisoners, the riskiest group of all.

Meanwhile, a bureaucratic sideshow in the Department of Justice threatened to make the situation even worse. In late 1982, the regional blood center in Marseilles, running short of blood, had petitioned the Justice Ministry to allow them to collect more often from the local prison. The ministry, which was responsible for the health of the prisoners, had restricted blood collections to three times a year for male prisoners (twice for female prisoners). The request from Marseilles drifted for a while in the ministry bureaucracy, until it landed on the desk of Justice Minister Myriam Ezratty. Ezratty knew nothing about risks of prisons and AIDS, nor, as the months passed and the evidence accumulated, had anyone bothered to inform her. She knew nothing about the Health Ministry's circular, of Garretta's decision, or of Noël's disturbing study. And so, when the time came to make her determination, she not only granted the request to Marseilles, but applied it to all prisons in France, increasing the limit on annual donations from each prisoner to five. Her edict went into effect on January 13, 1984. By 1985, as the AIDS epidemic crested and broke, collections in prisons reached an all-time high.

From his office at the prison hospital overlooking the *grand quartier* of the prison in Fresnes, a suburb of Paris, Dr. Pierre Espinoza thought little about the export of prison blood, even with collections taking place almost under his window. Espinoza had become director of the prison hospital as part of a national penal-medicine reform. Dedicated to the cause of prisoners' health, he was struck by the uniformity of his patients, who all seemed to be former drug addicts. Indeed, in recent years drug addicts had flooded French prisons. His task became more complicated in the spring of 1984, when he heard about AIDS and its relation to intravenous drug users. Deciding to conduct "a little public-health research," Espinoza surveyed the hospital's more than two hundred patients. He found that a dozen had come down with the disease, and sounded the alarm that prisons would become the next hotbed of AIDS.

At that point, Espinoza was not even thinking about blood; his only concern was to provide his patients with adequate care. Several months later, he received a visit from the local blood-center director, Dr. Najib Duedari. Duedari told him that a civilian patient had contracted AIDS. Working back through the patients' transfusion records, Duedari found that the infectious unit had come from a prisoner at Fresnes.

The two doctors decided to survey the prison donors, using every screening tool at their disposal, including medical exams, questionnaires, and hepatitis B tests. It would take weeks to do so, but at least this would give them a handle on the problem. Meanwhile, they conveyed their suspicions to Justice Minister Ezratty, and asked her to suspend prison collections. She asked Director General of Health Roux to attend a meeting of health, transfusion, and prison officials. Roux declined, dispatching a few subordinates instead, along with some disdainful handwritten instructions that the local transfusion-center directors "are adults" and should be able to work things out for themselves.

They could not. During the meeting, chaired by Ezratty, Espinoza said AIDS was running rampant in penitentiaries, and described the risks of giving prison blood to the general population. Roux's representative, Dr. Jean-Baptiste Brunet, an epidemiologist who became known as "Mr. AIDS" in France, agreed. Two other representatives from the Health Ministry worried more about causing a "rupture" in blood supplies. Ezratty limited her position to expressing the concern that suddenly halting the program would cause prisoners to panic. In the end, the officials decided to leave the program as it stood. They made one concession to safety, however: They requested the Health Ministry's director general's office to telephone regional blood centers with "appropriate recommendations" about the use of prison donors. It was never determined exactly how many centers were called, what they were told, or how they responded.

Ten days after the committee's decision, Espinoza and Duedari completed their survey. Two hundred sixty-four inmates had volunteered as donors. Of those men, more than 40 percent represented "populations at risk," including drug addicts, men who tested positive for hepatitis B, those who failed detailed questionnaires, and those who transferred and could not be traced. This represented an overwhelming proportion—more than sixty-seven times higher than from the "hottest" neighborhoods of Paris. The doctors mailed their results to Ezratty and

Roux. From this point forward, no one could pretend that blood drives in prison could possibly be safe. Meanwhile, throughout France, the "benevolent" collection of prison blood continued.

At Stanford University in California, Dr. Edgar G. Engleman had become fed up with his American colleagues. As director of the university hospital's blood bank, he had been following the AIDS debate since its inception. By now he had lost patience with the blood-banking establishment, with their false confidence and wait-and-see attitude about finding the perfect screening test. The time had come for action, not promises. Because of its long lag time, fatal consequences, and corrosive effects on public confidence in the blood supply, AIDS could not be dealt with like other diseases, by waiting until all the data were in. It was time to *do* something, even if that meant using a partial solution. Working at a research center gave Engleman access to a large number of reagents and equipment. He picked one procedure, the T4/T8 ratio test, which detected a white-cell abnormality typical among presymptomatic AIDS patients. The test was not perfect—it did not identify all cases and gave a small percentage of false positives—but he felt it would have to do. Each test cost his center $10. To offset the cost partially, he charged hospitals an extra $6 per pint.

To Engleman's surprise, the blood-banking establishment castigated him. How could he rely on such a test, they demanded, when the contamination of whole blood had not even been proved? Beyond that, the test was expensive, at least by the standards of the time. Ten dollars a pint might be affordable for a small blood bank like Stanford's but hardly for large ones like Irwin or New York. When Engleman submitted an abstract about his test to be presented at the 1983 annual meeting of the AABB, the organization rejected it.

Engleman's action and the controversy that surrounded it epitomized America's second year of dealing with blood-related AIDS. The Public Health Service announcement of March 1983 had marked the end of the first stage of the drama, in which widespread denial held sway. Now the nation entered a second period, characterized by confusion and half-steps. The AIDS virus had still not been identified. With insufficient science to go on, people were forced to make "decisions without data," as health-policy analyst Dr. Suzanne Gaynor has written, making educated guesses in a highly charged atmosphere of heightening market pressures and public concern. Everyone wrestled over what to tell the public, who might panic depending on the information. Unfortunately it turned out that just as in wartime, truth was an early casualty in the battle against AIDS.

Nowhere did this issue surface more painfully than in the National Hemophilia Foundation. Torn between their concern for hemophiliacs and their loyalty to the drug industry, the foundation's leaders stumbled through a series of missteps and miscommunication. On the one hand, they continued to press for safer clotting factors; now, for example, in addition to demanding the exclusion of gay donors, they insisted that the drug firms use the core antibody test, a costly proposition that the companies resisted. On the other hand, they urged hemophiliacs to keep using the clotting factors, despite growing evidence that the products were dangerous.

As the months went by, these assurances became irresponsible. In May 1983, for example, after discovering that a plasma donor came down with AIDS, Hyland recalled a lot of nearly two hundred vials of clotting factor. The foundation announced the withdrawal in its bulletin, but minimized the problem, saying, *"a recall action should not cause anxiety or changes in treatment programs"* (emphasis in original). The next month, Aledort stated his philosophy about Factor VIII use, when he spoke at an awards ceremony at Alpha: "My position is business as usual. There is no reason not to treat. There is no evidence that treatment *per se* is the cause of AIDS." A few months later, when Hyland and the Red Cross both announced product recalls, the foundation again urged its members to keep taking their medicine. Indeed, even after the Cutter company issued a massive recall in November, the foundation told its members to keep taking their clotting factors just as before.

Yet business as usual was becoming a perilous direction, as a massive contamination incident had already shown. In the fall of 1983, a transient gay man named Christopher Whitfield died of AIDS in Austin, Texas. The story made news as the city's first AIDS fatality. The name sounded familiar to the manager of a local plasma center who, looking through her records, found that her organization had purchased plasma from Whitfield forty-eight times in the previous year. He had no visible symptoms of AIDS when they examined him, and lied when asked if he belonged to a high-risk group.

Immediately she called Cutter Laboratories, which had a long-standing contract to buy plasma from the center. Cutter realized that most of the products had already been used, but recalled whatever they could right away. In the end, Cutter destroyed sixty-four thousand vials of factor—about 2 to 3 percent of the nation's entire annual supply. In a press announcement, Cutter spokesman Bud Modersbach said that, even though the company had taken this emergency action, "there is no evidence that AIDS is transmitted this way." Meanwhile, the NHF continued to assure the nation's twenty thousand hemophiliacs that clotting factors were basically safe.

It seemed that no one was telling hemophiliacs the truth—neither their mother organization nor the pharmaceutical firms with which they had traditionally enjoyed close relations. At one point Cutter issued a press release reassuring them that there were "no Cutter centers in New York, San Francisco, Los Angeles or Miami, where the vast majority of AIDS cases to date have been reported." Yet the company operated a center in Berkeley, just across the bay from San Francisco, no less popular among drug users and gays. They also ran plasma centers along the Mexican border and in a prison. Alpha Therapeutic, having touted their tough questionnaire policy, also advertised that they no longer collected plasma in high-risk areas. What they did not say was that when they shut down their San Francisco collection center, for example, they did not destroy the inventory of plasma. They shipped at least some to their laboratories to be made into Factor VIII.

The deception extended to the voluntary sector, whose Joint Task Force of the major blood-banking organizations had begun to issue authoritative advisories. In June 1983, the Joint Task Force addressed the question of directed donations. More and more members of the public, concerned about AIDS in the community blood supply, had been asking before surgery if their friends and family could give blood for them. The blood bankers opposed this, reasoning that being put in such a position might tempt a donor to bend the truth about his risk status, especially if the patient did not know his friend's sexual orientation. (Their motivation was not entirely selfless—keeping track of these separate donations would create a bookkeeping nightmare.) As part of this argument, the blood bankers issued a reassuring statement about the safety of the nation's blood supply, asserting that the AIDS risk from transfusion was something on the order of "one in a million." This number became a mantra for the blood bankers, who used it whenever the question of safety came up. They based it on a back-of-the-envelope calculation comparing the number of suspected transfusion-AIDS cases at the time—fewer than twenty—to the more than thirty million units of blood collected and distributed since the epidemic began.

Engleman, for one, felt the estimate misled. As he later explained to author Randy Shilts, the Joint Task Force only counted transfusion recipients with full-blown cases of the disease, not those with emerging infections. Nor did they account for the fact that the average recipient receives not one but three units of blood. Finally, they averaged the risk for the entire country. The more honest approach would be to provide

regional estimates, especially for the nation's hot spots. In such a calculation, Engleman said, the risk in San Francisco would be closer to one
in ten thousand, or even one in five thousand.

Meanwhile, Engleman's surrogate testing had proved its usefulness.
In one incident, he rejected a man who showed no apparent symptoms
but who eight months later was hospitalized with AIDS. He later
learned that the man had donated thirteen times to other blood banks
in the region, none of which had successfully screened him.

In the summer of 1983, just as Engleman was beginning his program, Dr. Herbert Perkins of the Irwin Blood Bank conducted a trial
with surrogate testing. Over a three-month period, he tested more than
eight thousand donors with the hepatitis B core antibody test. Then he
tabulated his results according to the ZIP code of the donors. He found
the strongest correlation not with ZIP codes representing the Castro,
the city's gay area, but with Chinatown. This made sense, since hepatitis B is endemic to Asia. It led him to conclude that the test was a useless screen against AIDS, since it correlated more closely with ethnic
group than sexual orientation.

CDC scientists scorned Perkins's experiment as amateurish and
vague. He had not asked the donors whether they were gay, but simply
tabulated the lab tests by ZIP code, a crude procedure at best. The epidemiologists continued to press for surrogate testing as the best available barrier against AIDS. They raised the issue at a meeting of the
FDA's Blood Products Advisory Committee in December, during which
a government representative suggested that the industry begin using
the tests. But the drug companies, having previously met to coordinate
a strategy, proposed an interim task force instead. "The general thrust
of the task force is to provide a delaying tactic . . . ," wrote Steven J.
Ojala, Cutter's representative, in a company memo. "This proposal
was one that had been agreed upon by all the fractionators the previous
evening. . . . It was generally agreed that core testing would eventually
become a requirement." Three months later, in an interim report, the
task force concluded that hepatitis B core testing was "not appropriate" for screening high-risk individuals.

The pressure for surrogate testing continued, especially in California, where people had heard about Engleman's testing and demanded
that their own blood banks take action as well. "We now had
patients . . . who were getting totally hysterical that we weren't doing
everything possible to make the blood supply safe," Perkins later testified. The head of the Veterans Administration Hospital in Palo Alto
wrote that he was no longer interested in buying blood from the Red
Cross San Jose chapter, which traditionally had supplied him, since

they did not use a surrogate test. Soon after that, the San Jose chapter, convinced that gay donors were slipping through the net, began using the hepatitis B core test, contrary to the Red Cross's national policy. At Irwin, Perkins had put the hepatitis B core test in place, even though he believed it to be ineffective. Soon the entire Bay Area, five blood banks in all, was screening its blood with surrogate tests. With the exception of two others, in Louisiana and Oklahoma, none of the nation's other blood banks had done so. Meanwhile, the Cutter and Alpha companies experimented with core tests, only to be told by the FDA that the companies could not print the claim on their labels that the testing made their products safer. The companies stopped testing.

Later, after the AIDS test became available, Engleman went back and re-evaluated samples he had saved of the nearly six hundred units his test had rejected. He found that only about 5 percent of the donors he rejected would have tested positive with the AIDS test. On the other hand, considering the volume of blood his center was handling, he had prevented thirty-three transfusion recipients from becoming infected.

The Factor VIII molecule is a big, ungainly entity, a sprawling glycoprotein so prone to clotting that with the slightest manipulation "it turns to glue," according to the technicians who isolated it. That tendency to congeal, of course, makes the medicine effective. It also explains why chemists did not even consider heat-treating the material when they began producing it, even though they had done so with albumin a generation before.

Dr. Edward Shanbrom led the team who developed Factor VIII at Hyland. Soon after they began producing the substance, he noticed that the lab workers developed jaundice after breathing the plasma mist in the cold rooms. Suspecting hepatitis, he conducted tests on the liver enzymes of experimental patients, which showed pathological changes. "It was obvious there was some kind of virus there," he said later. He did not consider withdrawing the product; like everyone who had seen its miraculous effects on hemophiliacs, he had no doubt that the benefits exceeded the risk. Yet he felt that, given that the product could spread a chronic disease, the company should try to minimize the risk. He notified his superiors to take preventive measures, such as closing its collection centers in hepatitis hot spots. They ignored his suggestion, and eventually he left Hyland.

Meanwhile, other methods were developed to control hepatitis. In the mid-1970s, as we have seen, the FDA mandated a series of increasingly sensitive hepatitis blood tests. The disease rates dropped for a while, as the tests screened out dangerous donors, but climbed again as a new form of the virus emerged. Called non-A, non-B hepatitis (and later renamed hepatitis C), it eluded their most rigorous laboratory screening. At this point several companies gave heat treatment another look, with little urgency. After all, even those hemophiliacs who developed the disease seemed to live long, active lives. Some doctors thought that once patients contracted the virus they developed an immunity or tolerance.

It was just about then that Shanbrom came back on the scene. An independent scientist now, he had developed a way to kill hepatitis by introducing a detergent to the fractionation. The detergent would break down the virus's lipid, or fat-based, outer shell, without which the virus cannot survive. As he proposed his formula from one company to another, "no one expressed interest, not a soul," he recalled. It later turned out that the AIDS virus also has a lipid outer shell. Indeed, years later a detergent-based treatment similar to his own became the preferred method of viral deactivation. If people had listened to Shanbrom at the time, the hemophilia-AIDS scourge might never have occurred.

Meanwhile, the fractionators moved forward with their own work on heat-treating Factor VIII to kill off pathogens. The secret eluded them until the Behringwerke Company of Germany developed a way to stabilize the Factor VIII protein with certain sugar-based chemicals before heating. The process was impractical, since it diminished the yield by as much as 90 percent, but the mere fact of its existence inspired the competitors to redouble their efforts. In March 1983, Baxter's Hyland Division received the first American patent to heat-treat Factor VIII. By early 1984, all the major fractionators had received FDA approval and began pasteurizing at least a portion of their Factor VIII for the removal of hepatitis. They had no proof that the process killed AIDS, since the causative agent had not been identified. Yet they thought, wrote Dr. Milton Mozen of Cutter Laboratories, "that *if* AIDS were caused by a virus and *if* this virus were heat labile, then heating the product *might* possibly prove beneficial in reducing the risk of infection."

The National Hemophilia Foundation hesitated to recommend the heat-treated products. After all, no one had conducted clinical trials, so the products' effectiveness remained unproved. Moreover, heating might alter the proteins, causing dangerous allergic reactions. It was not until October 1984, after the AIDS virus was identified and a team of scientists from the CDC and Cutter proved that heat treatment killed it, that the foundation urged that doctors "strongly consider" switching to the new products. Yet even then the old contaminated products remained. Rather than issuing an across-the-board recall, the FDA allowed the companies to phase in the new factor according to their individual schedules. The old tainted products circulated in the marketplace well into the following year.

If heating promised to solve the problem for hemophiliacs, it could do nothing for recipients of whole blood, since heating red cells would destroy them. The only defense was to screen out the pathogen *before* it reached the blood supply. Blood banks were doing that with increasingly stringent questionnaires, yet they were deadlocked, as we have seen, on the issue of surrogate testing.

It was science, finally, that ended the indecision. On April 23, 1984, at a highly publicized Washington press conference, Dr. Robert Gallo of the National Institutes of Health announced that he had discovered the virus that caused AIDS. It belonged to a family of pathogens called "retroviruses." He explained that, rather than kill their host cells directly, these furtive micro-organisms insert their own genetic code into the host's so that it reproduces more of the agent that infected it. The AIDS virus selectively invaded a critical component of the immune system called the CD4 helper cells. In time the viruses overwhelmed them, disabling the immune system. That mechanism explained both the lag time and the opportunistic nature of the disease. Standing with Gallo, Health and Human Services Secretary Margaret Heckler used the occasion to counter those who had criticized the Reagan administration's inaction. "Today we add another miracle to the long honor role of American medicine and science . . . ," she proclaimed. "Those who have disparaged this scientific search—those who have said we weren't doing enough—have not understood how sound, scientific research proceeds." She then predicted that a test to screen the blood supply with "one hundred percent certainty" would become widely available within six months.

By this time forty-nine Americans had been infected with AIDS through blood transfusions, and another forty-nine through clotting factors. Heckler's announcement mooted the arguments for surrogate testing, since a specific AIDS test was supposedly on the way. Contrary to Heckler's prediction, the test would not become available for nearly another year.

# CHAPTER 16

# "ALL OUR LOTS ARE CONTAMINATED"

There is nothing ingenious about the AIDS test. Adapted from an existing laboratory procedure, the ELISA test, as it is called (for "enzyme-linked immunosorbent assay"), involves filling a test tube with plastic beads coated with pieces of deactivated HIV. In order to test blood or plasma for the virus, the lab technician pours a sample over the beads. If the blood sample comes from a person with an infection, it will contain antibodies to the disease, which specifically adhere to the virus particles, or antigens, coating the beads. The technician incubates the mixture, washes it, and then adds an enzyme that turns yellow in the presence of the HIV antibodies. A deep yellow means that enough antibodies have stuck to the beads to infer the presence of the virus in the blood sample. A less dramatic enzyme reaction—a clear or pale yellow—means the sample is probably HIV-negative, since no antibodies have stuck to the beads. The technician does not just eyeball the solution, but measures the color density with a spectrophotometer. The test is cheap and easy, takes about three and a half hours, and produces reasonably specific results. In short, it is the ideal procedure for rapid blood screening on a national scale.

A couple of features limit the test. One is that, since the test establishes the presence or absence of antibodies and not the virus itself, it does not detect the earliest stage of a patient's infection. It takes time for the immune system to create antibodies to disease. This creates a troublesome "window" period of three weeks or so during which the disease remains undetectable. In these cases, the increasingly tough donor-exclusion policies limited the number of silent infections. The second limitation of ELISA is that a small number of people—the rate was 3 percent when the test was first commercialized—might test positive yet not have the AIDS virus. In order to check definitively for AIDS, doctors administer a more time-consuming and costly procedure called the "Western Blot" test, which, detecting one protein at a time, can specifically pinpoint the AIDS antibody.

Once Gallo had isolated the virus and developed the prototype test, the National Institutes of Health made it available to several pharmaceutical companies to adapt for industrial production. So it was only a matter of months before they grew enough of the virus, killed it, and produced and distributed millions of test kits.

The social dimensions proved less straightforward. Blood bankers worried that patients who tested positive for ELISA would automatically assume they had AIDS, even though it was only a rough screening test. They worried that good donors would stay away for fear of taking the test, or that high-risk donors would swamp the blood centers in order to learn their HIV status. They wondered what to tell donors who tested positive—had medical diagnosis and counseling suddenly become the blood bankers' business? These were troubling questions that required answers before collectors put the new tests in place. And so, through the early months of 1985, before the ELISA test had even been approved, blood bankers and public-health officials met to determine how to set the stage. It went without saying that they would not use any units that tested positive or even "indeterminate" for ELISA. Before notifying donors, however, they would retest the units with the more precise Western Blot procedure. In order to avoid making blood banks magnets for high-risk populations, the government set up free testing sites where people could have themselves tested anonymously.

On March 2, 1985, the government licensed the first ELISA test, produced by Abbott Laboratories. By summer, six brands of ELISA test had become available, and their use had become universal in American blood banks and plasma centers. At a meeting in August to evaluate the tests, Dr. Walter Dowdle, head of the CDC, described their performance as "just fantastic." Indeed, exuberance seemed the order of the day. The nation seemed to be making blood products safe again, with the heat treatment of clotting factors and testing of donors. In years ahead, the caseload would climb, as those who were infected reached the active stage of the disease, and the tragic dimensions of AIDS became clear. For the moment, however, the blood industry and health officials could savor their victory.

In Britain, as physicians watched the Americans introduce pasteurization, they agonized over which factors to prescribe. They had always perceived their own products as clean, donated as they were from middle-class volunteers, especially compared with American products, purchased from the filthy dwellers of Skid Row. Heat treatment suddenly shifted the paradigm. American clotting factors, once considered the most dangerous in the world, suddenly became the safest. As one specialist told author Virginia Berridge: "There was a terrible period from the end of 1984 until October 1985 [when routine AIDS testing began]. . . . Here we knew the epidemic was very small and we had a policy of educating people and turning people away. . . . The [British] pool was thought to be safe. . . . Our problem was—did you give US stuff where the pool was terrible or unheated British stuff from a safe donor pool. We gave at least two people unheated British stuff and they became infected. We didn't know that an unsafe pool was rendered safe by heating."

In December 1984, Britain's Haemophilia Society held an urgent meeting with Lord Glenarthur, the top official at the national health department. They demanded that the Blood Products Laboratory (BPL) at Elstree introduce heat-treatment technology, regardless of the cost. Glenarthur promised to do what he could. Meanwhile, they turned up the pressure by coercing the treatment physicians. They negotiated with American fractionators to increase their exports. Then, according to David Watters, the Haemophilia Society director: "We wrote a letter to clinicians saying that in ten days we were going to contact all our members and tell them to immediately stop using unheated products. The effect was dramatic. The BPL's order book dropped 75 percent overnight. The [American] pharmaceutical boys made a killing."

The Blood Products Laboratory introduced heat-treated clotting factors in the spring of 1985. They took to the end of the year to convert completely to pasteurization—more than a year later than the Americans. Meanwhile, the government kept slogging away at building a new fractionation plant at Elstree. Not until 1987 did they finish construction, years behind schedule. Even then, it supplied only a portion of the British demand, leaving American products to dominate the marketplace.

The British were less willing to open the door to the American ELISA test. When Abbott Laboratories applied for a British license in March 1985, their British competitor, the Burroughs-Wellcome company, was lagging. That circumstance made the government's response to the Abbott test suspicious. The British Department of Health and Social Security held up approval of the ELISA test for another five months, reportedly because it produced an unacceptable level of false positives. When the department finally acted on the AIDS tests, it gave the go-ahead to a Dutch firm called Organon and to Wellcome. According to the *Financial Times,* the delay gave the British firm "the chance to step into the business of producing diagnostic kits for AIDS, a market which could be worth 100–200 million pounds world-wide by the late 1980's and which is now dominated by the US companies, chiefly Abbott Laboratories." British health authorities maintained that it would have been "irresponsible" not to have seen how the American test performed with British doctors and donors. The British did not begin screening their blood until October 1985.

In Canada, Connaught Laboratories, with its antiquated equipment, could not hope to convert to heat-treating clotting factor. So the Red Cross decided that, as of December 1984, it would redirect its plasma to Cutter Laboratories in America, which was licensed to produce heat-treated Factor VIII. The Red Cross agreed to continue buying whatever untreated clotting factor remained in Connaught's manufacturing "pipeline," bowing to economic considerations (the unheated product was cheaper) and political pressures (the Ontario provincial government wanted to maintain the Toronto-based Connaught). The Red Cross also had its own inventory of unheated clotting factor, which it chose to distribute rather than destroy. The result was an unnecessary delay. Cutter had promised to deliver the safe clotting factors within five months of receiving the Canadian source plasma, but the Red Cross tacked on a two-month "transition" period so it could use up its own supplies. Thus, the agency distributed more than eleven million units of unheated clotting factor even though plenty of safe material was on hand.

The Canadians' performance with whole blood was no better. Like the French, they were slow to use questionnaires, fearing an infringement on their donors' rights and privacy. Later, when the ELISA became available, the Canadians were slow to adopt it as well, hampered by bureaucratic and budgetary delays. Even though virtually all U.S. blood banks were using the test by late March 1985, the Canadians did not do so until the following November. According to a study conducted by Dr. Donald Francis and Jack McDonald, a professor of social work at the University of Calgary, during the seven-month interim between the FDA's approval of the ELISA test and Canada's adoption of it, fifty-five transfusion recipients were infected by blood that could have been eliminated. Later, in a related scandal, thousands of Canadians would become infected with hepatitis C because of unnecessary screening delays in the collection of whole blood and plasma.

There was no such resistance to the ELISA test in Germany, nor did the Germans resist the heat-treatment technology, having quietly been phasing it in for well over a year. Unfortunately, the doctors did so without fully explaining the reasons to their patients. "They said, 'Here's a new product, a better one,' " recalled Werner Kalnins. But doctors never told him how dangerous the old product had been. "Meanwhile, I had the old vials at my house, at my parents' house. People kept taking the old stuff for months."

In Switzerland, an unsurpassed complacency held sway. This rich and famously clean nation had an internationally admired national Red Cross, whose Central Laboratory was a model of purity and innovation. Indeed, their combination of professionalism and enterprise was the reason Kellner approached them to become his partner in the Euroblood program with New York.

In 1985, a few weeks after the New York Blood Center began using the ELISA test, technicians noticed that a small proportion of the Swiss blood coming in through the Euroblood program tested positive. They destroyed the suspect units and telexed a warning to their partners in Switzerland.

The managers of the Swiss Central Laboratory were not ready to emerge from their pattern of denial. In the past couple of years, they had stalled on issuing donor questionnaires and in opening their borders to heat-treated products. Now that they had received the warning from New York, they began screening blood with the ELISA test—but only for export. It was not until May 1986 that the Central Laboratory screened all blood used domestically. "This was done by our beautiful Red Cross, a national monument beyond suspicion," Jacques Barillon, a lawyer for several infected hemophiliacs, later told *The New York Times*. "Distrusting the Swiss Red Cross was like saying that Mother Teresa starved children to death." Later Swiss courts would charge Dr. Alfred Haessig, the genial grandfather of the Swiss Red Cross's Central Laboratory, with criminal negligence, but he was eventually acquitted.

In the spring of 1983 in Japan, Dr. Takeshi Abe's blue-ribbon AIDS committee faced a decision. Baxter's international Travenol division had applied to sell its heat-treated clotting factors in Japan. It was the same product they had been selling for years, but now they had altered it with pasteurization. Abe's committee was asked to decide whether to grant a "partial modification" of the firm's existing license, whereby it could introduce the product within a few months, or insist on a full-scale relicensing procedure, which could take two years.

This would be no minor deliberation. The Japanese consumed more plasma products than any other nation—a full third of the world's annual production, from albumin to gamma globulin to clotting factors—and almost none of the products originated domestically. More than 90 percent came from the United States—some as finished products from the Baxter and Cutter companies, most as a combination of raw plasma and finished products sent by Alpha of California to its parent company, the Green Cross, to process and distribute. Even if the Japanese saw AIDS strictly as an American disease, they would have to suspect that it could travel to Japan in American plasma. The health minister, Atsuaki Gunji, said as much, in describing the imports to a Japanese reporter: "AIDS is a matter of serious concern."

The decision could have grave economic impacts as well. Baxter held only a minority position in a market that was dominated by the Green Cross, a multinational giant with assets greater than the Toyota car company. As sole possessor of the antiviral technology, Baxter could surpass Green Cross in sales.

The sad thing, industry observers recalled, was that this would never have happened if Dr. Naito, Green Cross's fabled "Old Locomotive," were still in charge. No one, they said, could "out-innovate" Dr. Naito, who had guided the firm with exemplary, even daring, leadership, however dark his past may have been. Staffing the company heavily with physicians, he remained up to date and attuned to the medical environment, equally ready to retreat or to plunge ahead. People remembered how he had wisely pulled out of the blood market in the wake of the hepatitis scandal some twenty years before. "One must also be ready to instantly withdraw from any area which is found to be potentially dangerous," he once said.

As time went on, however, and the pharmaceutical world became more complicated, Naito faced resistance to some of his ideas. In order to smooth over his dealings with government, he hired a former Health Ministry Bureau of Pharmaceutical Affairs chief named Renzo Matsushita as a vice-president. This practice, *amakudari* ("descent from the heavens"), in which former government bureaucrats assume positions in industry, occurs commonly in Japan. The arrangement helps both sides: The officials know they can look forward to a prosperous retirement, and because of their government expertise, they know where to go, whom to talk to, and which loyalties to call on for approval of new drugs.

After Naito died of cancer in 1982, a power struggle ensued between Matsushita and the company's scientific veterans. Matsushita won, after which the company sank into a period of stagnation. "The corporate culture distinctly changed," said a former executive of Alpha, Green Cross's American affiliate. "Naito was a Renaissance man; he was always far ahead of his time. Matsushita wasn't even qualified to hold Naito's shirt. When he took over, all the vision was lost." Now ruled by unimaginative bureaucrats, the company became slothful, ignoring opportunities Naito would have seized. For example, when scientists at Alpha developed their own heat-treatment process, they told the parent company and urged them to adopt it, only to be ignored. "The 'not-invented-here' syndrome existed in Green Cross to the *n*th degree," said the American executive. "They didn't want to give us credit for anything."

Indeed, even after AIDS became a clear and present danger, Green Cross executives seemed more interested in fostering confusion than in meeting the new challenge. One internal memo said that, since Green Cross plasma was imported from America, "there is no guarantee that the products are free of [AIDS]." Soon afterward, the same executive wrote, in a memo reportedly designed to reassure doctors, "The risk is nearly zero that [AIDS] will affect domestic patients via blood products."

It was at this point that Baxter came to Abe's committee with the product that could have crippled Green Cross. Abe was no stranger to the company, having been a longtime friend and technical adviser. In 1983, in fact, Green Cross donated $90,000 to a hemophilia foundation he was trying to establish. During the committee's first meeting, in June 1983, Health Minister Gunji urged Abe and his associates to grant Baxter the rapid "partial modification." He added that the task force should move as quickly as possible, even though Japanese companies would suffer "unavoidable" damage.

Seven days later, an unsigned memo reversed that position. Citing "suspicions regarding the effectiveness and safety of these blood products," the memo warned that an approval for Hemofil T (Baxter's heat-treated clotting factor) "would create an unfortunate precedent."

Immediately the process shifted into reverse. Abe's task force, still denying that AIDS had even landed in Japan, advised the ministry to withhold approval, and recommended the full licensing procedure instead. This meant that, for more than a year after safe products had been made available, the Japanese would allow only the importation of unsafe, unheated ones. Soon afterward, in a meeting with hemophiliacs, Abe stated: "Within the subcommittee I have heard that there are some who think the importation of [unheated] blood products should be stopped, but those are remarks from doctors who really don't know hemophilia."

The decision forced Baxter into a delay, during which Green Cross, urged on by Abe, rushed to develop its own technology. Despite the time lag between the two applications, the ministry approved both heat-treated products on the same day—July 1, 1985. Meanwhile, as we have seen, the use of clotting factors skyrocketed in Japan. In the nearly two years since the Baxter application, more than eighteen hundred Japanese hemophiliacs became infected with HIV.

Equally troubling, even after Green Cross obtained the ministry's approval for its heat-treated concentrate, the firm continued to distribute unheated products. Later the company submitted false reports as to when they had actually recalled the unheated clotting factor. Government investigations eventually revealed that the company had kept selling the dangerous material well into 1987.

Drs. Jean-Pierre Allain and Michel Garretta were as different as two colleagues could be. Allain, rumpled and professorial, cultivated the common touch. Garretta was debonair; tall, nattily dressed, with a handlebar mustache, he strove for a corporate image, driving a Jaguar to work at CNTS. Both disciples of Dr. Jean Pierre Soulier, the two men embodied the two component parts of transfusion medicine. Allain was the researcher-clinician, having spent years caring for boys with hemophilia. Garretta was the doctor-industrialist, having risen through the ranks of CNTS with time off for business school. Their differences seemed complementary at first, two sides of the same golden coin. But under the pressure of the AIDS epidemic their contrasting approaches made them antagonists, and threw them together into a drama that would traumatize the nation.

The "Affair of Contaminated Blood," or at least that portion of the scandal the French have labeled as such, began in a sense as Soulier prepared for his retirement. He knew that he could not appoint another director like himself—a scholarly researcher in the university model. Traditionally the center had been run as a research institution, changing the processes as investigators learned more. But these times called for pharmaceutical production. The organization would now require a manager, someone well versed in manufacture and quality assurance, an American-style boss. Late in 1983, Soulier announced that upon his retirement the following October he would appoint Garretta. Many opposed the choice—the man was too arrogant, they said, too domineering.

Garretta inherited a tough situation. The last years had been difficult for CNTS. The center, protected by a government monopoly, had always been run in a craftsmanlike way, but the blood world had changed around it, the multinational plasma companies surpassing the French in their quality and quantity of clotting concentrates. That was why, despite the official prohibition on importing commercial products, the French found it necessary to bring into the country anywhere from 10 to 30 percent of their Factor VIII, depending on the year. That was also why the Health Ministry gave CNTS the sole authority to authorize imports, with the goal of phasing them out. At the same time, the ministry had begun looking more critically at CNTS's perennial budget problems. Garretta's mandate, therefore, was an urgent one: Crank up the French production as rapidly as possible to achieve national self-sufficiency and turn a profit. His own aspirations set the goals even higher: a French plasma business based on nonprofit donors that could compete head to head with its commercial counterparts. By competing successfully in the global plasma marketplace, he hoped, the French ethos of *bénévolat, volontariat, anonymat* would become a beacon to the world.

The physical embodiment of this grandiose plan was the new fractionation plant in the Paris suburb of Les Ulis. Purposely built oversized, this new facility had been designed to handle plasma products for foreign customers as well as France, necessary to make CNTS a player on the world scene. The only problem was that the factory was out of date. At a time when other companies began to pasteurize their products, the planners of Les Ulis had made no provisions for heat-treatment technology.

Soulier had attempted to ameliorate that shortcoming. For two years he tried at least twenty different methods to pasteurize clotting factors. During this time Baxter's European affiliate, Travenol, approached CNTS with an offer to sell them Hemofil T. The French eventually bought small, research-sized quantities, but never committed to a major contract; the goal, after all, was to produce it themselves. Finally, despairing of finding a solution, Soulier negotiated a tentative agreement with the Immuno Company of Vienna, which had developed a pasteurization technology. Under the accord, Immuno would exchange its heat-treatment technology for CNTS's processes to make certain other products from plasma. The agreement was to be finalized in Munich in July 1984, when both sides planned to attend the annual meeting of the International Society of Blood Transfusion.

Soulier did not go to the meeting. Instead he sent Garretta, who had by now become deputy director in charge of manufacturing. After the conference, Garretta and Allain sat down with the Immuno representatives. Allain felt that he and Garretta held the inferior position. "For me it was simple," he later wrote. "We absolutely needed this technology . . . to deal with non-A, non-B hepatitis and potentially for the AIDS virus. . . ." On the other hand, the technology the French brought to the table was desirable but not critical. It was at this point, Allain later claimed, that the dark side of Garretta's management style emerged. Proud and defensive of CNTS's technologies and wary of giving away a piece of the marketplace, he attached condition after condition to the exchange. Finally, the Immuno director, exasperated by the minutiae, shouted that under Garretta's conditions he would be *giving* away the heat-treatment technology. Garretta blanched, hastily stuffed his documents in his briefcase, and left.

Ironically, even as Garretta was rejecting the technology another French fractionator came home from Munich with an entirely different position. Dr. Maurice Goudemand, director of the regional center in Lille, considered the research he had seen at the conference to represent "a genuine alarm cry." He wrote to Soulier proposing that the two centers collaborate on developing heat-treatment technology and distribute it free to all the centers in France. Soulier replied that he was retiring soon but would pass the suggestion on to his successor. Garretta never responded. The two centers had long been competitors, and he apparently felt they should proceed independently. Lille moved ahead on an emergency footing, and developed the technology by the end of the year.

In the fall of 1984, Garretta had taken over the directorship, and was ramping up the output of clotting factor, albeit unheated. Garretta and his colleagues had reason to underestimate the dangers they faced, since, despite the growth of AIDS in America, only two full-blown cases had appeared among French hemophiliacs. They assumed that converting to an entirely local plasma supply would make the clotting factors inherently safer—after all, they still believed in the sanctity of French donors.

By now the myth of *bénévolat* had begun to crumble. In November, for example, Dr. Jean Brunet reported that despite the small numbers in France the AIDS epidemic was "evolving rapidly." A researcher named Dr. Jacques Leibowitch at the Cochin Hospital in Paris had designed an experimental test for AIDS. Of the more than three thousand donors he tested at three different donor centers, eighteen turned out to be seropositive. (Seropositivity refers to the point at which a victim's immune system produces HIV antibodies, and hence registers positive on the ELISA test.) Six-tenths of one percent may not seem a large proportion on the face of it, but, as Brunet later wrote, considering the tremendous size of the plasma pools, if Leibowitch's proportions held true throughout Paris one would have to assume that all the plasma products were "currently infected."

Meanwhile, international consensus was swinging in favor of heat-treating all clotting factors. Everyone knew that the American National Hemophilia Foundation was recommending heat-treated products, and that the CDC considered them a necessity. The British journal *Lancet,* in a December 1984 editorial, stated that the time had come "to switch to heat-treated factor VIII concentrates." Though acknowledging that hemorrhage was still the primary cause of death among hemophiliacs, the editors wrote, "It would be indefensible to authorize the prescription and treatment at home of products known to be at risk when, apparently, safer products are available."

Allain, for one, needed no more convincing. His own studies had revealed that 45 percent of French hemophiliacs showed immunological signs of HIV infection, whether the products were French or American. Knowing this, he had continued to negotiate with Immuno. In early 1985, he wrote to Garretta that another chance had come to sign the agreement. He pleaded with the director to follow through quickly; to fail would mean a delay of anywhere from six to eighteen months, which would "discredit" the center in its fundamental responsibilities. Garretta signed. Soon afterward the Institut Pasteur concluded a study of eighteen "virgin" hemophiliacs who had only been using Hemofil T. None had become HIV-positive.

Knowing that it would take months to install the Immuno technology, Allain lobbied Garretta to import more of the American heat-treated products. But Garretta was facing pressures of his own. With an ongoing oversight into the center's chronically shaky finances, he could hardly throw away several million dollars' worth of unheated inventory. Nor could he ignore the Health Ministry's unrelenting pressure to achieve self-sufficiency. "Imports prove very costly . . . ," Roux had reminded him in a memo in early 1985. "It is regrettable that France, which is a developed country, is forced to import blood products." On the advice of a group of experts, Garretta agreed to import small amounts of the heated material—just enough for the hundred or so patients taking part in clinical studies. Any leftover material would be allocated to children under the age of four and to "virgin" hemophiliacs. He gave Allain the unpleasant job of drawing up the distribution lists and of saying no to doctors whose patients were not on it.

During this time Péron-Garvanoff, the jazz pianist with hemophilia, found himself thwarted in his efforts to gain more information. At a

party at Allain's, he had heard talk of "contamination" and "trouble."
He thought he heard someone mutter, "It's a mess." Yet his inquiries
brought no response. He began writing letters—dozens, later hundreds—that no one took seriously. He took his suspicions to the French
Hemophilia Association, whose representative "showed me to the
door," as he later recalled. Like its American counterpart, this association existed in a state of divided allegiance—its offices were located
upstairs from and furnished by CNTS, which also supplied the clotting
factor. Years later, at the notorious trial, Péron-Garvanoff confronted
his doctor. "Why didn't you *tell* me?" he said. Allain replied he felt he
owed his first loyalty to his superiors.

If Allain, the good soldier, appeared to stand in solidarity with his
colleagues, he fought them behind the center's closed doors. His relations with Garretta were falling apart. Discussion led to arguments,
arguments to blowups. Increasingly frustrated, he would commiserate
with his wife, Dr. Helen Lee, another CNTS researcher, showing her
some back-of-the-envelope calculations about the escalating risk. The
tension climaxed in a memorably awkward scene at a meeting of the
National Blood Transfusion Society in March 1985. Garretta was pre-
siding over two hundred notables who had gathered to discuss the
emerging epidemic. During a question-and-answer period, Lee raised
her hand. "Michel," she said, "do you realize that if we don't start
importing heated products we risk infecting ten to fifty hemophiliacs
each month?" Absolute silence. After the meeting Garretta repri-
manded her for not showing solidarity in public. Years later she admitted that she had privately criticized her husband as well for not
speaking out publicly. "He answered me that he wanted to continue
working from the inside to change things. . . ."

Even as they compromised the safety of hemophiliacs, Health Ministry officials committed another misjudgment that would endanger
thousands of ordinary French citizens. On February 11, 1985, the
American company Abbott applied for a license to market the ELISA
test in France. Just as in Britain, this application posed a problem for
the government, since the Institut Pasteur, a partially government-
owned research institution, was on the verge of producing its own
AIDS test. As early as the fall of 1983, Dr. Luc Montagnier had said
that he could produce such a test, and requested additional government funding. But the ministers ignored his requests for several
months. When he finally got funding and produced his ELAVIA test, as
he called it, the Americans were already marketing their product.

This was the second time the French had lost in the AIDS race to the
Americans, and it hurt. Montagnier's lab had been the first to isolate

the AIDS-related virus, back in early 1983. His researchers had found it in the swollen lymph node in the neck of an AIDS patient in Paris and called it LAV (lymphadenopathy-associated virus). Shortly after that, Dr. Robert Gallo, using a serum sample provided by the French, isolated a virus he called HTLV III (human T-cell lymphotropic virus). A bitter competition ensued as to which virus would be shown to cause AIDS. The rivalry culminated in an unseemly way when Gallo and Heckler touted their "discovery" of the AIDS virus at a press conference, as we have seen, while patronizingly doling out some credit to the French.

Now more than pride was at stake. The ELISA test represented a multimillion-dollar global market, and if Abbott seized a toehold in France, Pasteur would never be able to gain its rightful share. And so the National Public Health Laboratory (the French equivalent of the American FDA) delayed Abbott's application, notifying the company that they would require more data.

Like their British colleagues, the French had valid reason to proceed cautiously, since the Abbott test produced a slight false-positive rate that policymakers had to consider. Unlike the British, however, the French left an unmistakable record of their motives. The minutes of a May 9 meeting of the prime minister's Cabinet meeting spelled out their concern: "The moment the tests are authorized the French market will be largely captured by the American test. . . . [Therefore] the Cabinet of the prime minister requests . . . that the Abbott registration dossier be retained for some time by the National Public Health Laboratory." Another confidential government memo stated that the objective should be to guarantee Pasteur about 35 percent of the national market. Meanwhile, Dr. Robert Netter, director of the National Public Health Laboratory, had proposed a strategy. "Under the current circumstances it does not seem possible to delay registration any longer without risking a [charge of] abuse of power," he wrote. "I would therefore consider giving the Institut Pasteur an immediate registration and delaying Abbot's. . . ."

The Health Ministry acquiesced and approved Pasteur's AIDS test on June 21, 1985, a month before it granted Abbott's license. Pasteur would gain its foothold in France—at a price of several months' delay and thousands of untested donations.

In April 1985, Garretta and Allain journeyed to Atlanta for the first World Conference on AIDS. By now more than eleven thousand cases had been reported in the U.S. and Europe. (The epidemic's full fury had not yet been reported in Asia and Africa.) The delegates now definitively agreed that the virus, which traveled mainly through sexual contact, also spread though blood and blood products. Therefore, they recommended that nations immediately begin screening blood and plasma with new AIDS tests and virally deactivate their plasma products through heating.

Garretta, sobered by what he had heard, wrote to the Health Ministry stating the "absolute urgency" in preventing the disease among hemophiliacs and their families. Noting that about half the nation's four thousand hemophiliacs probably carried the virus, he projected an annual increase in cases of 10 to 25 percent. They had planned to convert to heat treatment in October; he now proposed to move the conversion date up to July. Even then, he noted, the three-month wait would mean "the death of five to ten hemophiliacs and a certain number of those close to them." Garretta proposed that during the interim CNTS start flying weekly shipments of Factor VIII to Immuno for custom-pasteurization.

Now that he had come up with a conversion plan, Garetta still faced the question of what to do with the inventory, only a portion of which could be treated in Vienna. The director of the Lille center had recalled all the old, contaminated product, an option that Garretta and his colleagues did not entertain. One plan, according to a May 7 memo, was to distribute "the entire stock of 'contaminated products' . . . before the substitute heated products are offered."

Later in the month, several of CNTS's department directors and their advisers met to formally consider their options. One of the participants had drawn up a chart estimating the likelihood that any of the remaining stocks at the center could still be untainted. Factoring the frequency of HIV rates in Paris in with the number of donors and the size of the plasma pools, he estimated that the chance of the stocks' *not* being contaminated probably was less than one in two thousand. In other words, he said, "all our lots are contaminated." Garretta mused that they all faced a "triple problem"—regulatory, moral, and financial. In the end they decided on a two-tiered classification. The quantities of safe product would be limited for a time, so not every hemophiliac would be able to obtain it. Therefore, whatever heat-treated material CNTS could import would first go to seronegative hemophiliacs—generally children and other "virgin" hemophiliacs who had not yet been exposed to contaminated factor. Meanwhile, they would give the old stock to patients who had tested positive for AIDS.

Garretta concluded the meeting by saying, "It is up to the supervisory authorities to accept their responsibilities and forbid us, where appropriate, from disposing of products with the financial consequences this entails."

That statement embodies the kind of buck-passing that held sway in government circles of the time. Receiving the report, Director General Roux chose not to "forbid" the Center from distributing tainted product. (Roux, it should be recalled, was the official who, confronted by the issue of blood collections in prisons, said the local officials were "adults" who could make the decision themselves.) Roux later claimed that he had lacked the power to interfere with Garretta, although that was not the case. Garretta claimed that he too had acted to the limits of his responsibilities—presenting the information, and waiting for his superiors' decision. Yet Garretta and his colleagues did not *need* Roux's intervention to avoid distributing the drugs. They could have followed the example in Lille and seen the patients as their first responsibility. After all, they were doctors.

Other doctors tried to remind them of that fact. In Toulouse, Professor M. Boneau objected to CNTS's system of distributing products according to seropositivity. It would be one thing if contaminated patients, "with an admirable courage and sense of solidarity," accepted the products to spare their HIV-negative comrades, but such was not the case. On the contrary—no one wanted the contaminated medicine. In a letter to Jean Ducos, president of the Consultative Commission, Boneau wrote, "My professional experience dictates to me that it is necessary to ban as of today the distribution of [all] non-heat-treated products." He pleaded for the "massive and transitory" importation of foreign products until France had developed an adequate supply. Later Ducos wrote Roux: "We know in fact that every day we are injecting blood products which will cause a conversion to HIV. . . . For how much AIDS will we therefore be responsible?"

The mandarins of the clotting factor chose not to listen. As the weekly flights began arriving from Vienna, Garretta noted, "The distribution of nonheated products should remain normal procedure as long as they remain in stock." Dr. Bahman Habibi, head of the clinical-applications department of CNTS, wrote to the distributors that any "insistent request" from HIV-positive hemophiliacs to obtain the new heat-treated material would have to be specifically approved. Moreover, "as a precaution," no one would be allowed to order more than a month's dosage at a time.

Péron-Garvanoff was one of those patients. In July 1985, he drove to CNTS headquarters in Paris and demanded they give him the heat-treated material. "There's only a little," the woman at the office apologetically told him, "and it's reserved for certain people."

By now the new technology was in place. In order to facilitate the switch to the new products, the government announced that the state health insurance would no longer cover nonheated clotting factors as of October. CNTS was now pasteurizing 100 percent of its clotting factors. Moreover, it had done so without having to discard a single contaminated unit—the nation's hemophiliacs had injected them all.

In May 1985, as we have seen, Pierre Espinoza and Najib Duedari had taken a survey of prison donors at Fresnes using questionnaires and surrogate testing. Despite the alarming nature of their findings and the research of others, the Health and Justice Ministries had failed to curtail the use of prison blood, awaiting more proof. Now that Espinoza and Duedari could test for the AIDS virus, they gathered the definitive information. They started by assembling 298 volunteer donors at the prison, screening out drug addicts, homosexuals, and people "with diverse medical conditions." Then they screened donors with the hepatitis B, hepatitis B core, and syphilis tests. Finally, they screened for HIV. Despite their rigorous winnowing of the donors, fifteen still tested positive for AIDS—nearly 7.5 percent of the "qualified" donors. In other words, in a population as contaminated as prisoners, no combination of screening and lab tests could prevent a large number of AIDS carriers from getting through. Indeed, the results of the combined screening showed that for various reasons more than half the donors were patently unacceptable. "Is it reasonable," wrote Espinoza in a letter to Justice Minister Ezratty and Director General Roux, "to pursue donations [where] 54% of the subjects belong to a population at risk?" Since screening could never be "viable" in prisons, he urged them to finally stop the collections.

The study *almost* triggered a reaction. After receiving Espinoza's report, Ezratty telephoned at least some of the regional prison authorities and told them not to take part in the collections, although she never sent formal notification. Roux also began to take notice. "Basically he's right," Roux said in a note attached to Espinoza's report. In September, in a policy document on blood screening, Roux's office included six paragraphs describing the dangers of prison blood and recommending the strictest possible measures to screen prisoners. Unfortunately, immediately after the document left Roux's office, other Health Ministry bureaucrats deleted the offending paragraphs.

Espinoza's colleagues encountered similar indifference. Duedari sent the data to Edmond Hervé, the minister of health, with some impassioned "reflections" on the dangers of prison blood. Dr. Luc Noël of Versailles sent his latest survey to Hervé's medical adviser, Dr. Claude Weisselberg. Neither Duedari nor Noël received a response.

To those who saw the dangers in prison collections, the government's nonchalance was incomprehensible. According to subsequent oversight investigations, it was almost as if health officials could not bring themselves to use the words "prison" and "AIDS" in the same sentence.

Perhaps the authorities were living in a dream world, in which no blood, even prison blood, could be assumed to be impure, as long as it bore the label *bénévolat*. If so, it was a world that few others inhabited. In the fall of 1985, Dr. Bahman Habibi, alerted by Espinoza's studies, sent a questionnaire to about thirty of his colleagues in other countries. The responses were "instructive," as he later recalled. Almost no one in the world collected from prisons. The Hungarians reported that they never bled prisoners; nor had the South African Red Cross. The correspondent from China explained that, though the state was obliged to gather blood from "all populations in cities and towns, including workers, farmers, businessmen, students, officials," prisoners had been excluded for the last forty years. The Finns, Australians, Canadians, and Americans had stopped using prison blood in the 1970s, when the hepatitis threat became clear. The Americans had stopped harvesting prison plasma for clotting factor by 1983, although some fractionators continued using it for other products for several years. The Israelis stopped collecting prison blood as soon as they began to hear about AIDS. A colleague from the National Blood Service in England wrote that no British blood centers recruited prison blood. Sympathizing with the French social concerns, he added: "A number of medical and prison officials are unhappy about this because they feel the prisoners are alienated from society and that it benefits them to contribute, in this way, to the community. That may be true but we are not in this business to carry out psychotherapy, after all."

That November, Habibi presented his survey to a meeting of the National Transfusion Society. He already had discussed it with the Consultative Commission, which, having seen Ezratty's and Noël's data, had recommended "with insistence" that all local blood collectors avoid prison blood. Now the society, along with the professional organization of transfusion-center directors, issued a joint letter warning against prison blood in the strongest possible terms. But no official circular or advisory ever followed. Nevertheless, the practice abated. Alerted by their own professional societies, almost all local blood bankers abandoned the prisons by late 1985 or early 1986, although at least three were still collecting in 1990.

Nineteen eighty-five should have been a time for rejoicing in America. After all, just three years after the first AIDS case had appeared in a hemophiliac, the nation had stopped the virus's advance through the blood system, with its innovative screening and viral-deactivation technology. Blood and its derivatives would become more expensive, but this was a reasonable price for a new era of security.

"I remember that Gorbachev got elected and the HIV test came out just about the same time," recalled Dr. Robert Westphal, who has held a variety of positions in the national and international Red Crosses. "And I thought, 'Gee, the whole world is getting better.' And then, a few months later, Rock Hudson died and everything went bad."

Rock Hudson's death, on October 2, 1985, marked a turning point in the public's perception of the disease. Once a mysterious gay ailment, it had now struck someone that people felt they knew. Along with humanizing the epidemic, however, Hudson's death seemed to magnify the public's irrational fears. As Randy Shilts has written: "When a movie star was diagnosed with the disease and the newspapers couldn't stop talking about it, the AIDS epidemic became palpable and the threat loomed everywhere. Suddenly there were children with AIDS who wanted to go to school, laborers with AIDS who wanted to work . . . and there was a threat to the nation's health that could no longer be ignored."

The late 1980s was a time of fearful perceptions, outrageous suggestions—for universal testing, for sending the promiscuous to prison. Commentator William F. Buckley proposed that the government should give everyone an HIV test and tattoo those who tested positive on their forearms and buttocks. "Our society is generally threatened," he claimed.

The hysteria swept over the "innocent" AIDS victims as well—transfusion patients and children with hemophilia. In Kokomo, Indiana, the local school district closed its doors to Ryan White, a fourteen-year-old hemophiliac who had been diagnosed as HIV-positive. White's courageous fight made him an international media celebrity, and the first public face of the "innocents" with AIDS. In Florida, the De Soto County school system barred three HIV-positive hemophiliac brothers named Ricky, Robert, and Randy Ray from elementary school. When the family took legal action, someone set fire to their house, driving them from town. They too became media heroes for the thousands who had been quietly suffering.

Inevitably, some of this anxiety became associated with the blood banks. Anytime a local AIDS case developed, the local television reporters would descend on the nearest blood banker for comment. Anti-AIDS measures were already in place, but the blood bankers, their credibility compromised, found that no amount of reassurance would help. In 1987, a poll conducted by the American Association of Blood

Banks found that 27 percent of the people they surveyed thought they could get AIDS merely by *giving* blood. Westphal recalled that his local television station in Vermont would use a drop of blood as a logo for AIDS stories, even in cases of sexual transmission. "You're scaring all our donors away," he told the station manager. The station changed the logo to a dirty syringe.

Yet the blood bankers must share blame for their loss of credibility. No one believed the "one-in-a-million" estimates anymore. Furthermore, even when they responded to the emergency, they did so inefficiently. A decade before, when the hepatitis B tests became available, blood bankers screened the new donations but not the blood and plasma in their inventory. Now, as they rushed to put the ELISA test in place, most blood bankers tested the incoming blood but continued to distribute the untested inventory. (The New York Blood Center, a notable exception, not only tested its inventory but sent teams to all forty or fifty hospitals it served and pulled the old blood off the shelves.)

As a result, a small number of tragedies occurred—inventory cases, as they came to be called. In Denver, for example, a young woman received two pints of blood during a hysterectomy in March 1985. The hospital had called the local blood center, which, having been testing for more than a week, had more than three hundred tested units on hand. But under their old first-in, first-out policy, the center sent out two untested pints, one of which infected the patient. "Our issue was reasonableness," said Denver attorney Peter Smith, who sued the blood bank on behalf of the woman. "Is it reasonable to have tested units sitting in storage while you have untested units going out the front door?" The jury thought not, and awarded the woman $5.5 million.

Another vestige of the old way of doing business was a complacency about record-keeping and procedures. The Reagan administration had slashed government services and inspections, particularly those in the FDA. In the mid-1980s, federal inspectors were visiting blood banks only on the average of once every two years. This *laissez-faire* attitude bred complacency. When, in response to AIDS, the FDA geared up inspections, officials found hundreds of violations, especially in the Red Cross. In almost every region, inspectors found that Red Cross blood banks had released blood that had failed the ELISA test, or neglected to notify the recipients of bad blood. Some of the violations were truly egregious. At the Los Angeles center, for example, inspectors found a "systematic failure" to follow standard testing procedures. In some cases, blood units that tested positive for AIDS or hepatitis "were then manufactured into products, labeled as non-reactive, released, and distributed for use." Burdened by the new tests and new responsibilities, the Red Cross seemed to be descending into chaos. In 1988, the FDA secured a special agreement in which the Red Cross promised to correct its long list of deficiencies—a promise that the Red Cross failed utterly to meet.

The drug companies could not afford to be so passive, since clotting factors were more widely contaminated and more intensively used. Most started exchange programs, trading new units for old. One hemophilia patient from New Jersey remembered an unsettling visit from his local hemophilia center when they learned he was still using unheated clotting factor: "A couple of guys came over the house with triple plastic bags and gloves on to take it all away. They had called me and said, *'Don't open that refrigerator.'* When they got here they asked me, 'Did you use any of this stuff?" And then I said, 'My God, I can't remember!' "

The drug companies pasteurized some of the returns, destroyed some, and sold some overseas. Some countries, such as Japan, had not yet approved the new heat-treated products, so the companies had no choice but to sell them unheated clotting factor. Yet the companies also knew that "dumping" could save money. Cutter addressed the issue in a series of memos in March 1986. One signed by an employee named Terry Johnson explained: "Our current policy for handling unscreened inventory is to clear the pipeline through normal sales. This also requires that we not distinguish between screened and unscreened lots for all our products except Koate [heat-treated Factor VIII]. . . . We need the unscreened inventory to meet our 1986 sales requirements. However, since there has been much pressure from our customers and governments to provide only screened material, I would like to request a review of our current policy—possibly on a product-by-product basis." The next day the company management decided to "move existing unscreened finished goods . . . before we move screened material." On the other hand, if "a foreign government wants only screened finished goods, we will comply, or if it is legally required."

This policy got the company embroiled in an ugly situation in Costa Rica. In 1985, Cutter had shipped at least one lot of unscreened Factor IX to the Costa Rican National Health Service, which in turn gave it to the country's hemophiliacs. Fifteen of them became infected with AIDS, which the Health Service alleged to have traced to those lots. That same year, the Armour Pharmaceutical Company suppressed a scientist's report that certain lots of its heat-treated factor had not been fully deactivated, and left them on the market for another two years. Some of those lots made their way to Canada, which was importing massive amounts of Factor VIII from Armour. The medicine infected six hemophiliacs in Vancouver, five of whom were children, all of whose doctors assumed it to be virus-free. (The incident was part of a larger episode in which Armour employed an inadequate heat-treating process despite the advice of a scientific consultant. Aside from the Canadian victims, the product gave HIV to at least six hemophiliacs in Great Britain, the Netherlands, and the United States before the company altered its process and took the old product off the market.) The company later paid each Canadian patient a $1.55-million settlement.

The Americans were not the only ones to export unsafe material. In the latter half of 1985, for example, the much-esteemed Institut Mérieux of France exported unheated plasma derivatives to several countries in southern Europe, the Middle East, and South America. (Even though the blood business in France was a government monopoly, a loophole in the law allowed Mérieux to import raw plasma from America and export the finished products.) Mérieux developed a heating technology, but after the competition. As a result, they continued to export unsafe products until early in 1986. Years later, when the French newspaper *Le Monde* exposed what they had done, Alain Mérieux, the company's CEO, admitted, "The company was a little slow in making its decision," and said he profoundly regretted the delay.

An Austrian company named Plasma Pharm Sera sent contaminated blood products to Portugal and Jordan. This incident was especially tragic, for it could have been avoided. The Portuguese Hemophilia Association had heard that the company had a mixed reputation and tried to warn the Health Ministry, which handled all imports. The bureaucrats ignored them and imported the Austrian clotting factor during the summer and fall of 1986. Meanwhile, the association had gotten hold of some samples, which tested HIV-positive. Again they pleaded with the ministry to stop. This time the officials listened, but took two and a half months to notify hospitals. By then, much of the material had already been injected. More than 120 people were infected in the incident, at least forty-six of whom have since died. Years later, when Portuguese newspapers uncovered the scandal, the Health Ministry blamed "health authorities in the country of origin," which had issued the appropriate certificates. Austrian health authorities suspected the approvals had been falsified, and shut down the company for its "murderous" behavior.

By the mid-1980s, Bruce Evatt and his colleagues at the CDC in Atlanta had learned quite a bit about AIDS in the blood supply. Using the AIDS test on archival samples, they found that the first case of AIDS in a hemophiliac probably occurred in 1978. Since then, the disease had climbed steeply—even more so than the case numbers would suggest. With the new tool of testing at their disposal, they no longer had to wait for "full-blown" AIDS cases to draw the growth curve, but could detect seropositivity. By 1985, when forty-eight hemophiliacs— or about .25 percent of all American hemophiliacs—had been diagnosed with AIDS, Evatt and his colleagues found that 74 percent of a sample group of Factor VIII users had become seropositive. This "striking and alarming" result, as he put it, bore ominous tidings: It meant that, within the next several years, thousands of hemophiliacs would develop full-blown AIDS.

Throughout the world, people began to see the epidemic's trajectory, as though a fog were lifting from the base of a mountain. In England, at a time when only three of the nation's fifty-two hundred hemophiliacs suffered from full-blown AIDS, scientists learned that an estimated 32 percent had seroconverted, which meant thousands of AIDS cases to come. In Canada, 55 percent of the hemophiliacs were thought to be positive. French scientists estimated that nearly half their hemophiliacs had seroconverted; in Denmark, 64 percent. Admittedly, the number of hemophilia-AIDS cases was small compared with the toll among other risk groups, especially gays; but people with hemophilia and their doctors now knew that an AIDS holocaust awaited them.

In France, Jean Péron-Garvanoff learned in the summer of 1985 that he had seroconverted. He never knew exactly when he had been tested; as a member of one of Allain's clinical studies, he had had blood drawn regularly. But he now knew why he had been denied the heat-treated factor when he had asked for it earlier that summer. In his mind, his doctor had considered him one of the expendables.

In Japan, Yasunori Akase learned that he was seropositive in 1986. Akase, a calligrapher who lived with his wife on an island near the city of Matsuyama, had no children of his own, but he loved kids just the same. That was what motivated him to participate so actively in working with hemophiliac kids. When AIDS was discovered, he could sense their reluctance to be tested for the disease, so he got himself tested to serve as an example. He did not withdraw, like many of his countrymen, on finding he was HIV-positive. Instead, he became a peer counselor for others, trying to free them from their feelings of humiliation. Later he became the first person in Japan to admit publicly he had AIDS.

Corey Dubin learned in 1985 that he had seroconverted, although he had suspected as much for years. As a severe hemophiliac and one of the nation's first Factor VIII recipients, he may have infused more than anyone else in the world, but "I decided maybe I could survive it," he said. He set off to Arizona for a while.

Dana Kuhn found it difficult to get tested. Doctors at the hospital were reluctant to test him, since, with only one dose of Factor VIII in his lifetime, his chances of an infection were vanishingly small. Finally, they acceded to his wishes. When they learned the results they were "appalled," he remembered. "They just couldn't believe it." Knowing he was married, they counseled him to begin using condoms immediately. But it was too late: His wife had become infected as well. She died a year later, leaving him with two children under the age of five.

Susie Quintana knew she was sick before she went for an AIDS evaluation. About a year after her surgery for the bullet wound, she began suffering abdominal pains, and then frequent bouts of diarrhea. Sores in her esophagus tormented her, and her mouth was invaded by a fungus called thrush. It was about two weeks before Christmas 1985 when her doctors finally diagnosed her with AIDS. After receiving the news, she went to her son's house. He bundled her into his pickup truck and sped off to Cortez, Colorado, where his father worked for the local gas company. They found him driving up East Main Street on the way to his dinner break. Susie was crying when they pulled up beside him. "Short," she said, using his nickname. "Short, they've given me AIDS."

# CHAPTER 17

# JUDGMENT

In 1988, Susie Quintana sued the United Blood Services for giving her blood that carried the AIDS virus. The transfusion had ruined her life, robbing her not only of health but of most human contact. When Quintana got AIDS, people in her town of Dolores, Colorado, did not know much about the disease, associating it with perversity and "queers." Now that one of their own had become infected, it was difficult for them to reach out in comfort; most folks avoided her. Her own husband found it impossible to touch her; they now used separate bathrooms and slept in separate beds. When a social worker eventually visited Quintana, he found her sitting alone, emaciated and hollow-eyed, the curtains pulled tight to seal out the day. He had never encountered a client who was so withdrawn and frightened, he said—virtually in a state of post-traumatic stress.

Advised by a local attorney to sue, Quintana took her case to the Denver law firm of Holland & Hart. In doing so, she became part of the "litigation explosion" of AIDS-related cases since the first transfusion-related case was filed in 1986. As the lag time in more AIDS cases expired and more people became sick, more of them decided to sue. By the end of the decade, dozens of damage cases had been filed; eventually the numbers would rise to nearly a thousand. Most hinged on the same basic premise—that in the years before the ELISA test became available the blood industry had been presented with several options for safety, which they ignored. That was what experts like Dr. Don

Francis had maintained, ever since the contentious CDC meeting of January 4, 1983, when he and his colleagues had suggested tougher donor screening and a surrogate AIDS test. On the other hand, the blood industry maintained that the infections took place before they had understood the pathogen they were dealing with. They acted out of concern for both the safety and the adequacy of the nation's blood supply. It was simple, they argued, to judge them in hindsight—simple, but unfair. Duncan Barr, the blood banks' leading attorney, once argued, "This was a tragedy that happened through no fault of anyone."

Quintana's attorneys, Bruce Jones and Maureen Witt, understood the complexity of the issues. They also faced legal obstacles. They knew about the "shield laws" that most states had enacted to protect the blood banks. Instituted after the Federal Trade Commission fiasco of the 1960s involving the Kansas City doctors and the Basses, these laws defined blood and its derivatives as a medical "service" rather than a "product." As such, blood, unlike other goods, did not carry an implied warranty, or strict liability. This meant that plaintiff's attorneys would have to prove negligence, which requires a higher standard of proof than simple damages. That led to yet another complication: If the lawyers wanted to prove negligence, they would have to show that the blood bank did not adhere to what is legally known as the "standard of care"—generally accepted practice at the time. This legal hurdle had caused almost all previous plaintiffs to lose.

Jones and Witt decided to puncture the precedent by showing that the standard itself was an example of negligence. They had wanted to review the whole scenario from the early 1980s, when the blood bankers considered and then rejected several options, such as core testing and the screening out of homosexuals. The judge ruled against them, however, and ordered them to stick to the specifics of the Quintana case. Hobbled from the beginning, they tried to show that the collection had been flawed, even by the blood industry's existing standards.

They structured their case around an interrogatory submitted by the donor of blood unit 12-308721—the pint that had infected Susie Quintana. Now an AIDS patient, the donor was allowed to conceal his identity by sending in handwritten answers to three dozen pages of questions. He drew a map of the schoolroom in which the blood had been collected and described every detail of the procedures he could remember. As to his understanding of the disease at the time, he wrote, "I did not know at the time that I donated blood that I could be infected with AIDS without manifesting symptoms. I don't think anyone did." He thought AIDS was most likely to occur "among Haitians,

intravenous drug users & homosexuals with over 1,000 different sexual partners." As for the effectiveness of the blood collectors' pamphlets and interviews, "The information provided at the blood drive
did not change or add to my knowledge of the disease."

Quintana's attorneys argued that even in following the industry
standard the blood center used ineffective techniques. They brought in
a communications professor who said that the blood bank had not used
"effective and proven communication principles" to discourage gays
from giving blood. A social scientist testifying for the blood bank contradicted him, saying it would have been "offensive and unthinkable"
at the time to ask specific and pointed sexual questions. Tom Asher, a
veteran of the plasma industry, testified that the blood bankers should
have known to screen out gay donors, promiscuous or not, since the
plasma collectors were already doing so by then.

Before the jury retired to deliberate, the judge gave them specific
instructions. They were only to consider whether United Blood Services had met the existing industry standard, he said, and further narrowed the standard to refer only to the blood banks, not the plasma
collectors. With such restrictive instructions, the decision was almost a
foregone conclusion. "We finally came to an accord that they met that
standard," said one of the jurors, with apparent regret. "It was mini-
mally adequate, but adequate."

Leaving the courthouse, a tearful Quintana said she knew the jury
faced "a tough decision. But it means AIDS patients will just have to
keep fighting. I fought this thing ever since I found out I had it and I'll
continue to fight it."

As hard as it had been for blood recipients to sue, it proved even
more so for infected hemophiliacs. In addition to shield laws, they
faced the problem of trying to show which company was responsible
for their condition, since most had used products from several. They
also found it almost impossible to recruit expert witnesses. Most of the
experts worked for the institutions the hemophiliacs accused—the
fractionators, the treatment centers, the government—so lawyers had
to search far and wide to find anyone to testify. At one point Dr. J. Garrott Allen came forward. It was Allen who had warned that pooled
plasma products were inherently dangerous, as part of his antihepatitis
campaign of the 1950s and '60s. Now he spoke as a prophet who had
seen his direst predictions come true. Testifying at a 1987 trial in Kentucky, he charged that the drug companies had known all along that
"no medical, economic or social reason could justify ever using . . .
pooled plasma or its concentrates. Large pools are highly profitable,
but they are medically bankrupt." Allen was elderly, however, and his

memory for dates and details was failing. The corporate attorneys picked him apart.

The only clear victory for a hemophiliac at the time was the case of Jason Christopher, an eleven-year-old mild hemophiliac who died of AIDS. His mother, Brenda Walls, sued the Armour company for negligence because it had not informed patients of the risk of using clotting concentrates. Jason had used cryoprecipitate for years, but took five doses of Armour's "Factorate" in 1983. The company argued that because he had used a mixture of products it was impossible to say which had made him sick. His doctors testified that, considering the timing of his infection and the clotting factor's tremendous concentration, the Factorate most likely had given him the infection. Their arguments satisfied the "reasonable medical probability" clause of Florida's liability laws, and the jury awarded the family $12 million. The company appealed, attributing the verdict to sympathy rather than evidence. Later the case was settled for an undisclosed amount.

Christopher, however, proved an exception. In most other cases, juries found that the drug companies had taken adequate precautions, especially since they had moved more quickly than the blood banks. That they had collected their products from high-risk neighborhoods rarely factored into the juries' decisions. The victims, confronted with legal barriers, expenses, and their own imminent mortality, settled quickly or quit.

During this time, the rising AIDS toll radicalized many in the hemophilia community. In 1992, Michael Rosenberg, a middle-aged hemophiliac with AIDS, formed a spin-off group called the Hemophilia/HIV Peer Association to challenge the National Hemophilia Foundation's legitimacy. This was a personal issue for Rosenberg, whose father had been the foundation's vice-president in the early 1960s. As far as he was concerned, the foundation had systematically lied to the hemophiliacs in assuring them of the safety of Factor VIII. Borrowing strategies from militant AIDS activists, the Peer Association disrupted meetings and staged street demonstrations. At one point, Rosenberg and a group sneaked into a testimonial dinner for Dr. Margaret Hillgartner, a prominent hemophilia doctor in New York. Seizing the microphone, he berated the participants for celebrating while patients to whom they had given AIDS were dying.

Rosenberg resembled the German actor Klaus Kinski, with his close-set features and edgy intensity. Born in Berlin, smuggled out as a baby at the onset of the war, he was raised in New York with a hemophiliac brother. Growing up with two "panicked and overprotective" parents, Michael and his brother responded like most little boys with hemophilia, becoming as rough and reckless as they possibly could. "There were all kinds of muscle bleeds, joint bleeds, ice, bed-rest bandages, and traction," he later recalled. As far as he was concerned, the fact that neither of them took concentrates until adulthood put the lie to "that bullshit" their doctors told them in the mid-1980s that they would die if they stopped using modern Factor VIII.

To Rosenberg, the worst betrayal was the one doctors had committed after the explosion of AIDS. During the hemophilia-related lawsuits, treatment providers such as Drs. Louis Aledort and Peter Levine testified as to the early state of knowledge about AIDS, and why they had made the choices they did. Rosenberg felt that, regardless of the specifics of their testimony, these doctors were testifying *for* the drug companies and *against* their own patients. His group published a "shame" list of the doctors who "betrayed" them, including Aledort, Levine, Shelby Dietrich, and others. In a wild exaggeration, Rosenberg referred to Aledort as the Josef Mengele of the hemophilia holocaust. Aledort, who had devoted his life to the care of hemophiliacs, was outraged and overwhelmed. "It is like a mob a scene out of the French revolution," he told a newspaper. "They want somebody's head to roll. . . . 'Let's knock off everybody who ever treated us.' Well, that won't solve anything."

Meanwhile, another upstart group had formed, called COTT, or the Committee of Ten Thousand—so named because, when it was founded in 1992, the organizers estimated that nearly half the nation's twenty thousand hemophiliacs had AIDS. Organized by an articulate Bostonian named Jonathan Wadleigh, along with Corey Dubin and Dana Kuhn, this group pursued the less theatrical tactics of lobbying for legal and legislative relief. For them it was a family matter as well: Dubin's father had organized the National Hemophilia Foundation's office in Los Angeles and eventually resigned in disillusionment. Now, as information emerged from the ongoing lawsuits, COTT members learned about the discussions that went on behind the foundation's reassuring messages, and the way money from the drug firms must have divided the organization's loyalty between its benefactors and patients. They fumed that virtually every other country whose hemophiliacs had suffered contamination had provided compensation to the victims. Only America—whose hemophiliacs endured the highest rates of all—had neglected to do so. Traveling to Capitol Hill, COTT members repeatedly testified, lobbying for an investigation into the events of the 1980s and for a plan to alleviate the suffering of the victims. Finally, as part of their agenda, COTT organized a class-action suit against the five major fractionators and the National Hemophilia Foundation. Recruiting attorneys from ten law firms around the country and nearly nine thousand patients, COTT prepared for what they hoped would become the most massive litigation in American jurisprudence.

In 1989, Yasunori Akase became the first person to file an AIDS-related lawsuit in Japan. His decision came during a period of growing activism among Japanese hemophiliacs. Ever since Abe's reassurances of years before, many had had the sense that they were being lied to. Even in a health system as proprietary as Japan's, news was emerging about the hemophiliacs' high HIV rates. They were also aroused by a series of legislative proposals requiring the public identification of all AIDS patients, with no consideration for privacy. People in hemophilia organizations retained law firms in Tokyo and Osaka, but they had trouble getting victims to come forward: The combined humiliation of hemophilia and AIDS was simply too much for people to bear openly.

It was at that point that the infected Akase volunteered—to give voice to those who felt intimidated into silence, he explained. On his first day in court in Osaka, that July, he gave a heartfelt statement about his predicament: "I simply cannot understand the ways of the Ministry of Health and Welfare. What they are saying is that [we] should die silently. I would like the government to know that even in death there is no relief for us. We are expected to die for no cause. Our funerals are held discreetly and a different diagnosis is indicated on our medical records. That is nothing but misery." That day, seven more victims filed suit. The following October, a handful of victims filed a parallel lawsuit in Tokyo. The victims felt such shame about their condition that the courts allowed them to file anonymously, setting a precedent in Japanese civil law. The judges referred to them only by number, and allowed them to testify from behind a screen.

The victims' case, like others throughout the world, hinged on the accusation of negligence. They argued that the drug companies and the agencies that regulated them knew about the dangers but decided neither to act nor to warn them. The government and drug companies replied, as they had elsewhere in the world, that until they could identify the virus they were powerless to stop it. As to why they waited to adopt heat-treated products, they argued that the efficacy of American heat treatment had not yet been proved. The trials in Osaka and Tokyo became an international tribunal, as lawyers from both sides flew in witnesses from the United States, the epicenter of blood-related AIDS and expertise. The plaintiffs brought in Drs. Bruce Evatt and Don Francis to bolster their scenarios of foreknowledge and negligence; defendants brought in Drs. Shelby Dietrich and David Aronson, formerly of the FDA, to describe the limited state of knowledge at the time.

The trials dragged on, bogged down by the government's and drug companies' claims that they had lost certain critical documents. Dozens and later hundreds of new plaintiffs signed on, as the original complainants weakened and died. "I don't want to die until I see the government and pharmaceutical companies take responsibility for causing this hell," declared a thirteen-year-old plaintiff. Akase succumbed in 1991. The gentle calligrapher who loved children had never meant to become a national example, featured in newspapers and television documentaries. All he had wanted, he told comrades and friends, was to lead a normal, inconspicuous life.

After Akase died, another hemophiliac became the public "face" of hemophiliacs with AIDS. Yoshiaki Ishida was not the combative type—he ran a record shop in Kyoto and wrote jokes for the local paper. In the late 1980s, however, he became quietly active in hemophilia rights. Later he chose to speak publicly. Testifying in court, he recalled Dr. Abe's false assurances that even if injected with AIDS-tainted concentrate virus only one in three thousand hemophiliacs would actually become infected. He gave a moving description of what life had been like since he learned he was HIV-positive: "I think my CD4 count [a measure of the strength of the patient's immune system] is like sand in an hourglass. Periodically, I am told by my doctors how much sand will be left. And each day I can watch the sand falling. When there are only a few grains of sand left, that would mean I am about to reach the state of death." He asked the court to act as quickly as possible before the sand in his hourglass ran out. He died in April 1995.

In Tokyo, a teenager, Reyuhei Kawada, rose to take Ishida's place. As a seven-year-old during the summer of 1983, Kawada had attended a camp for hemophiliacs where he learned to inject Factor VIII into a vein in the back of his hand—just after Abe's task force had decided to keep silent about AIDS. At age nine, Kawada became HIV-positive, but his mother could not bear to tell him for another year. His parents disagreed over whether he should litigate. His mother supported him, but his father said that, with no chance of winning, the boy should live out his remaining days in peace. His decision to move forward caused such dissension that his parents ultimately divorced. Now Kawada, his mother beside him, became the face of the "innocents" with AIDS, writing a book about his experiences, and speaking out at press conferences and demonstrations. At one protest at the Health Ministry in Tokyo, Kawada addressed more than a thousand participants. "If you bureaucrats are working for the people," he rhetorically demanded, "admit responsibility, apologize and take effective measures to help hemophiliacs suffering from HIV infection."

Until now, the government had resisted all approaches for a settlement in the ongiong trials, maintaining that neither they nor the drug firms had committed any wrong. But the mid-1990s had brought ferment to Japan, as several ministries became mired in scandal. With the ruling Liberal Party losing its majority, a series of new health ministers came into power. One, Churyo Mori, let it be known that if the courts saw fit to settle the cases he would be inclined to endorse the decision. In the fall of 1995, the Tokyo and Osaka courts jointly recommended a lump-sum settlement of $450,000 for each plaintiff, to be paid by the government and drug companies. Mori promoted this peacemaking effort, yet he hedged on one of the plaintiffs' key demands—a formal government apology. The trials crept on.

What finally brought closure to the scandal was the appointment of a new health minister named Naoto Kan. Kan, an energetic and charismatic activist, had come to the ministry vowing to reform it by "taking down the wall around the bureaucrats." One of his first acts was to order the release of all the ministry's previously "lost" documents. Suddenly the whole story became public—the details of the meetings, the evidence of the dangers, and government and industry officials' refusal to acknowledge them—almost all of which had been kept secret from the victims, their lawyers, and the public. "After seeing these documents it is apparent that the government will broadly accept responsibility," Kan stated at a press conference in February 1996. "This will have an impact on settlement procedures."

A week later, he met with two hundred plaintiffs and their families and delivered the apology they had sought for so long: "Representing the ministry, I make a heartfelt apology for inflicting heavy damage on the innocent patients. I also apologize for the belated recognition of the ministry's responsibility for the case. I understand the delay has tormented the victims."

The next month, the Tokyo and Osaka courts offered a new and more generous settlement, with provisions for continuing care. Now, in an emotional televised news conference, executives of the drug companies bowed before the plaintiffs in shame. "We would like to apologize from the bottom of our heart for the suffering of hemophiliac patients and their family members, who are unwitting victims of a terrible tragedy," Bayer's president Wolfgang Plischke told them. Bob Hurley, the president of Baxter's Japanese affiliate, offered a "sincere and deep" apology on behalf of Baxter employees worldwide. Green Cross's president Takehiko Kawano said, "We deeply regret that our products created a serious situation that resulted in pain and grief." His minimal statement outraged the victims. "Lives were at stake!" a woman berated him. "If it was your child, do you think a casual apology like that would be enough?" Kawano, visibly shaken, walked up to the woman, got down on his hands and knees, and bowed so deeply that his forehead touched the floor.

Justice had finally come for the victims, but not before one-third of the original plaintiffs had died. The following fall, authorities raided the offices of Green Cross and of Dr. Abe, removing crateloads of documents. Later they jailed the eighty-year-old hemophilia specialist, along with Renzo Matsushita and two other Green Cross executives, to await criminal charges of professional negligence leading to death.

By the time of their trial in 1992, Jean-Pierre Allain and Michel Garretta felt such mutual contempt that they did not speak to each other, even though they sat side by side. Then again, they had not spoken in a while. Allain had left CNTS in 1986. He had worked as a researcher for Abbott Laboratories in Chicago for a few years, and then accepted a position as professor of transfusion medicine at Cambridge University in England. Garretta had stayed at CNTS until publicity about his role made it impossible for him to work. He had found it necessary to have a bodyguard since 1989 when a group calling itself "Honor of France" blew up his Jaguar in the middle of the night. Then there had been the press allegations—the occasional leaking of CNTS memos that climaxed in a searing exposé by a doctor-journalist named Anne-Marie Casteret in 1991. Her revelations forced a government inquiry, which in turn fueled even more damaging publicity.

During those years, Jean Péron-Garvanoff had been suffering terribly. His half-brother Gabriel, also a jazz pianist, became sick in 1987 and died of AIDS the following year. His other half-brother, Christian, became sick in 1985. A "heroic fighter," as Jean portrayed him, he had said that his one wish was to stay alive long enough to bring justice to the doctors. Christian had filed a complaint, but the state prosecutor in Paris summarily dismissed it. Jean carried on, searching for a lawyer who would agree to take his case. None wanted to attack the revered French transfusion system. Across the political spectrum he searched, one attorney after another, until he met the lawyer for Jean-Marie Le Pen's far-right political party. Eager to embarrass the socialists in power, the lawyer filed the most serious charges that French law would allow him—"non-assistance to a person in danger." Meanwhile, a group of hemophiliacs brought suit against Allain and Garretta under an obscure statute forbidding "deception over the quality of a product." One of the attorneys, Sabine Paugam, who had wanted to try the doctors for murder, remarked that the charge was absurdly lenient— the equivalent of an indictment for selling spoiled mustard.

The trial took place in a civil court in the Palais de Justice in late June 1992. There, in the stifling classroom-sized chamber, sat the four defendants flanked by their attorneys—Garretta and Allain of CNTS; Roux, the director general of health; and Robert Netter, the director of the National Public Health Laboratory. During the breaks, activists would circulate with pamphlets denouncing the doctors. Péron-Garvanoff showed up every day, driving the fifty miles from the distant suburb of Nemours. He felt he had to bear witness for Christian, who had died four months earlier. Dressed in a suit and tie, he would sit listening intently and taking notes. Sometimes, when the testimony became too much to bear, he would hoarsely shout *"Assassins!"* before regaining his composure.

The testimony itself became a round-robin of finger-pointing. Garretta pushed the blame onto Roux, who he said should have stopped him from distributing the bad material. Garretta portrayed his own role as that of an expert, not a decision-maker. He had said as much in his conclusion to the infamous meeting of May 29, 1985, when he and his colleagues decided to continue distributing the unheated factor. The purpose of the meeting, he argued in retrospect, was not so much to arrive at a decision as to lay out all the possible options. One of those options—destroying the old stocks—would carry such grave financial consequences that he would need the approval of higher-ups. That was why, at the end of the meeting, Garretta concluded that it was up to the "supervisory authorities to . . . restrain us. . . ." Failing intervention, he distributed the stocks.

Roux argued that Garretta gave him credit for more power than he deserved. The now former director general of health described himself as only slightly more than a conduit between the blood centers and his superior, Deputy Health Minister Edmond Hervé; he sent information upward along with his advice, but could do little more. (According to this theory, decisions could be made either above or below Roux, but certainly not *by* him.) Roux testified that he had voiced his unhappiness over the nation's tardiness in obtaining safe products. Indeed, as early as February 1985, Roux had asked Hervé to set up a scientific commission to investigate CNTS, but the minister declined. As to whether he pressured Garretta not to import the heat-treated products, Roux gave a classic bureaucratic response: "I never refused importation but I did not insist on imports. Our ambition was self-sufficiency."

Allain pointed the finger at everyone but himself. He was astonished to have been charged with complicity, since he more than anyone had been pushing the center either to import the safe factor or to install the technology. His letter of January 1985 pleading with Garretta to sign with Immuno bore witness to that fact. (Ironically, the court saw his letter as evidence that he knew of the dangers but failed to respond adequately.) Defending himself against charges that he neglected to inform hemophiliacs when his studies began showing they were HIV-positive, he argued that so little was known about seropositivity at the time that informing patients would have needlessly scared them.

To one degree or another, all the defendants attempted to blame the French Hemophilia Association—a surrogate for the victims—which they said went along with the decision to give heat-treated factor to some patients and withhold it from others. The association countered that the doctors had deceived them from the beginning. The defendants all fell back on the broad theme of doubt, that so little was known about AIDS in the early years that they really had done the best they could. Yet even that general point came under attack. For example, the doctors had alleged that when they decided to give unheated clotting factor to HIV-positive patients they acted in the belief that a patient, once contaminated, could not be reinfected. Called to testify, Dr. Luc Montagnier contradicted them, arguing that, although that theory carried a certain weight at the time, it did not represent the broadest consensus. Actually, he recalled, scientists were uncertain about repeat contaminations—"but considering this doubt, fresh exposure to the virus should [have been] avoided" since multiple exposures could have hastened a patient's demise.

Like the ongoing trial of the Japanese specialists, the proceedings attracted international attention. Letters poured in from blood bankers and hemophilia specialists from the United States, Canada, Switzerland, and Eastern Europe. After the trial, the medical journal *Lancet* ran an editorial signed by thirty-seven scientists, mocking the courtroom as "Palais d'Injustice." From England, the Royal College of Pathologists issued a statement in support of Allain, saying that he had acted "at all times with a high level of professional competence and ethical propriety."

People compared the proceedings to other infamous French trials of the century, such as those of Alfred Dreyfus or the Vichy collaborators, depending on which side one took. Activists denounced the trial as a whitewash, designed to shield the Socialist Party's political appointees. A few months earlier, at a hematology conference in Paris, activists had rushed Dr. Bahman Habibi, handcuffed him to a table, and burst open a blood bag, covering him with its contents. Only later did he learn it had not been real blood, much less HIV-infected. Now, as they picketed the Palais de Justice, they demanded to know why Roux's superior Hervé had not been indicted, or his boss, Georgina Dufoix, minister of social affairs, or even their boss, Prime Minister Laurent Fabius. Hervé had made a grandiloquent gesture of offering himself for judgment, but French law forbids the prosecution of Cabinet officers for actions in performance of their duties. Meanwhile, Dufoix managed to inflame the issue even further when she stated, inanely, that she felt "guilty but not responsible" for the suffering of the victims.

In the end, the panel of judges found three of the four defendants guilty. Garretta, as the chief executor of the distribution policy, was found guilty of "deception over the quality of a product," and received a prison sentence of four years. Allain, whom the judges saw as Garretta's collaborator, and whom they faulted for not fully informing his patients, received four years with two years suspended. Roux received a two-year suspended sentence for "non-assistance to persons in danger." "He who abstains by indifference or lassitude is punishable as well as he who does it with intent to harm," the judges stated. Netter was acquitted.

Allain immediately filed an appeal. So did the prosecutors, who under French law can appeal a verdict if they find it insufficiently harsh. The next summer, they went through all the material again, with almost identical conclusions—Garretta went to jail for four years, and Allain for two. This time, Netter was found guilty of "non-assistance" in connection with the delay of the ELISA test and given a one-year suspended sentence. Roux, his health failing, received a four-year suspended sentence.

As cathartic as it may have been for the nation, the trial was a fiasco, little more than an exercise in national breast-beating. On the one hand, the charges were artificially limited: The court had arbitrarily limited the beginning of the "fraud" to March 21, 1985. They did this to ensure that the charges fell within the proscribed three-year statute of limitations of the particular charge *"tromperie sur la merchandise."* Without that restriction, the authorities might well have dated the doctors' culpability from January 1985, when Allain wrote his warning letter; or from the fall of 1984, when the world understood that heat treatment was desirable and the French knew their own stocks probably carried the virus; or to the previous summer, when they had walked out on the negotiations with Immuno. That would have shown the full context of their neglect. On the other hand, the doctors did far less damage than the French press repeatedly stated. By the time of the trial, nearly twelve hundred hemophiliacs had contracted AIDS, and the press attributed all cases to the doctors' neglect. In truth, however, most victims had seroconverted years before the doctors were made aware of the dangers, probably in the early 1980s. Their judgments possibly caused the unnecessary contamination of anywhere from 70 to 350 additional hemophiliacs—an unforgivable number, but not the holocaust that many had assumed.

More fundamentally, the trial ignored larger mistakes that caused the contamination of thousands. These involved two categories of negligence that were not even mentioned during the trials: the use of prison blood and the lack of donor screening.

Prison collections, as we have seen, continued well into the mid-1980s, despite all evidence that they should have stopped. In 1992, while the doctors' trial dominated the news, a team of investigators from the inspector general's offices of the Judicial Ministry and Social Affairs Ministry released a report in which they tried to assess the extent of prison-borne contamination. An actual count was impossible, they found, because the prison and civilian blood had been so thoroughly intermixed, so they tallied their numbers by using proportions. Factoring the quantities of prison and civilian blood collected with the relative rates of contamination, along with other complex coefficients, they realized that the damage from prison blood had been quite profound. In 1985 alone, for example—a year in which prisons provided only 0.37 percent of the nation's blood supply—penitentiaries accounted for 25 percent of all the contaminated blood in France. One could only guess at the number of resultant AIDS cases. Yet no one was tried for prolonging the policy in the face of dire warnings. Such revelations, as Espinoza later reflected, would not have been "convenient"—neither to the shell-shocked public, nor to the medical and political establishment, which would have seen dozens of its members carted off to jail.

Nor did anyone face charges for the years of collecting blood without donor screening—the subject of Roux's two "circulars" in 1983 and 1985. In the years before the ELISA test, sociological screening stood as the best defense against the disease, weeding out the most likely carriers. Many advanced nations used the technique with impressive results. In a retrospective study in San Francisco, for example, doctors at the Irwin Blood Bank found that in the years before the ELISA test their voluntary self-exclusion procedures eliminated 86 percent of the high-risk donors. In short, the practice prevented thousands of cases of transfusion-related AIDS.

Such screening would have been very effective in lowering the blood-borne AIDS rate in France. In 1993 Michel Setbon, a medical sociologist at Paris's City University, produced a comparative study of AIDS transfusion rates in France, Britain, and Sweden. All had similar AIDS rates among hemophiliacs—a clear indication that clotting-factor contamination was a global phenomenon. In contrast, the transfusion-related AIDS cases stemming from red-cell transfusions were far higher in France. Sweden and Britain had both been using donor-screening procedures before the AIDS test in 1985. When they eventually put the AIDS test in place, each nation found that about one in fifty thousand donors tested positive. This told them two things: that they probably had low AIDS rates to begin with, and that their sociological screening techniques had been effective. The French, in contrast, found that one in sixteen hundred donors tested positive. Even adjusting for differences in AIDS rates among the general population, Setbon's study meant that the French donor screening before the AIDS test was only about one-thirtieth as effective as that of Britain, which had a comparable population. From this information he extrapolated that thousands of French citizens had been contaminated from transfusions. This represented a massive example of denial and negligence, but amid the furor of the trial, he wrote, this explosive issue was "systematically forgotten."

In 1992, Susie Quintana's case came to court again. Her lawyers had appealed the initial decision, arguing that the judge made a mistake in not allowing them to challenge the industry standard. The appeals court agreed, and gave them the go-ahead to retry the case.

This time they came fully armed. Tom Asher, who had testified at the earlier trial about the standards in the plasma industry, was permitted to broaden his analysis to embrace the blood banks as well. Dr. Marcus Conant, the AIDS specialist from San Francisco, was permitted, as someone who had lived with the epidemic since its beginning, to say, "The entire blood industry was negligent. . . . Had they properly screened the blood and donor this woman would not have been infected by the AIDS virus. . . ." The most stunning witness, however, was Dr. Don Francis. Francis had never testified before (although he later bore witness at a trial in Japan), since the CDC had a policy of not allowing its experts to testify, presumably because it took too much time. But Francis had resigned in order to pursue AIDS-vaccine research, and now was free to speak as he pleased.

His testimony was devastating. Recalling the contentious CDC meeting of January 1983, he described how he had pounded his fist on the table, frustrated at the blood bankers' "inability to accept reality."

He talked about the early days of AIDS epidemiology, of knowing that "a few cases now would be a lot of cases later." He described the figurative sensation of sitting at the bend in the railroad track and feeling a rumbling, while the blood bankers denied that a train was on the way.

> I was frustrated because I saw a strategy here that we could do something to prevent a disease, and yet the target group who were going to have to do something about this in the field weren't willing to accept it. . . . If three things had been done with relatively low cost and great benefit, that is, educating of donors, direct confidential questioning of donors regarding at-risk behavior, and the use of a surrogate test . . . you would have built in a belt-and-suspenders system so that very few infectious units would have gotten through. . . .
>
> That's one of the saddest things about the blood banks: that not only did we present the data to them, but we really laid it on a silver platter what to do about it, in the least costly way.

Francis concluded that the blame must rest with the provider of the infected blood, United Blood Services, and with the blood-banking industry in general. If either of those parties had been vigilant, he said, "the blood would not have been transfused, and Mrs. Quintana would be healthy today."

Witnesses for the blood bank argued that Francis's views did not represent those of the federal government at the time, or even the CDC. Recounting the details of the period of uncertainty, they explained their legitimate doubts about the hepatitis B core test. As far as the exclusion of homosexuals was concerned, they thought at the time the disease affected only "fast-track" gays, whom the blood bankers thought they had excluded with questioning. Their testimony held little water with the jury, who merely by shifting their gaze from Francis to Quintana could see the personification of the warning given and the consequences of ignoring it.

Quintana's health was failing by now. The jury could see her declining by the day. "All I can say is that it's terrible," she said, during her brief testimony. "I wouldn't wish it on my worst enemy."

On the final day of her trial, Quintana was so sick that she had to be hospitalized. One of her attorneys, Maureen Witt, visited her that morning, and described her as "worse than I'd ever seen her." Witt then delivered an impassioned summation. She told the jury how Quintana, whose greatest ambition was nothing more than to be a fat grandma with a big enough lap to hold three of her grandchildren, was denied even that simple wish by an industry's carelessness. "One simple question, one drop of blood, one moment of care, that's all it would have taken. . . ."

As Witt was delivering her statement, Susie Quintana quietly passed away. When the news arrived in a note to the court, the judge sequestered the jury to shelter them from the new information. During a meeting in the judge's chambers, the blood bank's attorneys unsuccessfully motioned for a mistrial. The next day, the jury awarded Quintana $8.1 million in damages. It was only then that they learned of her death.

Immediately the blood bank's attorneys appealed, arguing that the moment Quintana died the case became a "wrongful-death" suit, which under Colorado law carries a limit of $250,000. Perhaps seeing the ghoulishness of that point, they quickly abandoned it and agreed to a settlement.

One would think that such a trial would bring closure, like the show trials in France and Japan. In those countries, rightly or wrongly, concretely or symbolically, mistakes were acknowledged and perpetrators brought to task. Yet no such closure would come to America. The Quintana case, although precedent-setting, was only one of many civil suits. Her verdict had established that the blood bankers had been negligent; yet later, in another pivotal case, a Washington, D.C., jury determined they had not. In still another case, a jury issued the bewildering decision that, though the industry as a whole had been negligent in its standards, the Red Cross, which followed them, had been scrupulous and correct. The verdicts in hemophilia cases were similarly mixed.

In 1995, the National Academy of Sciences' Institute of Medicine released an analysis of the events surrounding the HIV contamination of blood products. About twenty thousand Americans had contracted HIV through blood and its derivatives, according to CDC figures at the time—more than eight thousand hemophiliacs and twelve thousand blood recipients. (For perspective, blood-products cases constituted about 3.4 percent of the total AIDS cases in America, which stood at 581,429 in 1996.) Prompted by COTT's relentless lobbying, the institute's staff had spent two years interviewing scores of people involved in the issue in the early 1980s—from industry, medicine, and government—and examined thousands of documents. In the institute's version of events, no clear-cut villain emerged. What they did find was an overall failure, a diffuse trail of missed opportunities and insufficient leadership. They agreed, for example, with Evatt and Francis that after the CDC meeting of January 1983 the blood industry should have excluded gay donors and started surrogate testing. Although no one could have known with certainty the results of these changes, both of which were eventually adopted, starting them earlier would have reduced the AIDS cases. In this regard the committee faulted the federal government for failing to take leadership and speak to the industry with a single, clear voice.

The government was not alone in its deficiencies. The institute faulted the blood bankers for acting too conservatively and for sending overly reassuring messages to the public. They blamed the fractionators for not beginning their research into heat treatment earlier. They also blamed the National Hemophilia Foundation, whose financial ties to the fractionators led to an inevitable conflict of interest, causing them to downplay the risks of clotting factor. Admittedly, this was a time of uncertainty, but even in uncertain times, the institute concluded, those who deal with a resource as dangerous as blood must constantly be ready to minimize the threat. They must never feel comfortable with the *status quo*. As Donna Shalala, secretary of health and human services, later said: "I believe the IOM report shows that our entire public health system missed opportunities to intervene and save lives."

The report might have brought closure if it had received any serious publicity, but the American news media paid scant attention. The blood industry, meanwhile, protested. Speaking at a Senate hearing that fall, James Reilly, president of the plasma industry's trade group, ABRA, complained, "Many of the IOM's findings and conclusions are without foundation and are incorrect. . . ." He stated that it was the "lack of information" that prevented the fractionators from making prescient decisions, and nothing more. Dr. Thomas Zuck, past president of the American Association of Blood Banks, co-authored a paper to correct many of the institute's "misconceptions" about blood banks. He pointed out that, given the information at hand, blood bankers did all they could. "In reality, it was not so much a lack of leadership . . . that made decision-making difficult, but rather a lack of solid data regarding the risks to the blood supply in the early years of the epidemic."

Zuck's argument makes sense when placed in an international context. At every stage of the AIDS epidemic—from openly revealing the link to the blood supply, to putting in place donor-deferral procedures, to virally deactivating plasma products and using the ELISA test—the Americans acted more quickly than any others. One could argue that the epidemic first surfaced in America, so naturally provoked an earlier reaction. Yet one must also concede that at each stage of knowledge the Americans set the standard for response. Still, given the scope of the tragedy—at least ten thousand hemophiliacs who seroconverted and twelve thousand transfusion recipients who became ill—one must ask whether the American blood establishment could have acted more effectively.

Part of the problem was that AIDS changed the entire paradigm of blood banking, and, to a degree, of public health. In previous days, when a problem arose, doctors could wait for definitive data, taking steps as technology became available, and meanwhile issue calming messages to the public. That was what had been done with hepatitis. Unlike hepatitis, however, AIDS was not a virus that people could live with. Every AIDS infection was fatal (although new treatments are changing that). Furthermore, as a retrovirus with a long lag time, it presented a greater threat than assumed, always a step ahead of the researchers. The heightened dangers called for a new paradigm, in which blood bankers would begin to take actions based on approximate data and to share doubts with the public more honestly. The record is striking in how often during those years people privately expressed grave concern yet voiced traditional confidence to the public, as with the "one-in-a-million" statistic. After ELISA screening began, a national survey found that one in four hundred donors tested positive—not to be confused with AIDS cases *per se,* but a clear indication of a significantly higher carrier rate. Years later, James McPherson, director of the Council of Community Blood Banks (recently renamed America's Blood Centers) and the Red Cross's regulatory affairs head at the time, would ruefully admit they had made a mistake: "When we were touting the risk as one in a million, we hadn't a clue. It was a stupid thing to say, but we were afraid. We had this desire to reassure the public. We should have just been honest and said, 'We don't really know what's going on.' If it started a panic, well, so be it." The result, said McPherson, is that blood bankers' credibility became "a mile wide and an inch deep."

Francis, who sounded the alarm, argued that some well-directed panic might have helped. What if the blood banks had issued an emergency call for female donors, he argued, the lowest risk group for AIDS? Goodness knows, he said, the media were primed: If given the information forthrightly, they might have triggered a mobilization, akin to the great wartime blood drives. No one knows whether the idea would have worked, but then again, no one attempted to find out. The blood bankers, intimidated by the potential for bigotry and fear, and worried about the traditional problems of maintaining a supply, moved more cautiously than they should have.

The drug companies, as we have seen, moved quickly on the issue of high-risk-donor exclusion. They frequently were less than forthright, however. In 1982, for example, an attorney for Cutter Laboratories wrote, "It appears to me to be advisable to include an AIDS warning in our literature. . . . Litigation is inevitable, and we must demonstrate diligence in passing whatever we do know to physicians who prescribe the product." Yet none of the companies affixed warning labels to their products for at least another year. Furthermore, each time the drug companies had to issue a recall, they would assure the public that the AIDS risk was minimal, even when they lacked a basis for assurance. Finally, the companies misled by exaggerating whatever positive measures they took, as when they advertised the closing of high-risk plasma centers while saving the inventory or continuing to operate others in the vicinity. The result was that hemophiliacs and their doctors continued to use a dangerous product.

Once AIDS became an acknowledged reality, the drug firms quickly developed screening tests and virus-killing technologies. For this they deserve praise: They preceded not only the blood bankers, but virtually everyone else in the world. Yet they need not have waited so long to react. If they had heeded the warnings of Factor VIII pioneer Ed Shanbrom and others who issued early warnings about viruses, they would have abandoned the high-risk collection centers sooner and worked more intensively on virus-killing technology. They knew that virtually all the hemophiliacs had become infected with hepatitis. If they learned anything from the hepatitis epidemic, it should have been that they faced a *general* problem with viruses, not only with hepatitis in particular.

In the early 1940s, as Dr. Edwin Cohn prepared albumin for the military, he refused to release the substance until his men could find a way to pasteurize the product, even during the wartime emergency. Twenty years later, when Shanbrom and his colleagues isolated Factor VIII, they assumed it impossible to heat-treat the concentrate. Balancing the anticipated good against the risks, they (and the government, which approved the product) made a calculated decision to release the dangerous concentrate. One cannot help wondering how Cohn the perfectionist would have responded.

Shanbrom himself addressed this issue at a 1996 conference of AIDS victims in Japan. He had not taken the podium in order to defend himself; indeed, the moderator had introduced him as the man whose advice the industry should have taken. Nevertheless, Shanbrom lowered his head in a long, ceremonial bow. "I would like to officially and openly apologize for the pain and suffering to all the hemophiliacs and their families," he said. "While we attempted to do good we also did harm. For this I apologize."

Tragedy takes the form of many realities. There is the reality of fact, with its numbers and dates; the legal reality of statutes and precedents; the scientific reality of microbes and their control. There is also the emotional reality, equally substantive, yet far more difficult to define and comprehend. This, the most painful and enduring of the realities, can defy all attempts to soothe it. It has a language of its own, in which words like "betrayal" and even "murder" become common, words that to an outsider may seem harsh and misplaced.

Every once in a while—sometimes during a trial or demonstration—this reality would reveal itself. One occasion was the annual meeting of the National Hemophilia Foundation in Indianapolis in 1993. Meetings like this used to have an atmosphere of genial camaraderie—almost like extended-family reunions—among the foundation, its chapters, the patients, and the doctors. Even the drug companies would attend, setting up "infusion suites" in which patients could treat themselves if they needed some clotting factor, or parents could inject it into their children. At this meeting, however, the atmosphere was different, poisoned by the rising toll of illness and suspicion. The drug companies were nowhere to be seen; nor, for that matter, were some of the prominent doctors.

The business of the meetings usually involved electing officials, going over the budgets, attending seminars, and hearing about new developments in therapy. This time AIDS dominated the agenda. One of the function rooms had been set aside as a memorial hall, with individual handcrafted panels from the AIDS Quilt hanging in rows. Because hemophilia strikes only males, and because this generation was prematurely dying, the room was a mosaic of lost adolescence, of boys who would never grow into men. The quilts bore images of basketballs and dogs, computers and cars. "You are surrounded by love," read one with a circle of fingers pointed at a young man. Another read: "Loved and missed but never forgotten." Another bore the words of an anonymous poet:

> *Do not stand at my grave and weep*
> *I am not there, I do not sleep.*
> *I am a thousand winds that blow*
> *I am the diamond glints on snow*
> *I am the sunlight on ripened grains*
> *I am the gentle autumn's rains. . . .*
> *Do not stand at my grave and cry*
> *I am not there. I did not die.*

The air in the convention center was electric with resentment. Signs of impending rebellion were everywhere. Children in the hallways sat lettering posters under their parents' approving gaze: "I was betrayed by the NHF"—and this at an NHF convention. The tension finally exploded during a "town meeting" to discuss a compensation package the foundation had attempted to negotiate. A few of the fractionators had made an offer that many found insultingly low. To the audience this was but another example of the supine posture their organization had assumed in dealing with the drug firms. Now, as members stood up to comment, the meeting transmuted itself from a policy discussion to a revival-like recitation.

"It pains me to say this, because my father was the vice-president of the NHF," shouted Rosenberg. "But the foundation has become the handmaiden to the industry!" He then announced that the class-action suit the rebels had filed would name the Foundation as one of the defendants.

People started cheering. "You guys are heroes—heroes!" a woman screamed.

Jean White, Ryan White's mother, stood up in the audience and said that she had received no support from the foundation—"no support at all"—when the school district so ignorantly castigated her son.

A burly man stood up and hollered: "I don't give a shit about the compensation. What are the chances of putting these criminals in jail? I'll give you everything I've got. I'll sell my house, I'll sell my business—just *get* those sonofabitches!"

Rosenberg replied: "Do you really believe the only doctors who should be in jail are the ones in France?" Shouts, wild cheers.

Katherine Royer stepped to the podium. Hers had been one of the early wrongful-death cases; she had lost her son to multiple infections, and later lost her lawsuit as well. She talked about the betrayal she felt when one of the doctors on the foundation's medical-advisory board testified as an expert witness for the drug company. "After the trial I confronted him with what he did. And do you know what he said? He said, 'Don't worry, Mrs. Royer, I donated my fee to charity.' And I said, '*I'm* the charity! Damn you all, don't you know what's going on?' "

Royer was so upset that she began to collapse. Members of the audience rushed up to catch her, surrounding her, holding her, leading her away. "You have nothing to be sorry about," said Michael Druck, an activist from New York. "*They're* the ones who should be ashamed."

Later a group of activists staged a demonstration outside the convention center, which TV crews had rushed over to cover. Some of the demonstrators wore black cloaks and death masks. Others, like Rosenberg, had no need to do so: Everyone could tell, merely by looking at him, that he only had a couple of months to live. At this point in his life, only anger sustained him. They waved placards: "HEMOPHILIA HOLOCAUST." They chanted: "My loss, their profit!" and "Shame! Shame! Shame! Shame!"

Rosenberg hobbled between the marchers and the media, a participant and a commentator at once. "This is the '*et tu Brute*,'" he proclaimed. "This is the history of shame and betrayal. These were the people who were supposed to *protect* us."

Corey Dubin was there, his hulking presence belying his HIV status. He would take over as the spokesman after Rosenberg died. Ricky Ray's and Ryan White's mothers were marching, ordinary women who had achieved a kind of sainthood through their sons' martyrdom. "I still have two living sons who are HIV-positive," said Mrs. Ray. "I promised them I would keep doing this until the day that I die." Brenda Walls circled with the marchers, remembering her son Jason Christopher. Even though she had won her case, she told everyone within earshot that she would continue to fight. As she joined in with the chorus—"What did they know, when did they know it?"—she fought back tears, alternately chanting and biting her lip. Suddenly she let fly with an outburst: "*These people killed my son!*"

There was a pause as everyone, shocked by the accusation, seemed to physically absorb what the woman had said. It may seem hard for outsiders to understand, for logic dictates otherwise, but these parents and thousands of others felt directly responsible for what had happened to their children, because they had administered the fatal injections. Guilt was devouring them. As one mother wrote years after this incident: "Today, I got up to shower and dress and while brushing my hair looked in the mirror and saw for the umpteenth time the person who injected the lethal dosage into the veins of my hemophiliac son. . . . My mind often harasses me, trying to pinpoint which treatment that was."

What could be more devastating to a parent? That was why Brenda Walls's cry was so stunning—in the flash of a moment she redirected the blame, relieving those present of their survivor guilt. Suddenly the specifics did not matter anymore. None of the details—the dates, the memos, the policies, the advisories—counted in light of the blinding realization that these parents were as innocent as their children. All shades of gray would henceforth disappear. It was them against the drug firms and their lackeys, mothers reunited in victimhood with their sons.

If the moment embodied an emotional reality, it represented a symbolic one as well. For something had gone terribly wrong with the technology, with the centuries-old dream of mastering blood. The dream had always been about healing; it had been so from the macabre experiments of Denis and Yudin to the brilliant advances of Landsteiner, Lewisohn, and Cohn. Over the years, the technology had restored life to millions. Starting with a small quantity of material (only 5 percent of Americans regularly donated), the global blood industry had come to the point where it could process and package nearly a score of injectable substances that were saving lives all over the world—a modern version of the fishes and loaves. What better gift could one possibly give? What better restorative than plasma or red cells could a doctor administer to a hemorrhaging patient? What greater blessing could a mother bestow than to inject clotting factor into a child with hemophilia? Yet, if blood from another was a gift, there was no denying it was also a commodity and embodied the properties and contradictions of both. "Give the gift of life," the Red Cross would implore, and as people did so the organization made money. Mothers had infused the gift of life into their sons, but found they had injected sickness instead. So much had been promised; no wonder the blood scandals that erupted around the world resounded so loudly with cries of broken faith and betrayal.

It is unlikely that the demonstrators considered such reflections. Like other protesters in other locales, they knew that certain people had failed to protect them and that they or their loved ones were wounded or dead. And so the group took up her cry, almost as a mantra, a self-hypnotic chant. Openly sobbing now, Walls stumbled forward, chanting, affirming, *insisting* with fellow mourners: "They killed our sons! They killed our sons! They killed our sons! *They killed our sons!*"

# EPILOGUE

# BLOOD IN A POST-AIDS SOCIETY

The blood supply is safer now. Sobered by the experience of AIDS, the blood industry has put in place a battery of tests and a long list of intrusive donor-deferral questions. The American Association of Blood Banks, for example, publishes a uniform donor questionnaire that, among other questions, asks men if they've ever had sex, "even once," with another man since 1977; and asks women if they have had sex with a male who has had sex with another male. One donor, a sixty six-year-old man who had given eleven gallons of blood over the years, complained to a blood-banking newsletter, "In recent years my blood center has made the blood donation process so ugly and repugnant that I quit donating." Indeed, the industry has lost donors, plaguing it with continuing supply problems. But the blood is safer now than it has ever been, even before AIDS. Most estimates put the odds of contracting HIV through transfusions at about one in 450,000, which, considering that most patients receive multiple units, makes the risk about one in 90,000. Other diseases are also being carefully screened, with the chance of contracting transfusion-related hepatitis B at one in 63,000 per unit, and that for hepatitis C at less than one in 100,000 per unit—all of which indicates a reasonably safe supply, at least for the time being.

Hemophiliacs face lower risks than ever before, with plasma concentrates subjected to a variety of treatments. In addition to heat treatment, fractionators may also use the "solvent-detergent" method

345

developed by scientists at the New York Blood Center, which, similar to Shanbrom's idea, kills the virus by destroying its outer coating. Some fractionators employ "monoclonal" purification, in which antibodies derived from mouse cells with an affinity for Factor VIII are used to "fish-hook" the Factor VIII molecule selectively, leaving contaminants and viruses behind in the residuum. The result is an extremely pure concentrate. Most recently, drug companies have begun to produce artificial clotting factors from genetically engineered animal cells into which they have inserted the genetic sequence to produce human Factor VIII. As a result of these technologies, the contamination rate among "virgin" hemophiliacs has dropped to practically nil. The hemophiliacs dying of AIDS infections today will probably be the last to be so tragically affected.

This knowledge came at a terrible price. During the blood-related AIDS epidemic, at least half of America's hemophilia population—more than eight thousand people—and twelve thousand other transfusion recipients contracted the virus. (There were hundreds of thousands of hepatitis C victims during and after this period as well.) Worldwide, more than forty thousand hemophiliacs have tested positive for HIV. A recent study based on World Health Organization statistics throws a more focused light on the blood-related AIDS cases in Europe, where more than five thousand people have developed the disease. The numbers show that the countries with the highest rates among their hemophiliacs (with the exception of France) were those that relied on imported products—such as Germany, for example, where nearly half the nation's hemophiliacs contracted HIV. In contrast, Belgium, which used mainly cryoprecipitates from local donations, suffered a rate of only 7 percent. One cannot deny that during the period of uncertainty about AIDS a more cautious use of Factor VIII would have saved lives. Meanwhile, two countries stood out as having the highest rates among red-blood-cell recipients: Romania, whose health system in the 1980s was a dangerous wreck, and France, with its history of hubris and denial.

Most estimates of viruses in blood products leave out Africa and parts of the Third World, where data collection is scarce and unreliable. Blood-bank consultants returning from Africa describe a lack of the most basic equipment, such as refrigeration and disposable needles. Centers that administer ELISA tests can often afford to do so only sparingly, testing samples every Friday, for example, so that infected blood that comes in any other day gets through. Many blood banks throughout the Third World require patients to replace blood by recruiting family and friends. "Most, of course, utilize professional paid donors who hang around the hospital," observed an International Red Cross consultant after touring Latin America. In Pakistan, according to the same Red Cross consultant, "little if any testing is done, and blood transfusions . . . can be arranged for and provided out of pharmacies on the street." "In some parts of Southern Asia," wrote another International Red Cross official, "not only [are] donors in the marketplace . . . ready to sell their services, but blood bags can be found lying in the sun on the small shop desk so that anybody can establish an 'instant blood bank.' " Under such conditions, it is little wonder that more than 40 percent of the AIDS cases in Pakistan have been contracted through transfusion, more than 10 percent of the cases throughout Africa, and that doctors in India estimate that 95 percent of their blood is not safe.

The situation in poor nations is cause for alarm, but it is clear that the wealthier nations have learned from their experience with AIDS. Most have reinvented their blood-banking infrastructures to make them more accountable.

The French brought their disparate transfusion centers under the control of a French Transfusion Agency, which, answering directly to the minister of health, makes all the policy and import decisions—no more "multiple baronies" for them. Shaken by their previous negligence, the French have conducted the world's most ambitious effort to trace and compensate victims of tainted blood products. As for the doctors convicted in the infamous AIDS scandal, they served out their terms and returned to private life—Allain as a humble professor in Cambridge (he was the only one of the defendants whose legal bills were not covered by the government); Garretta as a telephone employment recruiter, his career in ruins, his name in disgrace. Meanwhile, French prosecutors, still unsatisfied with the original sentences, mounted an ongoing investigation of poisoning charges against the doctors and indicted three former ministers above them for manslaughter.

The British established a centralized National Blood Authority to coordinate the resources of the various regions. They also completed a new fractionation facility that now meets 70 percent of the nation's needs.

The Canadians, astonished at the mistakes of the peak AIDS years, convened a commission to produce a definitive account. For four years, the commission, chaired by jurist Horace Krever, met and took testimony, only to have its final report frozen for several months while the Red Cross sued to prevent its release. (Connaught was out of the picture by then, having abandoned its clotting-factor business in 1987.) In the end, the Canadian government took the blood program away from the Red Cross, allowing the organization to recruit donors but nothing more. The Red Cross's leaders, insulted by the demotion, decided to leave the blood business altogether.

In Japan, the government set a goal of eliminating all imported blood products, which the nation achieved for Factor VIII, although it still imports albumin and gamma globulin. The most dramatic post-AIDS changes in Japan came with the fall of the once powerful Green Cross. The company's behavior had proved so scandalous that shareholders sued and hospitals boycotted its products, forcing repeated lay-offs and downsizing. In 1998, the firm was taken over by a smaller drug company called Yoshimoto Pharmaceutical Industries. The new company bore neither the Green Cross name nor its logo—a melancholy end to Dr. Naito's dream. More happily, at the same time that Green Cross disappeared so too did the stigma of hemophilia and AIDS in Japan: In the fall of 1996, an HIV-positive hemophiliac named Satoru Ienishi was elected to Japan's Diet.

The fractionators learned from the crisis as well. In addition to implementing tougher screening and disinfection procedures, they agreed, under congressional pressure, to lower their pool sizes from 100,000 donors or more to a maximum of 60,000 in order to lower the risk of contamination.

Twenty-two countries, including most of Europe, Canada, and even Thailand, have established funds to compensate their infected hemophiliacs. The United States, which experienced the worst of the hemophilia holocaust, has not done so. The American hemophiliacs who filed their class-action suit had hoped to bring closure to the issue, and perhaps even a measure of justice, but a panel of judges ruled that if the suit went ahead the damages could be large enough to "hurl the industry into bankruptcy." Reasoning that no single civil jury should hold the "fate of an industry in the palm of its hand," the judges decertified the class, ruling that if the hemophiliacs wanted to sue they would have to do so individually—a finding that provoked hundreds of new lawsuits, which will entangle the American court system for years. Meanwhile, several hemophiliacs and their lawyers uncovered enough evidence of foreknowledge and neglect to negotiate a settlement with the industry. The agreement, worth a total of $640 million, would give each of the hemophiliacs in the group an estimated $100,000—an amount that may seem generous on the face of it but would drain away after a few years' purchases of Factor VIII. The government has been considering a compensation package for infected hemophiliacs and transfusion recipients (named the Ricky Ray act, after the young

Florida martyr) since 1995, but at this writing the bill continues to languish in Congress. Meanwhile, an estimated two HIV-infected hemophiliacs in America die every day.

Though slow on compensation, the Americans did enact new procedures to prevent an AIDS-like crisis from happening again. Besides setting up additional surveillance centers, the government gave overall responsibility for blood products to a new, centralized Blood Safety Committee. "Blood safety must never again be a secondary issue," Donna Shalala, secretary of Health and Human Services, told a congressional hearing in 1995. "I am elevating it to the highest level of the department." The FDA, stung by criticism of overly cozy relations with blood bankers and fractionators, set strict guidelines for its Blood Products Advisory Committee, ruling out potential conflicts of interest, and stipulating that the committee must only deliberate over safety and efficacy—not over cost effectiveness, as in the past. The committee would now include consumers as well, one of the first of whom was the outspoken hemophilia activist Corey Dubin. Blood bankers also became more sensitive to risk: When the first hepatitis C test was put in place in 1990, the blood bankers put a freeze on the blood in their inventories until they could go back and test it.

Some of the old ways occasionally persist. Despite the new testing and processes, the industry still makes mistakes; human error and sloppy procedures have forced consent decrees and blood-product recalls at blood banks as notable as the New York Blood Center, the Red Cross in Los Angeles, and United Blood Services in Tucson, Arizona, and at industry leaders such as Centeon (formerly Armour) and Alpha Therapeutic. Moreover, officials have sometimes vacillated about the values to consider when formulating new policies. In the summer of 1995, for example, the government's Blood Products Advisory Committee debated whether to institute a new procedure to close the AIDS-testing window (the period of time after a person contracts the HIV virus but before his immune system produces detectable antibodies) called the p24 antigen test. Although increasingly sensitive versions of the ELISA test had narrowed the window to less than a month, the new test could restrict it to a mere ten days, directly and indirectly preventing an estimated sixty-eight infections a year. At a cost of $2–4 a test, the procedure would mean a national expenditure of about $30 million, or nearly $500,000 for each infection prevented. Based on that analysis, the committee voted not to use the procedure, over the objections of Dubin and a few others. "I was absolutely stunned by the vote and felt a real sense of history repeating itself," Dubin told a congressional hearing. Appalled at the committee's actions, Congressman Christopher Shays wrote to FDA Commissioner David A. Kessler, reminding him that "timidity in confronting the AIDS threat has already exacted a tragic toll," and demanding that he reverse the committee's "illogical" decision. Kessler not only required the new test, but demanded the resignation of most of the committee members.

If AIDS brought a new era in vigilance, it also ushered in a new period in blood economics. The battery of new tests raised the price of each unit by anywhere from $25 to $35, a significant portion of the unit's current retail price of $150 to $200. This, in addition to health considerations, motivated doctors to use less blood whenever possible, or to allow patients to set aside blood for themselves. Beyond that, tough new federal inspections were forcing blood banks to spend more money on training employees and on quality control.

No organization suffered under the new economic order as much as the American Red Cross. After decades of wealth and glory, the agency crashed, both in its balance sheet and in public opinion. Having failed to live up to its 1988 agreement with the FDA, the agency was slapped with a court-ordered consent decree, under which it must make a series of improvements under a rigidly enforced schedule. The Red Cross had hired a new and famous president—former Labor Secretary Elizabeth Hanford Dole—to reverse its sagging fortunes. Shortly after taking office, she announced a "transformation" campaign to cut personnel, centralize laboratories, and revamp the computer system, a seven-year program that cost $287 million. And at the same time that the agency's finances were failing, it faced an unexpected health crisis. In 1994, a longtime Red Cross donor died of a rare malady known as Creutzfeldt-Jacob disease, or CJD. This poorly understood affliction, also known as mad-cow disease, can linger for decades in the victim before creating spongelike holes in the brain, ultimately leading to dementia and death. (A veritable panic erupted in Europe when British beef was named as a possible disease vector.) No transfusion-related cases of CJD had ever been documented, but in the lingering wake of the AIDS epidemic, the Red Cross quarantined all the plasma pools to which the man had contributed. When, after months of deliberation, the FDA ordered the material's destruction, it cost the Red Cross several million dollars. Subsequent withdrawals have cost the agency a total of $130 million.

These losses caused a hemorrhage in the agency's biomedical-services division, which lost $113 million in 1995 alone. The Red

Cross tried to compensate by expanding its business, offering loss-leader prices to hospitals in regions where it had not sold blood before, while raising prices in old areas to compensate. The tactics and layoffs eroded morale; some Red Cross employees in Springfield, Missouri, became so disenchanted that they broke off and formed their own community blood bank. The Red Cross responded by suing the employees for allegedly stealing their donor lists. That and other incidents made it clear that, without the common enemy of AIDS, the alliance among the blood bankers had frayed, and the rivals returned to business as usual—feuds, slander, and cutthroat competition—contributing to periodic blood shortages in several regions.

A new economic era unfolded in Europe, where markets shifted dramatically. As Europe became one economic community, the rules about blood products changed. Now officially classified as pharmaceuticals, blood products could freely cross European borders, defying traditional efforts to keep them out. With American and German products entering France, for example, the French monopoly on clotting factors crumbled: Of the seven original fractionators in France, only two have continued to operate. Other small fractionators, such as Sweden's Kabi and Denmark's Novo Nordisk, found it increasingly difficult to survive. Even two of the industry's giants—French-owned Armour and German-owned Behringwerke—consolidated their assets in a new company in Pennsylvania called Centeon. Thus, more and more of the world's plasma products have become controlled by a shrinking number of global corporations.

The former Communist nations experienced wrenching changes in their blood economies, for as the governments fell so too did the old systems of "pressure recruitment." Donations across Eastern Europe plummeted, and impoverished residents flocked to for-profit plasmapheresis centers brought in by Western entrepreneurs. In Romania, for example, a German firm called UB Plasma set up a center in an industrial neighborhood of Bucharest, paying 12 German marks ($8) per donation, and not bothering with screening tests. They had collected one and a half tons of plasma before the Romanian Health Ministry froze the operation. For six months the plasma sat in a refrigerated warehouse, only to go to Germany when a reactionary regime came into office and allowed the German businessmen to ship it out. Once in Germany, the plasma became part of a broader scandal when it was revealed that, even with their German collections, UB Plasma had only sporadically tested for AIDS. The company sent thousands of units of blood products to hospitals in Germany (including an American military hospital), and to at least half a dozen foreign countries. In 1995, authorities closed the company and sentenced its chief officials to prison.

At the age of seventy-three, Dr. Ryoichi Naito conducted his most daring experiment. He had been traveling abroad several years earlier when he read a news article about an unusual liquid. The fluid, called Fluosol, belongs to a class of substances known as perfluorocarbons, of which Teflon is a member, and can hold large quantities of dissolved oxygen. He had read about an experiment in which Dr. Leland Clark at the University of Cincinnati had submerged a mouse in a jar of Fluosol as he bubbled oxygen through the liquid. The rodent survived for more than an hour. Naito was fascinated by the potential of this liquid, which he thought he could use as a substitute for red blood cells.

The Old Locomotive immediately set to work, visiting America to learn more about the material, hosting numerous symposia, and gearing up his own research in Japan. Within a few years, he developed a formula called Fluosol-DA, which he injected into cats, rabbits, monkeys, and dogs. Generally the animals did well, although the dogs suffered a rapid but temporary blood-pressure drop—signs of an allergic reaction. Then he sent a quantity to researchers in Germany, who injected more than a liter into each of seven accident victims who had been pronounced brain-dead and were being sustained on life support. The liquid circulated freely through their bodies, with no signs of organ damage in subsequent autopsies. The next step, he knew, would be to inject Fluosol into a healthy human being. He debated the question with a surgeon friend at the Kobe University Medical College, who warned him that the kind of blood-pressure reduction they had seen in the dogs would kill a human being. For months they argued over whom to use as the first human subject.

"I was deeply troubled," Naito wrote in his memoirs. "If I were to ask somebody other than me to receive the infusion and if he were to die . . . I would be the killer. Tremendous effort and vast amounts of money spent for the study would instantly come to naught and I, as the president of the company, would be fully responsible."

In the end Naito's arguments won out—he felt he was the only person on whom they should experiment. He kept his decision secret until the night before the trial, when he told his wife and made out his will.

The experiment took place on February 8, 1979. Naito lay on a hospital bed, sprouting catheters and electrodes, surrounded by worried doctors and colleagues. The doctors infused one milliliter of Fluosol and waited half an hour for a reaction. When the instruments showed nothing abnormal, they added a few milliliters more, until they had infused twenty in all. After twenty-four hours without a reaction, the doctors sent Naito home, perfectly healthy. Later a dozen more executives volunteered, and none experienced untoward reactions.

Naito's experiment opened the way for a cascade of research into a long-standing dream—a substitute for red cells with none of their disadvantages. A synthetic oxygen-carrier such as Fluosol would never have to be tested for compatible blood types or disease. The need for refrigeration would vanish. Military units and emergency vehicles could always keep a ready supply on hand. Hospitals would always be fully stocked. Gone would be the days of seasonal shortages and public appeals—blood and all its components would finally become simple pharmaceuticals. Naito and his colleagues moved forward enthusiastically, as did some American colleagues. In the long run, however, although the liquid seemed safe, large infusions would leave the patient oxygen-starved, like the long-term plasma recipients in World War II. It seemed that Fluosol could not transport enough oxygen to sustain patients.

The next generation of research took scientists into the heart of the system—the oxygen-carrying pigment in the red cells. Hemoglobin is a complex of proteins, a four-part structure twisted into the shape of several three-dimensional pretzels, each of which holds an iron atom in its core. This protein complex, housed within the red cell, readily absorbs and releases oxygen. Over time, however, the red cell deteriorates. The hemoglobin pigment leaks out and breaks down into its subunits—fragments that not only fail to carry oxygen, but can prove damaging if they accumulate in the kidneys. The problem, then, was to find ways to keep these subunits intact.

Over the past fifteen years, scientists, largely funded by the military, have been trying to recapture, repair, or synthesize the hemoglobin. One company, for example, called Biopure, removed hemoglobin from cow's blood and purified it sufficiently to avoid allergic reactions. Having conducted small-scale tests before the Persian Gulf War, Biopure lobbied the military to ship their product to Saudi Arabia. There, just as with the invasion of Normandy, the military had been projecting casualties and amassing blood products. (This time, however, their projections were wrong. Having estimated a daily need of six thousand units of red cells to meet the needs of fifteen hundred casualties, the military used a total of two thousand pints, most of which they administered to wounded Iraqis.) The military declined, worried that the foreign protein might cause kidney damage in the already dehydrated GIs.

The Pentagon's reluctance did not prove too discouraging, and by now several firms have prepared hemoglobin substitutes safe enough for human testing. What most have in common is a process in which they stabilize the pigment so it remains intact outside the protective membrane of the red cells. One process, for, example, developed by Baxter, involves collecting expired blood from the blood banks, breaking open the cells, and using a chemical "stitch" to rejoin the hemoglobin subunits. Another method, developed by a company called Somatogen, involves altering the genetic coding of bacteria, inducing them to produce a customized hemoglobin with strong enough crosslinks to survive outside the protection of the red cell. With these products and others now in advanced testing, the chances are high that a substitute for hemoglobin will become available within the next several years.

On that day, another of blood's mysteries will have been solved. The history of blood use has always been one of deconstruction, from plasma to albumin, to antibodies and clotting agents—and artificial hemoglobin would represent one more step. Yet it would not be the final step. For all the optimism of these companies and their backers, the hemoglobin substitutes have a much shorter life span than red cells—lasting two or three days versus several weeks—which will make them short-term substitutes at best, to be followed up by red-cell infusions. Furthermore, the products may prove too expensive for most of the world.

The need to collect blood will remain. Therefore, it will be increasingly important to remain sensitive to risks—for blood, as the most human commodity, is as vulnerable as the humans who carry it. Although certain diseases have been controlled in the blood system, new ones stand poised to invade.

CJD, as mentioned, poses a new threat. Even though the disorder has never been shown to be passed by human blood, blood services in several countries now screen donors for any family history of the disease or exposure through certain transplanted tissue. In Britain, the "hypothetical" chance of blood-borne transmission has prompted health authorities to avoid using domestic plasma for pharmaceutical production and to resume using imports instead. In America, dozens of recalls and withdrawals have occurred, resulting in industry-wide losses of hundreds of millions of dollars. The massive withholding of suspected CJD-tainted material contributed to national shortages of immune globulin, vital to tens of thousands of people. The shortage caused a problem of such magnitude that Congressman Christopher Shays held hearings in the spring of 1998 to determine if the industry was hoarding the material or shipping too much overseas. Multiple causes actually contributed—increasing demand, factory shutdowns while companies retooled to meet necessary health standards, foreign market demands, and the need to avoid even the theoretical risk of spreading disease. In short, the crisis arose from the complexities of marketing blood products in a post-AIDS society.

Other exotic diseases threaten as ecological disruption, human migrations, and global jet travel bring foreign pathogens closer to home. After the Persian Gulf War, when seven returning soldiers came down with leishmania, a Middle Eastern parasite, blood banks temporarily banned a million veterans from donating blood. Chagas' disease, caused by a red-blood-cell parasite endemic to Latin America, has surfaced in the United States as a result of travel and immigration. The disease can escape detection for decades—a fact that so worries some blood centers that they exclude donors who have traveled to Latin America, regardless of whether they show signs of infection.

Even a disease that has been screened from the blood supply can later surface to cause a quiet epidemic. In 1990 and 1992, American blood bankers introduced a series of rigorous tests to screen out donors who carried hepatitis C, dramatically reducing its prevalence in the blood supply. Yet with a lag time of decades, the disease's true morbidity is only now becoming known. More than four million Americans are thought to have contracted the virus, several hundred thousand through transfusions. The blood-related death toll could eventually surpass that of AIDS. Indeed, no matter how well scientists think they understand blood, they can never assume the safety of all its components. Even immune globulins, always assumed incapable of carrying pathogens, have in certain cases transmitted hepatitis C. The inevitable conclusion: No matter how safe we collectively make our blood supplies, new dangers will appear—and only in retrospect will they be fully understood.

The resource and its risks will always be with us, and it is up to humanity to view them realistically. It would be tempting to try to find a universal system for managing blood, one infrastructure that works more effectively than the rest. But in examining the tainted-blood tragedies of the 1980s, it becomes clear that no system was immune from mistakes, whether capitalist or socialist, monolithic or decentralized. Countries that emerged from the crisis with relatively low blood-borne disease rates had a few simple, common elements: diligent people in charge who fostered rapid response, open communications, and close control over the source of their supplies. Safety is a matter of practice, not ideology.

When I started researching this book, several blood bankers I interviewed urged me not to portray blood as a commodity, but only as a gift given freely by donors. The implicit message was that products derived from paid plasma donors are not only risky but also inherently immoral. That thinking, however, got the blood bankers of many nations in trouble: Viewing blood in the context of national pride, they failed to act quickly when blood products became tainted. On the other hand, viewing blood as nothing more than a commodity led to the abuses of the 1970s, when "vampire" collectors set up in the Third World.

Piet Hagen, who described the global blood system in his book *Blood: Gift or Merchandise,* represented a prevailing dichotomous view in his title. To him, to scholars like Titmuss, and to many non-profit blood bankers, blood can be either a commodity or a gift—corrupted or moral, tainted or pure. In truth, blood is both gift *and* merchandise, and only by accepting the dual nature of blood products will humanity use them with sufficient care and intelligence. Blood is a precious and dangerous medicine. We must be careful how we use it.

# NOTES

*The research for this book involved travel in nine countries over a period of five years, and the gathering of thousands of documents and hundreds of interviews. For the sake of clarity and brevity, I have included only those references that directly apply to the material cited. Whenever possible, I have included all references and archives so that readers may follow. In some instances, however, documents were provided by individuals who must remain anonymous. Although these documents are genuine and verified as such, it may not be possible to find them in a public archive.*

## PREFACE

x      SELLS FOR ABOUT $13: Price quotes courtesy of the American Petroleum Institute, circa March 1998.

WOULD SELL FOR MORE THAN $20,000: This approximation is based on a roughly $60-per-pint cost of collecting whole blood (including labor, equipment, and other costs) prior to testing and component separation. The precise cost is highly controversial and the subject of much study among blood bankers. Interviews with Drs. James P. AuBuchon, Walter Dzik, and Robert Westphal.

xi      $18.5 BILLION PER YEAR: This rough estimate combines the value of plasma products and whole blood. The annual plasma-products revenue is about $5 billion per year (American Blood Resources Association). Whole blood is not usually assigned monetary value, since it is handled on a non-commercial basis. Nevertheless, assigning it value is instructive. World-wide, approximately 91.6 million pints are collected annually. (Ennio C. Rossi, Toby L. Simon, Gerald S. Moss, *Principles of Transfusion Medicine*, 2d ed. [Baltimore: Williams and Wilkins, 1995], p. 918), which at today's retail value of $150 and upward per pint (industry reviews) is worth about $13.5 billion. Adding the value of plasma products yields a conservative estimate of $18.5 billion.

## CHAPTER 1: THE BLOOD OF A GENTLE CALF

3    HE SUFFERED "PHRENSIES": J. Denis. "An extract of a letter . . . ," *Philosophical Transactions* 2 (Nov. 10, 1667): 617–24.

"MANY PEOPLE OF QUALITY . . .": Ibid., p. 620.

5    SEVENTEENTH-CENTURY MEDICINE: Fielding H. Garrison, *An Introduction to the History of Medicine* (Philadelphia, London: W. B. Saunders, 1929), pp. 236–309; Will and Ariel Durant, *The Story of Civilization: The Age of Louis XIV* (New York: MJF Books, 1963), pp. 522–30.

DENIS WAS A SOMBER-LOOKING MAN: Personal observation, portrait of Denis at the Faculty of Medicine in Paris.

BORN TO A MODEST FAMILY: Jean-Jacques Peuméry, "Les Origines de la transfusion sanguine: II," *Clio Medica* 9, no. 3 (1974): 215–18.

DENIS WATCHED FOR SIGNS: Denis, "Extract," pp. 617–24.

6    "WE OBSERVED A PLENTIFUL SWEAT . . .": Ibid., p. 621.

7    GREAT CHANGE WAS SWEEPING EUROPE: Durant, *Civilization,* pp. 495–98.

"WE ARE DRIVEN TO WONDER": Garrison, *History of Medicine,* p. 220

8    ". . . THERE ARE NO PORES!": Earle Hackett, *Blood* (New York: Saturday Review Press, 1973), p. 91.

THE EXPERIMENTAL PHILOSOPHY CLUB: Years later, the Experimental Philosophy Club moved to London and became the Royal Society. Both organizations published *Philosophical Transactions,* a journal that was the premier learned chronicle of its time, and is still produced today.

LOWER WAS BORN INTO A FARMING FAMILY: John H. Talbott, *A Biographical History of Medicine* (New York, London: Grune & Stratton, 1970), pp. 151–53.

9    A SERIES OF EXPERIMENTS STARTING IN 1665: Merril W. Hollingsworth, "Blood Transfusion by Richard Lower in 1665," *Annals of Medical History* 10 (1928): 213–25; Geoffrey Keynes, ed., *Blood Transfusion* (Bristol: John Wright & Sons, 1941; London: Simpkin Marshall, 1949), p. 9.

"SPECTACULAR NEW EXPERIMENT . . .": Hollingsworth, "Blood Transfusion." This reference is a translation of a portion of Lower's seminal work, *Tractatus de Corde,* which was written in Latin. In it, Lower used the old Julian calendar in referring to his experiments, even though most of Europe had converted to the modern Gregorian calendar. Thus, the dates he gives in the text do not coincide with what scholars have determined to be the actual years.

". . . STOCKED WITH THE BLOOD OF A *COWARDLY* DOG": [Robert] Boyle, "Trials proposed by Mr. Boyle to Dr. Lower . . . ,": *Philosophical Transactions* 1 (Feb. 11, 1666): 385–88.

10    "A PRETTY EXPERIMENT . . .": Samuel Pepys, *Diary of Samuel Pepys,* Nov. 14, 1666, cited in Hollingsworth, "Blood Transfusion," p. 224.

NINETEEN TRANSFUSIONS AMONG DOGS: Peuméry, "Origines," p. 152.

HIS MOST DARING IDEA: J. Denis, "Concerning a new way of curing . . . ," *Philosophical Transactions* 2 (July 22, 1667): 489–504.

11    A SIXTEEN-YEAR-OLD PATIENT: Ibid., pp. 501–3.

A BURLY, FORTY-FIVE-YEAR-OLD LABORER: Ibid., pp. 503–4.

12    LOWER PUBLISHED A VITUPERATIVE RESPONSE: Hollingsworth, "Blood Transfusion."

12      LOWER AND KING PAID 20 SHILLINGS: Samuel Pepys, *Diary of Samuel Pepys,* Nov. 21, 1666, cited in Hollingsworth, "Blood Transfusion," p. 225.

THE TECHNIQUE SPREAD: Bernard J. Ficarra, "The Evolution of Blood Transfusion," *Annals of Medical History,* 3rd ser., no. 4 (1942): 305–6; Peuméry, "Origines," pp. 227–32.

THE FRENCH INTELLIGENTSIA WERE HIGHLY POLITICAL: Peuméry, "Origines," pp. 215–50; Francis R. Packard, "The Physicians of Paris Versus Those of Montpelier," *Annals of Medical History* 4 (1922): 357–75.

13      A SERIES OF PAMPHLETS TRASHING DENIS: Peuméry, "Origines," pp. 225–50.

THERE STOOD ANTOINE MAUROY: J. Denis, "An extract of a printed letter . . . ," *Philosophical Transactions* 3 (June 15, 1668): 710–13.

14      WOULD HE "EXERCISE THE CHARITY": Ibid., p. 711.

THE NEWS WAS "BRUITED ABROAD": Ibid., pp. 710–13.

AMONG THE ODDEST REVERSALS IN JUDICIAL HISTORY: "An extract of the sentence, given at the Châtelet . . . ," *Philosophical Transactions* 3 (June 15, 1668): 713–15.

15      ANTOINE WAS DYING: Antoine would have died in any case. If his wife had not poisoned him and Denis had succeeded in getting blood into the patient, Antoine's immune system would have reacted so violently to his third exposure to foreign blood protein that he would never have survived.

THE FRENCH PARLIAMENT OFFICIALLY BANNED ALL TRANSFUSIONS: Ibid., p. 715.

"ORIGINAL . . . OF LOVE": "A letter written by an intelligent and worthy English man from Paris . . . ," *Philosophical Transactions* 4 (Dec. 13, 1669): 1075–77. Dr. Byron Myhre, Chief of Clinical Pathology at the UCLA–Harbor Medical Center, suggested the fever-malaria explanation and provided history about the febrile treatment of malaria (personal interview).

## CHAPTER 2: "THERE IS NO REMEDY AS MIRACULOUS AS BLEEDING"

17      DOCTORS WOULD SEASONALLY "BREATHE A VEIN": Gilbert R. Siegworth, "Bloodletting over the Centuries," *New York State Journal of Medicine,* Dec. 1980, p. 2024.

18      A PATIENT BEING BLED FROM THE FOOT AND NECK: Fielding H. Garrison, "The History of Bloodletting," *New York Medical Journal* 97 (1913): 432–37, 498–501.

"BLEED IN THE ACUTE AFFECTIONS . . .": Ibid., p. 434.

HE RECOMMENDED BLOODLETTING FOR A WIDE ARRAY OF MALADIES: Ibid., p. 435.

AVICENNA OF PERSIA: O. Cameron Gruner, *A Treatise on the Canon of Medicine of Avicenna* (London: Luzac, 1930), pp. 501–8; John H. Talbott, *A Biographical History of Medicine* (New York, London: Grune & Shatton, 1970), pp. 20–21.

19      "BLEEDING THE BODY PURGES IN DISGUISE . . .": Garrison, "History," p. 435.

19    A WHOLE STRATUM OF BLOOD WASTE: Peter Bowron, "Bloodstained Mementos of Medieval Medicine," *History Today,* Oct. 1988, pp. 4–5.
TALMUDIC AUTHORS: Fred Rosner, "Bloodletting in Talmudic Times," *Bulletin of the New York Academy of Medicine* 62 (1986): pp. 935–46.
BLOODLETTING CALENDAR: Garrison, "History," p. 498.
PATIN EFFUSED: Francis R. Packard, "Guy Patin and the Medical Profession in Paris in the Seventeenth Century," *Annals of Medical History* 14 (1922): 366; Garrison, "History," p. 499.

20    HE WAS A WICKED FLEMISH RASCAL: Packard, "Guy Patin," p. 363.
"WE ALSO BLEED . . . CHILDREN . . .": Packard, "Guy Patin," p. 232.
BLEEDING DOES THE BODY NO GOOD: There are a few exceptions to this rule, even in modern times. One condition that responds to bloodletting is erythrocytosis, an excess of red blood cells; another is a metabolic malfunction called "hemachromatosis," in which the body absorbs too much iron.
TYPICAL APOLOGIA: Packard, "Guy Patin," p. 232.
"TALL, WITHERED EXECUTIONER . . .": Garrison, "History," p. 499.

21    "SIRE, WE TALK TOGETHER . . .": Will and Ariel Durant, *The Story of Civilization: The Age of Louis XIV* (New York: MJF Books, 1963), p. 121.
AN IMPRESSIVE ARRAY OF HARDWARE: Audrey Davis and Tony Appel, *Bloodletting Instruments in the National Museum of History and Technology* (Washington, D.C.: Smithsonian Institution Press, 1979).

22    *LOECE,* "TO HEAL": Ibid., p. 35; Gilbert R. Siegworth, "Bloodletting over the Centuries," pp. 2022–28.
"THE MOUTH OF THE WOMB . . .": Siegworth, "Bloodletting," pp. 2026–27.
"PRINCE OF BLEEDERS": Paul J. Schmidt, "Transfusion in the Eighteenth and Nineteenth Centuries," *New England Journal of Medicine* 279 (Dec. 12, 1968): 1319.
PATIENTS DID NOT SUFFER FROM DISEASES LIKE "TUBERCULOSIS": One notable exception was smallpox. By the revolutionary era, doctors had not only identified it as a disease but come up with an inoculation as well. By pricking the pustule of an infected patient and sticking the same needle into someone else, they effectively conferred immunity. George Washington had his troops treated in this manner—the first mass inoculation in military history.

23    ALEXANDER FORBES: J. Worth Estes, "Patterns of Drug Usage in Colonial America," *New York State Journal of Medicine* 87 (Jan. 1987): 37–45; additional information on colonial-era medicine from interview with Dr. J. Worth Estes.
"HE DIED . . .": Paul J. Schmidt and James E. Changus, "The Bloodletters of Florida," *Journal of the Florida Medical Association,* Aug. 1980, pp. 743–47.

24    "LET ANYONE WHO DESIRES . . .": William Pepper, "Benjamin Rush: An Address Delivered Before the American Medical Association at Its Annual Meeting in, R.I., June 1889," *Journal of the American Medical Association* (hereafter *JAMA*) 14 (April 26, 1890): 593–601.
SCENES MEDIEVAL AND SURREAL: Detailed accounts of the plague and Rush's role were obtained from John H. Powell, *Bring Out Your Dead: The Great Plague of Yellow Fever in Philadelphia in 1793* (New York:

Time, Inc., 1965); J. Worth Estes, "Introduction: The Yellow Fever Syndrome and Its Treatment in Philadelphia, 1793," in J. Worth Estes and Billy G. Smith, eds., *A Melancholy Scene of Devastation* (Canton, Mass.: Science History Publications, 1997), pp. 1–17; interview with Dr. J. Worth Estes.

25    ". . . A DAY OF TRIUMPH . . .": L. H. Butterfield, ed., *Letters of Benjamin Rush*, vol. 2, 1793–1813 (Princeton, N.J.: Princeton University Press, 1951), p. 663.

"AT FIRST I FOUND . . .": Butterfield, ed., *Rush Letters*, p. 695.

"YOU HAVE NOTHING BUT A YELLOW FEVER": Powell, *Bring Out Your Dead*, p. 22

26    "I TREAT MY PATIENTS . . .": Ibid., p. 130.

"NEVER DID I EXPERIENCE . . .": Pepper, "Benjamin Rush," p. 599.

27    *"THE TIMES ARE OMINOUS . . .":* Paul F. Lambert, "Benjamin Rush: Physician in Politics," *Oklahoma State Medical Association* 65 (1972): 218–24.

"THE PENNSYLVANIA HIPPOCRATES": Nicholas E. Davies, Garland H. Davies, and Elizabeth D. Sanders, "William Cobbett, Benjamin Rush, and the Death of General Washington," *JAMA* 249 (Feb. 18, 1983): 914.

"OLD LIGHTNING ROD": Ibid., p. 913.

"I VERILY BELIEVE": Ibid., p. 914.

"A MAN BORN TO BE USEFUL . . .": Ibid.

"AND SO IS A MOSQUITO . . .": Ibid.

A JURY FOUND COBBETT GUILTY OF SLANDER: Ibid.

". . . *GENERAL WASHINGTON WAS EXPIRING* . . .": Ibid.; see also "The Medical History of George Washington (1732–1799)," *Mayo Clinic Proceedings Staff Meetings* 17 (1942).

WASHINGTON INSISTED: Worthington C. Ford, *The Writings of George Washington*, vol. 14 (New York: G. P. Putnam's Sons, 1893), pp. 246–49.

"OLD PEOPLE CANNOT BEAR BLEEDING . . .": Davies et al., "Cobbett, Rush," p. 914; for a detailed discussion see also John P. Carroll and Mary W. Ashworth, *George Washington*, vol. 7 (New York: Charles Scribner's Sons, 1957), pp. 617–17.

28    "DON'T YOU THINK . . .": Davies et al., "Cobbett, Rush," p. 914.

"I HAVE BEEN MUCH AMUSED . . .": Butterfield, ed., *Rush Letters*, p. 1211.

"BLEED *AD LIBITUM* . . .": J. Henry Clark, "Bloodletting in View of the Peculiarities of the Present Age," *Medical and Surgical Reporter* 9 (April 1958): 231.

FRENCH DOCTORS IMPORTED 41.5 MILLION LEECHES: Garrison, "History," p. 500.

"A GREAT DRAWER OF BLOOD . . .": Ibid.

28–29    GIOVANNI RASORI: Ibid.

29    THE *MUDFOG PAPERS*: Quoted in A. L. Pahor, "Charles Dickens and the Ear, Nose, and Throat," *Archives of Otolaryngology* 105, no. 1 (Jan. 1979): 1–5.

"I ALTOGETHER GAVE UP BLOOD-LETTING . . .": John H. Warner, "Therapeutic and the Edinburgh Bloodletting Controversy: Two Perspectives on the Medical of Science in the Mid-Nineteenth Century," *Medical History* 24 (1980): 241–58.

PIERRE-CHARLES-ALEXANDRE LOUIS: Walter R. Steiner, "Dr. Pierre-Charles Louis, a Distinguished Parisian Teacher of American Medical Students," *Annals of Medical History* 2 (1940): 451–60; see also Garrison, "History," pp. 500–501.

30 ". . . WE THOUGHT WE WERE RIGHT . . .": J. Worth Estes, "George Washington and the Doctors: Treating America's First Superhero," *Medical Heritage,* Jan.–Feb. 1985, pp. 43–57.

## CHAPTER 3: A STRANGE AGGLUTINATION

31 "*ALORS,* HOW CAN I DO THAT? . . .": T. W. Clarke, "The Birth of Transfusion," *Journal of History of Medicine,* Summer 1948, pp. 337–38.
". . . MY PATIENTS ARE ONLY MY DOGS AND CATS": Carrel related the incident to his roommate, another doctor, who later committed the story to paper. That account, plus Carrel's subsequent letters, provide a detailed description of the episode.

33 NEW YORK BECAME HOME: J. Hirsh and B. Doherty, *The First Hundred Years of the Mount Sinai Hospital of New York* (New York: Random House, 1952), pp. 92–103.
CARREL WAS A SHORT, INTELLIGENT, OPINIONATED MAN: R. J. Bing, "Carrel: A Personal Reminiscence," *JAMA* 250 (Dec. 23/30, 1983): 3297–98; W. S. Edwards and P. D. Edwards, *Alexis Carrel: Visionary Surgeon* (Springfield, Ill.: Charles C. Thomas, 1974).

35 WORLD-CLASS CENTER FOR MEDICAL RESEARCH: On the early history of Rockefeller University, see Edwards and Edwards, *Alexis Carrel,* pp. 38–42. (Until 1954 it was known as the Rockefeller Institute.)
A SIGHT THAT WAS "PITIFUL INDEED": Clarke, "Birth of Transfusion," p. 338.

36 "YOU'D BETTER TURN IT OFF . . .": L. G. Walker, "Carrel's Direct Transfusion of a Five Day Old Infant," *Surgery, Gynecology & Obstetrics* 137 (Sept. 1973): 494–96.
"I SHALL ALWAYS FEEL YOU SAVED MY LIFE . . .": Ibid., p. 496. Mary Robinson Lambert, the girl whom Dr. Carrel transfused, lived to be a happy and productive young woman. Years later, when she turned twenty-one, the Lamberts threw a dinner party at which Carrel was an honored guest. Mary went on to perform social work at the Neurological Institute in New York; she died of a sudden brain hemorrhage at the age of thirty-four.
"*SOUVENIR PRECIEUX*": Ibid., p. 496.
"REMARKABLE SURGICAL SUCCESS": Ibid., p. 495.
". . . THEY ARE TOO IGNORANT IN FRANCE . . .": Edwards and Edwards, *Alexis Carrel,* p. 61.

37 BLUNDELL INJECTED TWELVE TO FOURTEEN OUNCES: Harold W. Jones and G. Mackmull, "The Influence of James Blundell on the Development of Blood Transfusion," *Annals of Medical History* 10 (1928): 242–48; Geoffrey Keynes, ed., *Blood Transfusion* (Bristol: John Wright & Sons, 1941; London: Simpkin Marshall, 1949), p. 21.
"STRONG AS A BULL": Ibid., p. 245.
INVENTED A VARIETY OF INSTRUMENTS: V. Mueller & Co., Chicago, "Blood Transfusion Outfits" (advertisement); B. J. Ficarra, "The Evolution

of Blood Transfusion," *Annals of Medical History,* 3rd ser. 4 (1942): 302–23; J. H. Aveling, "On Immediate Transfusion," *Transactions of the Obstetrical Society of London* 6 (1865): 130.

37     DR. J. H. AVELING INVENTED AN APPARATUS: Aveling, "On Immediate Transfusion," p. 132.

"THE MENTAL IMPROVEMENT WAS NOT . . .": J. H. Aveling, "A Successful Case of Immediate Transfusion," *Lancet,* Aug. 3, 1872, pp. 147–48.

38     DR. ALFRED HIGGINSON: A. Higginson, "Report of Seven Cases of Transfusion of Blood, with a Description of the Instrument Invented by the Author," *Liverpool Medical and Chirurgical Journal* 1 (1857): 102–10.

TWO TRANSFUSIONS DURING THE CIVIL WAR: W. J. Kuhns, "Historical Milestones: Blood Transfusion in the Civil War," *Transfusion* 5 (Jan.–Feb. 1965): 92–94.

MILK TRANSFUSIONS: C. E. J. Jennings, *On Transfusion of Blood and Saline Fluids* (London: Baillière, Tindall, and Cox, 1888), pp. 106–15.

GESELLIUS FOUND THAT 56 PERCENT ENDED IN DEATH: Fritz Schiff, *Selected Contributions to the Literature of Blood Groups and Immunology,* vol. 4, pt. 2, *Blood Groups and Their Areas of Application* (Fort Knox, Ky.: United States Army Medical Research Laboratory, 1971) (translation of *Die Blutgruppen und ihre Anwendungsgebiete* [Berlin: Julius Springer, 1933]).

EMINENT PHYSICIANS DENOUNCED TRANSFUSION: Ibid., p. 180.

KARL LANDSTEINER: Biographical material from P. Speiser, F. Smekal, and G. Smekal, *Karl Landsteiner,* trans. R. Rickett (Vienna: Verlag Bruder Hollinek, 1975); M. W. Chase, "Notes About Dr. Karl Landsteiner," unpublished memo, n.d. (courtesy of Rockefeller Archive Center); P. Rous, "Karl Landsteiner," *Obituary Notices of Fellows of the Royal Society* 5 (March 1947): 71–124.

39–40     "A PECULIAR REGULARITY . . .": Karl Landsteiner, "Über Agglutinationserscheinungen normalen menschlichen Blutes," *Wiener Klinische Wochenschrift Klinische* 14 (1901): 1132–34, trans. in *Transfusion* 1 (Jan.–Feb. 1961): 5–8.

40     THEY LABELED THIS "AB.": Several years after Landsteiner's discovery, an American and a Czech researcher working without knowledge of each other or Landsteiner "discovered" the four blood groups as well. In 1907, the Czech, J. Jansky, named the groups I, II, III, IV, in order of their frequency. In 1910, the American, W. L. Moss, named them IV, III, II, I. The multiple naming systems caused considerable confusion until, at an international blood congress in 1937, the medical community settled on A, B, AB, and O.)

41     DR. GEORGE WASHINGTON CRILE: G. Crile, "The Technique of Direct Transfusion of Blood," *Annals of Surgery* 46 (Sept. 1907): 329–32.

A HANDSOME $500 FEE: Interview with Dr. Richard E. Rosenfield, Mount Sinai Hospital, New York.

"PSYCHIC FACTOR . . .": Crile, "Technique," p. 330.

". . . I MARVEL AT OUR RECKLESSNESS . . .": Bertram M. Bernheim, *Adventure in Blood Transfusion* (New York: Smith & Durrell, 1942), p. 83.

42 "SLOWLY, DELIBERATELY . . .": Ibid., pp. 15–16.

43 "ONE NEVER KNEW HOW MUCH BLOOD . . .": Reuben Ottenburg, "Reminiscences of the History of Blood Transfusion," *Journal of the Mount Sinai Hospital* 4 (1938): 268.
"HIGHLY EDUCATED, CULTURED WOMAN . . .": Bernheim, *Adventure,* pp. 73–74.

44 UNGER AND LINDEMAN FOUND THEMSELVES COMPETING: L. J. Unger, "Blood Transfusion—1914 Model," *Haematologia* 6 (1972): 47–57; Richard E. Rosenfield, "Early Twentieth Century Origins of Modern Blood Transfusion Therapy," *Mount Sinai Journal of Medicine,* 1974, pp. 626–35; interview with Dr. Richard E. Rosenfield.

45 OTTENBERG REDUCED THE ACCIDENT RATE TO ZERO: Reuben Ottenberg and David J. Kaliski, "Accidents in Transfusion," *JAMA* 61 (Dec. 13, 1913): 2138–40.
"A HEMORRHAGE BEGAN THIS MORNING . . .": R. Massie, *Nicholas and Alexandra* (New York: Bantam Books, 1967), p. 113.

46 HISTORY'S MOST FAMOUS HEMOPHILIA CARRIER: Ibid., pp. 146–52.
"GOD HAS SEEN YOUR TEARS . . .": Ibid., p. 186.
THE MONK WOULD TELL STORIES TO THE BOY: Doctors treat hemophilia with substances that promote coagulation, which did not exist in the tsar's time. It may also be true, though, that calming an agitated patient during an episode can help abate the bleeding.
"BE MORE AUTOCRATIC": Massie, *Nicholas and Alexandra,* p. 330.

47 "HE LIKES OUR FRIEND": Ibid., p. 344.
"NAME PROTOPOPOV": Ibid., p. 352.
"OUR FRIEND'S OPINIONS": Ibid., p. 353.
"FATE INTRODUCED HEMOPHILIA": Ibid., p. x.
A MODEL "OF LUXURY, CLEANLINESS AND ORDER . . .": Letter from Dr. Turner to Miss Benedict, Nov. 3, 1939, Archives of the Mount Sinai Medical Center.
A HOSPITAL STRICTLY FOR STRIVERS: Interview with Dr. Richard E. Rosenfield.

48 DR. RICHARD LEWISOHN: Interview with Dr. Richard E. Rosenfield; "Dr. Richard Lewisohn, 86, Dies; Discovered Blood Preservative," obituary, *New York Times,* Aug. 13, 1961.
"MUST WE ACCEPT THIS COAGULATION TIME . . . ?": R. Lewisohn, "The Development of the Technique of Blood Transfusion Since 1907," *Journal of the Mount Sinai Hospital* 10 (Jan.–Feb. 1944): 605–22.

49 "NOBODY HAD EVER FOLLOWED . . .": Ibid., p. 612.
"TECHNIC VERY SIMPLE . . .": Archives of Mount Sinai Medical Center, Historian's office files, Lewisohn file.
"HOW BEAUTIFULLY THIS DRUG RESPONDS . . .": Lewisohn, "Development, p. 612. Two other researchers came upon sodium citrate virtually at the same time—Albert Hustin of Belgium and Luis Agote of Brazil. Lewisohn receives most credit for the discovery because he precisely determined the safest concentration.

50 ". . . IT WAS ALMOST AS IF THE SUN . . .": Bernheim, *Adventure,* pp. 139–40.

## CHAPTER 4: BLOOD ON THE HOOF

53   OLIVER AND THREE CO-WORKERS RUSHED TO THE HOSPITAL: British
     Red Cross Society, *Report of the Blood Transfusion Service for the Year
     Ended Dec. 31st, 1926* (London: Petley & Co. Printers, n.d.), pp. 5–9. For
     a full account of the service's early years, see ibid., pp. 5–21; Geoffrey
     Keynes, ed., *Blood Transfusion* (Bristol: John Wright & Sons, 1941; Lon-
     don: Simplein Marshall, 1949), pp. 347–60). Additional information on
     Oliver's blood service from interview with Dr. Allen Waters.

54   HOSPITALS CALLED HIM 428 TIMES: British Red Cross, *Report*, p. 4.
     "[THE] WORK IS CONSTANT . . .": Ibid., p. 18.
     "CLERGYMEN ARE VERY COURTEOUS . . .": Ibid., p. 13.

55   "TRANSFUSION NATURALLY PROVIDED . . .": Geoffrey Keynes, *The Gates
     of Memory* (Oxford: Clarendon Press, 1981), p. 144.
     ". . . MY SUPERIORS WERE AFRAID . . .": Ibid., p. 189.
     "IT WAS SUPREMELY NOBLE . . .": British Red Cross Society, *Blood Trans-
     fusion Service Quarterly Circular,*" no. 13, Oct. 1936, p. 8.

56   "HOW MUCH BLOOD IS THERE . . .": British Red Cross Society, *Blood
     Transfusion Service Quarterly Circular,* no. 1, Oct. 1933, p. 3.
     "AN OCCASIONAL OBSTACLE . . .": British Red Cross Society, *Blood Trans-
     fusion Service Quarterly Circular,* no. 4, July 1934, p. 3.
     ". . . SHE PROMPTLY FAINTED . . .": Ibid.
     MORE THAN THREE THOUSAND CALLS PER YEAR: British Red Cross
     Society, *Blood Transfusion Service Quarterly Circular,* no. 16, July 1937,
     pp. 5, 8.

57   "READY TO BE SENT TO ANY HOSPITAL IN TOKYO . . .": Letter from Dr.
     H. Ijima to P. L. Oliver, Sept. 8, 1936, reproduced in British Red Cross
     Society, *Blood Transfusion Service Quarterly Circular,* no. 13, Oct. 1936.

57–58  "OUR 1,001ST WAY . . .": *New York Times,* Feb. 11, 1923.

58   MORE THAN 150: The rising popularity of selling blood can be seen in the
     following *New York Times* articles: "Ask Set Price for Blood," Oct. 22,
     1933; "150 Students at Michigan Give Blood to Pay Their Way," April 15,
     1925; "Sold Blood for Education," Jan. 28, 1924, "Sells His Blood to
     Wed," Sept. 20, 1923; "Yale Lists Blood Selling," Oct. 17, 1925; "Blood
     Donors Establish New and Lucrative Trade," Aug. 10, 1924.
     SOME MEN DANGEROUSLY DEPLETED THEMSELVES: "Giver of Blood
     Dies on Way to Find Work," *New York Times,* Dec. 14, 1924. Percy Oliver
     disdained the situation in New York. In a 1929 address to the Institute of
     Hygiene in London, he said: "Our service does not claim to be a 'Band of
     Heroes' but we can at least claim that we have saved London
     from . . . such scandals as apparently exist today in New York."
     HE KNEW IT WAS TIME TO LEAVE: P. Speiser, F. Smekal, and G. Smekal,
     *Karl Landsteiner,* trans. R. Rickett (Vienna: Verlag Bruder Hollinek,
     1975), p. 58.

59   ". . . FOR A MAN LIKE LANDSTEINER . . .": Letter from Dr. Storm Van
     Leeuwen, Pharmaco-Therapeutisch Instituut to Dr. Flexner, May 7, 1921,
     courtesy of Rockefeller Archive Center.
     "I WOULD LIKE A LITTLE COTTAGE BY THE SEA . . .": Speiser et al., *Land-
     steiner,* p. 61.

59     ". . . I AM JUST WORKING . . .": Ibid., p. 72.
       ". . . THE MASTER OF DEATH": E. Bendiner, "Karl Landsteiner: Dissector
       of the Blood," *Hospital Practice,* March 30, 1991, p. 102.

60     ALL DONORS WERE REQUIRED TO REGISTER: D. Stetten, "The Blood
       Transfusion Betterment Association of New York City," *JAMA* 110 (1938):
       1248–52.
       "KEEP HIMSELF IN GOOD PHYSICAL CONDITION . . .": Blood Transfusion
       Betterment Association Incorporated, "Information and Instructions to
       Blood Donors of the Blood Transfusion Betterment Association Incorpo-
       rated," 1929, courtesy of Rockefeller Archive Center.
       THE ASSOCIATION DISPATCHED A POLICEMAN: Blood Transfusion Bet-
       terment Association minutes, March 26, 1937, p. 3, courtesy of Rocke-
       feller Archive Center.
       ANSWERING MORE THAN NINE THOUSAND CALLS PER YEAR: Stetten,
       "Betterment Association," p. 1251. For an overview of the bureau's prin-
       ciples and practices, see pp. 1248–52.
       DONORS RECEIVED $35: Blood Transfusion Betterment Association, min-
       utes, Nov. 17, 1931, pp. 1–2; Betterment Association, "Information and
       Instructions," both courtesy of Rockefeller Archive Center.

61     IN ONE FAMOUS CASE: J. Thorwald, *Crime and Science: The New Fron-
       tier in Criminology,* trans. R. and C. Winston (New York: Harcourt, Brace,
       & World, 1967).
       IN RUSSIA: W. Schneider, "Chance and Social Setting in the Application of
       the Discovery of Blood Groups," *Bulletin of the History of Medicine* 57
       (1983): 553.
       GERMANY ACCEPTED SEROLOGICAL TESTS: Ibid., p. 553.
       A NOTORIOUS BABY-SWITCHING INCIDENT: The incident is described in
       Chicago *Daily Tribune,* July 19–26, July 28–31, Aug. 1, 9, 12, 19, 1930;
       June 9, 1931; Schneider, "Chance and Social Setting," pp. 552–53; A.
       Weiner, "On the Usefulness of Blood Grouping in Medicolegal Cases
       Involving Blood Relationship," *Journal of Immunology* 24 (1933): 450.

62     "THE LAWS OF HEREDITY": "Shuffled Babies Howl as Science Toils on
       Puzzle," Chicago *Daily Tribune,* July 23, 1930.

63     "ROUND AND HARD": Ibid.
       "OOCH!": Ibid.
       "I WOULDN'T CALL THESE FINGERPRINTS": "Doctors Decide Infants Are
       in Right Homes," Chicago *Daily Tribune,* July 25, 1930.
       "I'M SICK OF THIS": Ibid.
       "NOBODY IS GOING TO TAKE": "Bambergers Steal March by Baptism,"
       *New York Times,* July 28, 1930.

64     "I'LL SUE": Ibid.
       "I DON'T KNOW NOTHING": "Baby Mixup Is Still a Mixup to Mr.
       Watkins," Chicago *Daily Tribune,* July 26, 1930.
       "HYSTERICAL CONDITION": "Bambergers Steal March."
       "THIS CAN BE SOLVED": "Bambergers to Clinch Claim of Baby by Bap-
       tism," Chicago *Daily Tribune,* July 28, 1930.
       "TO SETTLE THE QUESTION": Ibid.

65     CHARLIE CHAPLIN: T. Huff, *Charlie Chaplin* (New York: Henry Schu-
       man, 1951), pp. 283–85; D. Robinson, *Chaplin: His Life and Art* (New
       York: McGraw-Hill, 1985), pp. 518–29.

65    CALIFORNIA DID NOT DO SO UNTIL 1953: Robinson, *Chaplin*, p. 528.

66    SERGE YUDIN CONDUCTED: S. S. Yudin, "Transfusion of Stored Cadaver
      Blood," *Lancet*, Aug. 14, 1937, pp. 361–66; Serge Judine, *La Transfusion
      du sang de cadavre à l'homme* (Paris: Editeurs Masson, 1933). (Yudin and
      Judine are alternate spellings.)

      "PHYSICIANS FORWARD": Interview with Dr. Vitaly Korotich.

      DR. ALEXANDER BOGDANOV: Biographical information from F. Schiff,
      *Selected Contributions to the Literature of Blood Groups and Immunol-
      ogy*, vol. 4, pt. 2, *Blood Groups and Their Areas of Application* (Fort
      Knox, Ky.: United States Army Medical Research Laboratory, 1971), pp.
      224ff; Y. N. Tokarev, I. Y. Maltseva, and G. D. Gloveli, "A. A. Bogdanov
      k 115-letiju so dnia rozhdenia," *Gematologiia i transfuziologiia*, Dec.
      1988, pp. 51–55.

      "HOW IS IT THAT OUR MEDICINE . . .": A. Bogdanov, *Red Star: The First
      Bolshevik Utopia*, trans. C. Rougle (Bloomington: Indiana University
      Press, 1984), pp. 85–86.

67    "PHYSIOLOGICAL COLLECTIVISM": Schiff, *Selected Contributions*, p.
      224.

      HE LINGERED WITH UREMIA: Ibid.; interview with Dr. Steven Tahan,
      Harvard Medical School.

      HE FINALLY DIED: Tokarev et al., "A. A. Bogdanov," pp. 51–55.

68    SHAMOV CONDUCTED A SERIES OF EXPERIMENTS: W. N. Shamov, "The
      Transfusion of Stored Cadaver Blood," *Lancet*, Aug. 7, 1937, pp. 306–9.

      SHAMOV PRESENTED HIS FINDINGS: Judine, *Transfusion*, pp. 7–10.

68–69  "I WAS CALLED OUT TO THE RECEIVING ROOM . . .": Yudin, "Transfu-
      sion," p. 361. For a complete account of Yudin's experiments, see Judine,
      *Transfusion*.

70    BY 1938: C. R. Drew, "The Role of Soviet Investigators in the Develop-
      ment of the Blood Bank," *American Review of Soviet Medicine* 1 (April
      1944): 360–69.

      "THE DANGER OF IMPURITIES . . .": Schiff, *Selected Contributions*, p.
      194.

      "BRITISH TEMPERAMENT . . .": British Red Cross Society, *Blood Transfu-
      sion Service Quarterly Circular*, no. 20, July 1938, p. 5.

      TWO DOCTORS IN THE CHICAGO AREA: Donald F. Farmer, "Transfu-
      sions of Cadaver Blood: A Contribution to the History of Blood Transfu-
      sions," *Bulletin of the American Association of Blood Banks*, June 1960,
      pp. 229–34. For experiments in India, see also G. N. Vyas, U. L. Munver,
      D. S. Salgaonkar, and N. M. Purandare, "Human Cadaver Blood for
      Transfusion," *Transfusion* 8 (July–Aug. 1968): 250–53.

      THE LEAD INVESTIGATOR WAS DR. JACK KEVORKIAN: Jack Kevorkian
      and Glenn W. Bylsma, "Transfusion of Postmortem Human Blood," *Amer-
      ican Journal of Clinical Pathology* 35 (May 1961): 413–19; Jack
      Kevorkian and John J. Marra, "Transfusion of Human Corpse Blood
      Without Additives," *Transfusion* 4 (March–April 1964): 112–17.

70–71  "IT WOULD BE PRESUMPTUOUS": Kevorkian and Bylsma, "Transfusion of
      Postmortem Human Blood," p. 418.

71    THEY INITIATED BLOOD STORAGE ON A NATIONAL SCALE: " 'Canned
      Blood' Adds Transfusion Values," *New York Times*, Nov. 28, 1937.

      IN 1937 ALONE: Ibid.

71 BERNARD FANTUS: Bernard Fantus, "The Therapy of the Cook County Hospital," *JAMA* 109 (1937): 128–31; M. Telischi, "Evolution of Cook County Hospital Blood Bank," *Transfusion* 14 (Nov.–Dec. 1974): 623–28.

## CHAPTER 5: PRELUDE TO A BLOOD BATH

72 DR. HANS SERELMAN: "Says Transfusion Can't Alter Race," *New York Times,* Oct. 20, 1935.

73 "TYPICAL JEWISH POSTURE": R. Proctor, *Racial Hygiene: Medicine Under the Nazis* (Cambridge, Mass.: Harvard University Press, 1988): 150.
"BEST NORDIC HEAD": Ibid., p. 95; Pauline M. H. Mazumdar, "Blood and Soil: The Serology of the Aryan Racial State," *Bulletin of the History of Medicine* 64 (1990): 209.
RECHE HAD CARVED OUT A NOTABLE CAREER: Proctor, *Racial Hygiene,* pp. 150–51.
LUDWIG HIRSZFELD: For biographical information, see Frank R. Camp, Jr., Ellis A. Fuller, and Kenneth I. Tobias, "Ludwik Hirszfeld: Physician, Scientist, Teacher (1884–1954): Relevancy of His Studies to Karl Landsteiner, Military Medicine, and Other Areas," *Military Medicine,* Feb. 1978, pp. 115–19; William H. Schneider, "Chance and Social Setting in the Application of the Discovery of Blood Groups," *Bulletin of the History of Medicine* 57 (1983): 545–62.

74 THE HIRSZFELDS COLLECTED ABOUT EIGHT THOUSAND BLOOD SAMPLES: Ludwik Hirschfeld and Hanka Hirschfeld [alternate spelling], "Serological Differences Between the Blood of Different Races," *Lancet,* Oct. 18, 1919, pp. 675–79.

75 "THE SEROLOGICAL FORMULA . . .": Ibid., p. 678.
OTTO RECHE AND HIS COLLEAGUES: A thorough account of Reche's serological research can be found in Mazumdar, "Blood and Soil," pp. 187–219, and in Schneider, "Chance and Social Setting," pp. 558–62.
"BEARERS OF POLISH NAMES . . .": Ludwig Hirszfeld, trans. *Constitutional Serology and Blood Group Research* (Fort Knox, Ky.: U.S. Army Medical Research Laboratory, n.d.), p. 154.
"BETTER TO RETAIL TRADE": Schneider, "Chance and Social Setting," p. 556.

76 "I WISH TO SEPARATE MYSELF . . .": Ibid., p. 561.
TWO "AGGLUTINATION POLES": Mazumdar, "Blood and Soil," p. 190.
THE ARYANS MUST CONQUER AND RESETTLE: As compelling as the blood-group research seemed, the Nazis never used it to carry out policy: It was simply too cumbersome to use against individuals, since any blood type can appear in any race. After the Germans conquered Poland, they set out to resettle it with "pure" Germans, in a program called Blood and Soil (Blut und Boden). The Reich's resettlement centers, which screened more than a million applicants for colonizing Polish lands, used the old-fashioned techniques of examining family trees and physical characteristics to determine who was a true Aryan. Nevertheless, Reche's work in serography illustrates the degree to which the Nazis perverted scholarship and science.

76      NUREMBERG BLOOD PROTECTION LAWS: An account of the laws and the
        German medical establishment's complicity can be found in Proctor,
        *Racial Hygiene,* pp. 65–94; and in Robert Jay Lifton, *The Nazi Doctors:
        Medical Killing and the Psychology of Genocide* (New York: Basic Books:
        1986).

77      "A FEW DAYS AGO . . .": Letter from Dr. Herman Nielsen to Dr. Karl Land-
        steiner, April 15, 1933, courtesy of Rockefeller Archive Center.
        THE MAN "HAS BEEN IMPRISONED . . .": Letter from Dr. Wilhelm
        Dressler to Dr. Karl Landsteiner, December 27, 1938, courtesy of Rocke-
        feller Archive Center.
        THE NUMBER OF JEWISH DOCTORS IN GERMANY: William Coleman,
        "The Physician in Nazi Germany," *Bulletin of the History of Medicine* 60
        (1986): 236.
        "NO MAN OF GERMAN BLOOD . . .": Proctor, *Racial Hygiene,* p. 91.

78      A TOLERANCE FOR QUACKERY: Coleman, "Physician," pp. 236–37.
        NORMAN BETHUNE: For information about Bethune's life and writings,
        see Roderick Stewart, *The Mind of Norman Bethune* (Westport, Conn.:
        Lawrence Hill, 1977), which contains an extensive collection of his let-
        ters; *New York Times,* Dec. 25, 1936; March 7, 1937.

79      "ALL VARIETIES AND KINDS": Stewart, *Norman Bethune,* p. 51.
        "WE COLLECT": Ibid., p. 57.
        "OUR NIGHT WORK": Ibid., p. 56.
        "A PRICK OF THE FINGER": Ibid., p. 52.
        "A GRAND COUNTRY": Ibid., p. 58.
        *"NADA":* Ibid.

80      "PLANS ARE UNDERWAY": Ibid., p. 57.
        "SIGN OF AN ENGAGEMENT": Ibid., pp. 63–64.

80–81   "DON'T EVER GET INVOLVED": John Gerasi, *The Premature Antifascists*
        (New York: Praeger, 1986), p. 105.

81      "THE INCESSANT STREAM": Stewart, *Norman Bethune,* p. 62.
        "WHAT WAS THE CRIME": Ibid., p. 63.
        AT LEAST 60 PERCENT OF BLOOD RECIPIENTS DIED: R. S. Saxton, "The
        Madrid Blood Transfusion Institute," *Lancet,* Sept. 4, 1937, pp. 606–7.

82      "IN THE WHOLE OF BARCELONA . . .": J. W. Cortada, ed., *A City in War:
        American Views on Barcelona and the Civil War, 1936–39* (Wilmington,
        Del.: Scholarly Resources, 1985), p. 132.
        LIKE THE RUSSIANS, HE REFRIGERATED THE BLOOD: F. D. Jorda, "The
        Barcelona Blood-Transfusion Service," *Lancet,* April 1, 1939, pp. 773–76.
        See also "Spain Bottles Blood," *New York Times,* June 20, 1937; P. H.
        Mitchiner and E. M. Cowell, "The Air-Raid: A Series of Articles on Med-
        ical Organisation and Surgical Practice in Air Attack," *Lancet,* Jan. 28,
        1939, pp. 228–31.

83      "A GREAT ADVANCE ON ANY SYSTEM . . .": "Stored Blood," *Lancet,* April
        1, 1939, p. 231.
        AN ESTIMATED NINE THOUSAND LITERS OF BLOOD: Jorda, "Barcelona
        Service," p. 773.

84      "NOT TO STORE BLOOD FOR LARGE-SCALE TREATMENT . . .": Alastair
        H. B. Masson, *History of the Blood Transfusion Service in Edinburgh*
        (Edinburgh and South East Scotland Blood Transfusion Association, n.d.),
        p. 24.

84      A COMBINED TOTAL OF EIGHT PINTS: Ibid.

85      "AN ATTRACTIVE WOMAN": Evelyn Irons, "The Undergraduate," in Pauline Adams, ed., *Janet Maria Vaughan, 1899–1993: A Memorial Tribute* (Dame Janet Vaughan Memorial Fund, n.d.); see also Max Blythe, "Dame Janet Vaughan DBE FRS in Interview with Max Blythe," Wheatley, Oxon (Oct. 1987), Royal College of Physicians and Oxford Polytechnic Medical Sciences Videoarchive, interviewer's copy; interviews with Drs. Patrick Mollison and Helen Dodsworth.

     "TOO STUPID": Helen Dodsworth, "Dame Janet Vaughan," unpublished ms., p. 31.

     ". . . THE ONLY BLOOD THAT WAS SHED [DURING] MUNICH . . .": Max Blythe, "Interview"; G. A. Elliott, R. G. Macfarlane, and J. M. Vaughan, see also "The Use of Stored Blood for Transfusion," *Lancet,* Feb. 18, 1939, pp. 384–87.

86      "THEY MUST ADMINISTER BLOOD . . .": J. M. Vaughan, "War Wounds and Air Raid Casualties," *British Medical Journal,* May 6, 1939, pp. 933–36.

     "THE CHILDREN USED TO GRUMBLE . . .": Max Blythe, "Interview."

     ". . . SAID I WAS PRETTY NAUGHTY . . .": Ibid.

     ". . . WE SHOULD LIKE TO BE ASSURED . . .": Masson, *History,* p. 26.

87      "START BLEEDING": Max Blythe, "Interview."

## CHAPTER 6: WAR BEGINS

88      A DISTINCTLY IMPROVISED FEEL: Interview with Dr. Patrick L. Mollison; Committee of Privy Council for Medical Research, *Medical Research in War* (London: His Majesty's Stationery Office, 1947), pp. 184–87.

89      "HE DIDN'T THINK . . .": Max Blythe, "Interview."

     ". . . 'HOW LIKE JANET TO SET UP IN A BAR!' ": Max Blythe, "Interview."

     "THE GYM WAS TURNED INTO A LABORATORY . . .": Interview with Dr. Patrick L. Mollison.

     "IT WAS GRATIFYING . . .": Committee of Privy Council, *Medical Research,* p. 184.

     ". . . SATISFIED WITH THE RESULTS": W. d'A. Maycock, "Blood Transfusion in the B.E.F.," *British Medical Journal,* Oct. 5, 1940, p. 467.

90      "ALL LONDON WAS BURNING": Constantine FitzGibbon, *The Winter of the Bombs* (New York: W. W. Norton, 1958), p. 45.

     "THE DELIVERIES ALL ARRIVED ON TIME . . .": Committee of Privy Council, *Medical Research,* p. 187.

91      DOCTORS LABORED UNDER NEAR-COMBAT CONDITIONS: J. M. Vaughan, "The Transfusion of Blood and Blood Derivatives Under Emergency Conditions," *JAMA* 123 (Dec. 18, 1943): 1020–25.

     "IT IS NOW WIDELY HELD . . .": Ibid., p. 1021.

91–92      "I WENT BACK TO HAVE A SECOND LOOK . . .": Max Blythe, "Interview."

92      LONDON WAS CONSUMING VAST QUANTITIES OF BLOOD: Committee of Privy Council, *Medical Research,* p. 188.

92–93      "IN SMALL EUTHANISTIC INSTITUTIONS . . .": W. S. Edwards and P. D. Edwards, *Alexis Carrel: Visionary Surgeon* (Springfield, Ill.: Charles C. Thomas, 1974), p. 100. For Carrel's friendship with Lindbergh, see "Paying Tribute to Dr. Carrel, Lindbergh Recalls His Days in the Lab," *American Medical News,* Aug. 20, 1973; J. D. Newton, *Uncommon Friend* (San Diego, New York, London: Harcourt Brace Jovanovich, 1987).

94    JOHN ELLIOTT: Biographical information from William DeKleine, *The History of the American National Red Cross,* vol. 33-B, *Early History of Red Cross Participation in Civilian Blood Donor Services and in the Blood Procurement Program for the Army and Navy* (Red Cross internal monograph) (Washington, D.C.: American National Red Cross, 1950), pp. 9–28; (courtesy of American Red Cross); interview with Dr. Paul Schmidt.

95    "[THE] RESPONSE OF THE PATIENT . . .": W. L. Tatum, J. Elliot, and N. Nesset, "A Technique for the Preparation of a Substitute for Whole Blood Adaptable for Use During War Conditions," *Military Surgeon,* Dec. 1939, pp. 481–89. For other references on early work with plasma, see M. M. Strumia and J. J. McGraw, "The Development of Plasma Preparations for Transfusions," paper read at American College of Physicians meeting, Boston, April 24, 1941; J. Elliott, "A Preliminary Report of a New Method of Blood Transfusion," *Southern Medicine and Surgery,* Dec. 1936, pp. 643–45; John Elliot, Walter L. Tatum, and George F. Busby, "Blood Plasma," *Military Surgeon,* Feb. 1941, pp. 118–28; DeKleine, *History of American Red Cross.*
      "PLASMA FOR BRITAIN": DeWitt Stetten, "The Blood Plasma for Great Britain Project," *Bulletin of the New York Academy of Medicine* 17 (Jan. 1941): 27–38.

96    DR. CHARLES DREW: Biographical information from Patrick P. Craft, "Charles Drew: Dispelling the Myth," *Southern Medical Journal* 85 (Dec. 1992): 1236–40, 1246; C. E. Wynes, *Charles Richard Drew: The Man and the Myth* (Urbana and Chicago: University of Illinois Press, 1988); W. Montague Cobb, "Charles Richard Drew, M.D., 1904–1950," *Journal of the National Medical Association* 42 (July 1950): 238–46.

96–97  "IN AMERICAN SURGERY . . .": Craft, "Charles Drew," p. 1238.

97    UNDER DREW, PLASMA FOR BRITAIN: Stetten, "Blood Plasma Project."
      ONE BATCH THAT TESTED CLEAN: Meeting of the Plasma Committee Blood Transfusion Betterment Association, "A Report on the Contaminated Plasma Discovered in England," Nov. 27, 1940, letter to Mr. George C. Smith from M. M. Davidson, Director, Blood Transfusion Betterment Association—Blood Plasma Division, Dec. 12, 1940. Both documents courtesy of Rockefeller Archive Center.
      "WHEN WE BEGAN THIS WORK . . .": Stetten, "Blood Plasma Project," p. 33.

98    ". . . OUR MAJOR TROUBLES HAVE VANISHED": Ibid., p. 37.
      NEARLY FIFTEEN THOUSAND PEOPLE: Report, Blood Transfusion Betterment Association. Jan. 31, 1941, p. 11, courtesy of Rockefeller Archive Center.
      THEY ACCEPTED BLOOD FROM NEGRO DONORS: Stetten, "Blood Plasma Project," p. 34.

99    ITS LEADERS THOUGHT IT BEST NOT TO COLLECT AFRICAN-AMERICAN BLOOD: "American National Red Cross Chronology of the Development of Blood Donor Service," American Red Cross internal doc., p. 3; Robert H. Fletcher, *The History of the American National Red Cross,* vol. 32A, *An Administrative History of the Blood Donor Service, American Red Cross*

*during the Second World War,* Red Cross internal monograph (Washington, D.C.: American National Red Cross, 1950), pp. 134–36, courtesy of American National Red Cross.

99      ON APRIL 1, 1950: Wynes, *Charles Richard Drew,* pp. 1–3, 103–5; Craft, "Charles Drew," pp. 1239–40.

100     "IS THAT DR. DREW?": Wynes, *Charles Richard Drew,* p. 104.
         PRESIDENT HARRY TRUMAN: "Text of Address by the President on Civil Rights," *New York Times,* June 14, 1952.
         DICK GREGORY TOLD THE STORY: Wynes, *Charles Richard Drew,* p. 107.
         "ALL THE BLOOD IN THE WORLD": Wynes, *Charles Richard Drew,* p. 106.

## CHAPTER 7: BLOOD CRACKS LIKE OIL

101     EDWIN J. COHN: Biographical information about Cohn, his work, and his times from interviews with Dr. John Ashworth, Dr. John T. Edsall, Dr. Henry Isliker, Dr. Douglas Surgenor; Louis K. Diamond, "Edwin J. Cohn Memorial Lecture: The Fulfillment of his Prophecy," *Vox Sanguinis* 20 (1971): 433–40; John T. Edsall, "Edwin Joseph Cohn, 1892–1953: A Biographical Memoir," reprinted from *Biographical Memoirs,* vol. 35 (New York: Columbia University Press, 1961), pp. 47–82; John Tilletson Edsall, "Transcripts of Interviews Sponsored by the Oral History Committee, Harvard Medical School, 1990–1991," Boston, 1992; G. Scatchard, "Edwin J. Cohn Lecture: Edwin J. Cohn and Protein Chemistry," *Vox Sanguinis* 17 (1969): 37–44; James Tullis, "Edwin Cohn: The Man and His Science," transcript of lecture, Nov. 1, 1990.

102     GREATEST MOBILIZATION OF SCIENTISTS TO DATE: For a comprehensive history of American science during the war and the Office of Scientific Research and Development, see J. P. Baxter 3rd, *Scientists Against Time* (Boston: Little, Brown, 1946).

103     "IN THE INTEREST OF CLEAR THINKING . . .": "Abstract of Minutes of Meeting, December 2, 1940, Office of Scientific Research and Development, Subcommittee of Blood Substitutes," in Subcommittee of Blood Substitutes and Subcommittee of Blood Procurement, Committee on Transfusions, *Bulletin on Blood Substitutes* (Washington, D.C.: National Research Council Division of Medical Sciences), vol. 1, May 31, 1940–Dec. 6, 1945, p. 12.

104     "THE PREPARATION OF LARGE AMOUNTS . . .": E. J. Cohn, "The History of Plasma Fractionation," *Advances in Military Medicine* 1 (1948): 368. This article, pp. 364–443, serves as a definitive reference on the development of fractionation.

105     "WE MADE CONNECTIONS . . .": Interview with Dr. Sam Gibson.
         "ALL OF THESE PATIENTS . . .": "Conference on Albumin: Minutes of the Meeting January 5th, 1942," in *Committee on Medical Research Bulletin on Blood Substitutes,* vol. 1, p. 151. For a full description of Ravdin's experience with albumin at Pearl Harbor, see Isidor S. Ravdin, "The Reminiscences of Isidor Ravdin," transcript of interviews conducted by S. Benson for the Oral History Research Office of Columbia University, 1955–62. For information on the condition of the wounded at Pearl Har-

bor, see "Report on Air Raid Attack by Japanese . . ." memo, U.S. Naval
Hospital, Pearl Harbor, Dec. 19, 1941, National Archives.

105     EIGHTY-SEVEN PEOPLE HAD RECEIVED ALBUMIN: "Conference on Albumin: Minutes January 5th, 1942," p. 154.

106     ". . . 'PACK YOUR BAGS . . .' ": Interview with Dr. James Lesh.

". . . PLASMA IS PRACTICALLY MIRACULOUS . . .": "As We Go To Press,"
*Red Cross Courier,* Feb. 1943, courtesy of American Red Cross.

107     "WHEN [SOLDIERS] KNOW . . .": "Plasma Paragraphs," *Red Cross Courier,*
Jan. 1944, courtesy of American Red Cross.

"WOMEN AT WAR!": P. J. Schmidt, "Tampa Chartered First Blood Bank,"
*Sunland Tribune: Journal of the Tampa Historical Society* 16 (Nov. 1990):
54–60.

108     A PREVIOUSLY FILLED BOTTLE: D. B. Kendrick, *Memoirs of a Twentieth-
Century Army Surgeon* (Manhattan, Kans: Sunflower University Press,
1992), p. 86.

"THOSE RECEIVING TRANSFUSIONS . . .": R. Fletcher, *The History of the
American National Red Cross,* vol. 33, *An Administrative History of the
Blood Donor Service, American Red Cross during the Second World War,* p.
134, courtesy of American Red Cross. On the issue of "Negro donors,"
see also pp. 135–37.

"THE PREJUDICE AGAINST NEGRO BLOOD . . .": "Blood and Prejudice,"
*New York Times.* June 14, 1942.

109     "MOST MEN OF THE WHITE RACE . . .": Fletcher, *History of the American
National Red Cross,* p. 136.

THE RED CROSS BANNED "BLACK" BLOOD: Ibid., p. 135.

AT THE SCHLITZ BREWING COMPANY: "Tapping America's Vein-Power,"
*Red Cross Courier,* Feb. 1944, courtesy of American Red Cross.

109–10     STREETCAR EMPLOYEES IN ST. PAUL, MINNESOTA: "Plasma Paragraphs,"
*Red Cross Courier,* April 1944.

110     "PLEASE, I WANT TO GIVE BLOOD . . .": Ibid.

"I CHANGED MY MIND": "Plasma Paragraphs," *Red Cross Courier,* Nov.
1943, courtesy of American Red Cross.

THEY WOULD COLLECT BLOOD IN THIRTY-FIVE CITIES: "Partnership
with the Wounded: The Blood Donor Service in Wartime," American Red
Cross internal doc. no. 494.2, Dec. 4, 1946, courtesy of American Red
Cross.

THE FOLLOWING YEAR: The increasing demand kept changing the target
throughout 1943. In March 1942, the military asked for nine hundred
thousand pints by July 1943. By January 1943, however, the military had
increased the demand to more than three million pints by July and then to
four million pints for the calendar year. Much to its credit, the Red Cross
had by the end of 1943 collected for processing some five million pints.
See Fletcher, *History of the American National Red Cross,* pp. 51–100.

NO LABORATORY WAS MORE THAN TWENTY-FOUR HOURS AWAY FROM
ITS SOURCE: "Wartime Blood Program," American Red Cross internal
doc. no. 81834, n.d., p. 62.

"THREW THE PACKAGE OUT A THIRD STORY WINDOW . . .": William
DeKleine, W., *The History of the American National Red Cross,* vol. 33-B,
*Early History of Red Cross Participation in Civilian Blood Donor Services
and in the Blood Procurement Program for the Army and Navy,* Red Cross

internal monograph (Washington, D.C.: American National Red Cross, 1950), p. 38, courtesy of American National Red Cross.

110      NAVY SUBJECTED THE KITS: Douglas B. Kendrick, *Blood Program in World War II* (Washington, D.C.: Office of the Surgeon General, Department of the Army, 1989), p. 168. This book is the definitive history of the wartime blood program.

111      THE PROCESS COULD TAKE FIFTEEN MINUTES OR MORE: Interviews with Dr. Sam Gibson, Dr. William Crosby, Dr. John Ashworth.
"THE CONCENTRATED ATTENTION . . .": Edsall, "Edwin Joseph Cohn," p. 66.

112      ". . . THERE WAS JUST DR. COHN BEATING ON US . . .": Interview with John Ashworth.
"THE YOUNG MAN BEING TOLD . . .": Diamond, "Edwin J. Cohn Memorial Lecture," p. 435.
"THE GOOD DOCTOR NEVER TWITCHED": John Lear, "You May Be Drafted to Give Blood," *Collier's,* Mar. 10, 1951, p. 58.
"IN ANY SCIENTIFIC GATHERING": Lear, "You May Be Drafted," p. 12.
THREE BOUND VOLUMES: Committee on Transfusions, *Bulletin on Blood Substitutes,* vols. 1–3.

113      A CHARACTERISTIC MEETING: "Conference on the Preparation of Normal Human Serum Albumin: Department of Chemistry, Harvard Medical School, June 6, 1942," in ibid., vol. 1, pp. 258–72.

114      ". . . WE OUGHT TO BE DOING OVER 25,000 . . .": Committee on Medical Research, *Bulletin on Blood Substitutes,* July 19, 1943, p. 758.
"LET US ALL TRY . . .": Ibid., p. 784.
TOTOFUSIN: G. Brock, "German Blood Substitutes," letter, *Lancet,* Dec. 6, 1941, p. 716.
PERISTON: "Conference of the Albumin and By-Products Group of the Sub-Committee of Blood Substitutes of the National Research Council, July 28, 1943," Sub-Committee on Blood Substitutes, in *Bulletin on Blood Substitutes,* vol. 2, p. 785; "Meeting of the Subcommittee on Blood Substitutes, Appendix C and D," Committee on Medical Research *Bulletin on Blood Substitutes,* pp. 871–75.
"SHOOTING OF FOUR PEOPLE . . .": *Trials of War Criminals Before the Nuremberg Military Tribunals Under Control Council Law No. 10* (Washington, D.C.: U.S. Government Printing Office 1952), vol. 1, p. 167.

115      THEY CONTINUED TO USE DONORS-ON-THE-HOOF: For an understanding of the status of German wartime transfusion, see Howard E. Snyder, "Inspections of German Hospitals," in Medical Department, United States Army, *Surgery in World War II: Activities of the Surgical Consultants* (Washington, D.C.: Office of the Surgeon General, Department of the Army, 1962), vol. 1, pp. 457–60; "Organization of Blood Donors in Germany," letter, *JAMA* 105 (Aug. 24, 1935): 610–11; "German Views on Blood-Transfusion," *Lancet,* Nov. 1, 1941, p. 533; F. Holle, "Die Technik der Bluttransfusion im Felde" ["The Technique of Blood Transfusion Under Battle Conditions"], *Zent. F. Chirurgie* 69 (June 13, 1942): 984–91, trans. and commentary in *Bulletin of War Medicine,* Jan. 1941, p. 276; F. Nöller, "Die Bluttransfusion unter besonderer Berücksichtigung der Blutkonservierung und des Trockenblutes" ["Blood Transfusions,

with Special Reference to Stored Blood and Dried Blood"], *Bruns' Beiträe z. klin Chirurgi* 173, no. 1 (May 18, 1942): 73–128, trans. and commentary in *Bulletin of War Medicine,* Jan. 1941, pp. 276–77; K. Lang and H. Schwiegk, "Erfahrungen mit . . . Serumkonserve und mit Plasma also Bluter satzmittel" ["Observations upon the Value of Serum and Plasma as Blood Substitutes"], *Deut. Millitärazt* 7 (June 1942): 379–84, trans. and commentary in *Bulletin of War Medicine,* March 1943, pp. 392–93.

115   "SENSELESS RACE THEORIES . . .": "Aryan Blood Demand Handicaps Nazi Wounded," Associated Press, March 1, 1942.

"WAR SURGERY CAN BE VERY GRIM . . .": E. R. Churchill, *Surgeon to Soldiers* (Philadelphia, Toronto: J. B. Lippincott, 1972), p. 379.

ALLIED SOLDIERS REPORTED GRUESOME EXPERIMENTS: G. Daws, *Prisoners of the Japanese: POWS of World War II in the Pacific* (New York: William Morrow, 1994), pp. 258–59.

THE DUTCH SET UP AN UNDERGROUND SYSTEM: Interviews with Dr. J. A. Loos, Head of Transfusion Department, and Dr. Johannes J. van Loghem, former Head of Transfusion Service, at the Dutch Red Cross; J. Spaander, "Dutch Blood-Transfusion Service During the German Occupation," *Lancet,* April 12, 1947, 494–95; G. G. A. Mastenbroek, *Hoe van het Een het Ander kwam* (Amsterdam: Nederlands Produktielaboratorium voor Bloedtransfusieapparatuur en Infusievloeistoffen BV Emmercompascuum, 1985) (booklet produced by Netherlands Red Cross Blood Processing Laboratory).

116   "THE GERMANS HAD NO INTEREST . . .": Interview with Dr. van Loghem.

THE SOVIET USE OF CADAVER BLOOD: Elmer L. DeGowin, "Report on Proposal to Use Cadaver Blood as a Source of Serum Albumin for the Armed Forces," app. C of "Meeting of the Subcommittee on Blood Substitutes, February 24, 1943," in Subcommittee on Blood Substitutes, *Bulletin on Blood Substitutes,* vol. 2, p. 608–9.

117   ". . . CADAVERS CAN SERVE THEIR COUNTRY ONLY ONCE": Ibid., p. 609.

"WHEN WAR BROKE OUT . . .": A. Bagdasarov, "Blood Transfusion in the U.S.S.R.," *British Medical Journal,* Oct. 17, 1942, p. 445.

"PETROV'S SOLUTION . . .": L. T. Blum, "Transfusion of Blood and Blood Substitutes in the USSR," *American Review of Soviet Medicine* 2 (Feb. 1945): 276–77; I. P. Petrov, P. N. Veselkin, M. L. Dernovzkaya et al., "The Comparative Value of Three Blood Substitutes," trans., *American Review of Soviet Medicine* 1 (June 1944): 450–55.

118   "FROM EARLY MORNING . . .": Bagdasarov, "Blood Transfusion," p. 446.

"THE USE OF THE WOMAN'S NAME . . .": W. Pennfield, "The British-American-Canadian Mission to the U.S.S.R.," *Canadian Medical Association Journal* 49 (1943): 455–61.

120   DR. CHARLES JANEWAY OF HARVARD REPORTED: "Abstract of Minutes of Meeting, April 19th, 1941," *Subcommittee on Blood Substitutes, Bulletin on Blood Substitutes,* vol. 1, p. 15.

"NO SIGNIFICANT REACTION . . .": "Abstract of Minutes of Meeting, March 10, 1942, Office of Scientific Research and Development, Committee on Blood Substitutes," in ibid., p. 183.

A SIXTY-TWO-YEAR-OLD MAN REACTED: "Conference on Bovine Albumin, Minutes of the Meeting, July 16, 1942," in ibid., pp. 308–14.

120 "SUDDENLY AND UNPREDICTABLY" DIED: "Conference on Albumin Testing: Minutes of the Meeting, October 19, 1942," in ibid., pp. 373–79.

121 "THE WHOLE GROUP HAD URGED . . .": Ibid.
TWO AND A HALF MILLION PACKAGES OF DRIED PLASMA: Albumin figures from "Conference of the Albumin and By-Products Group, Meeting of November 17, 1943," Subcommittee on Blood Substitutes, *Bulletin on Blood Substitutes,* vol. 2, p. 942. Plasma figures derived from "Production Report of Bleedings from Blood Centers, December 31, 1943," American Red Cross, in Fletcher, *History of the American National Red Cross,* p. 219.

## CHAPTER 8: BLOOD AT THE FRONT

122 DR. EDWARD "PETE" CHURCHILL: Biographical information from Edward D. Churchill, *Surgeon to Soldiers* (Philadelphia, Toronto: J. B. Lippincott, 1972); "Edward Delos Churchill" (obituary), *Harvard University Gazette* 69, no. 37 (June 7, 1974); interview with Richard Wolf, Francis A. Countway Library of Medicine, Harvard University Medical School, Boston.
THE MILITARY DISPATCHED A CADRE OF MEDICAL CONSULTANTS: For a full report of the consultants' activities, see Medical Department, United States Army, *Surgery in World War II: Activities of the Surgical Consultants* (Washington, D.C.: Office of the Surgeon General, Department of the Army, 1962), vol. 1.

123–24 "A SURGEON WOULD SAY . . .": Churchill, *Surgeon,* p. 180.
124 ". . . AS ONE DOES AN AUTOMOBILE RADIATOR": Ibid., p. 40.
125 "THE CORPSMEN HAD DONATED . . .": Ibid., p. 37.
"WE HAVE HAD ALL TYPES . . .": Ibid., pp. 113–14.
126 THE BRITISH HAD SET UP A LAYERED ORGANIZATION: L. E. H. Whitby, "The British Army Blood Transfusion Service," *JAMA* 124 (Jan. 12, 1944): 421–24; L. E. H. Whitby, "Transfusion in Peace and War," *Lancet,* Jan. 6, 1945, pp. 6332–34.
"WHEN I VISITED . . .": Churchill, *Surgeon,* pp. 49–51.
127 ON MARCH 24, 1943: E. Churchill, Headquarters North African Theatre of Operations, United States Army, Office of the Surgeon, "Memorandum on Whole Blood Transfusions," March 24, 1943, courtesy of Countway Library of Medicine.
"SIGNIFICANT PROPORTION OF THE WOUNDED": E. Churchill, North African Theatre of Operations, Office of the Surgeon, "Memorandum on Whole Blood Transfusion," April 16, 1943, courtesy of Countway Library of Medicine.
SURGEON GENERAL NORMAN T. KIRK: Biographical information from interviews with General Douglas M. Kendrick, M.D., Dr. William Crosby; "Norman Thomas Kirk," U.S. War Department biographical update, Oct. 25, 1946.
"A HUGE VESTED INTEREST . . .": Churchill, *Surgeon,* p. 47.
DOUGLAS B. KENDRICK, M.D.: Biographical information from Douglas B. Kendrick, *Memoirs of a Twentieth-Century Army Surgeon* (Manhattan,

Kans.: Sunflower University Press, 1992); W. Crosby, "World War II's War Within a War: What Delayed the Delivery of Whole Blood to Overseas Combat Troops?," *MD,* Dec. 1991, pp. 35–37.

128    "HE TURNED ME DOWN COLD . . .": Interview with General Douglas M. Kendrick, M.D.

CHURCHILL AND HIS MEN IMPROVISED: E. Churchill, North African Theatre of Operations, Office of the Surgeon, "Whole Blood Transfusion," April 16, 1943, courtesy of Countway Library of Medicine.

"YOU MUST BREAK THE STORY . . .": Churchill, *Surgeon,* p. 51.

128–29    "CHURCHILL HAD BALLS . . .": Interview with Dr. William Crosby.

129    "IT IS QUITE APPARENT . . .": Letter from Douglas B. Kendrick, War Department, Office of the Surgeon General, to Dr. Edward D. Churchill, Nov. 1, 1943, courtesy of Countway Library.

"WE WORKED *AROUND* KIRK . . .": Interview with General Kendrick; see also Crosby, "War Within a War."

A SECOND BLOOD BANK WAS SET UP: Kendrick, *Blood Program in World War II,* pp. 431–32.

THE SCENE WAS CHAOTIC: For a detailed description of the Anzio operation, see Howard E. Snyder, "Fifth U.S. Army," in Medical Department, United States Army, *Surgery in World War II: Activities of the Surgical Consultants* (Washington, D.C.: Office of the Surgeon General, Department of the Army, 1962), vol. 1, pp. 345–49, 417–19; R. L. Bauchspies, "The Courageous Medics of Anzio," *Military Medicine* 122 (Jan. 1958): 53–65; memo from Lt. Col. Samuel A. Hanser, Medical Corps, Subject "Blood Bank," Feb. 4, 1944, courtesy of the Countway Library of Medicine.

130    FLOWN OUT ON A DESIGNATED "BLOOD PLANE": Kendrick. *Blood Program in World War II,* pp. 417–20.

"TEMPORARY STRUCTURE OF ROUGH HEWN BOARDS AND BEAMS": Kendrick, *Memoirs,* p. 131.

"THEY MADE ME 'DICTATOR OF BLOOD' . . .": E. Benhamou, "Allocution de Monsieur le Professeur Benhamou de l'Académie de Médicin, Président de la Société de Transfusion Sanguine," in *Médicin-Général Jean Julliard 1902–1960: Allocutions et Notices* (Paris: Masson, 1961), p. 17.

131    "MOTHER HOUSE" ESTABLISHED IN ALGIERS: For a description of the blood-collection efforts of the Free French in North Africa, see E. Benhamou, "Notes pour servir à l'histoire de la transfusion sanguine dans l'armée française de 1942 à 1945 à partir de l'Afrique du Nord," *Revue des corps de santé,* no. 7 (special), 1966, pp. 859–62; Y. Burguet, "Le Centre de transfusion sanguine de Fés (Maroc) (1943–1958)," *Revue des corps de santé,* no. 7 (special), 1966, pp. 863–73; J. Blomet, "La Transfusion-réanimation aux armées (O.R.T. 1 et 2)," *Revue des corps de santé,* no. 7 (special), 1966, pp. 875–85.

"EXPLAINING THE PURPOSE . . .": Benhamou "Notes pour servir," p. 860.

132    THE CHARGES WERE FALSE: Even as his friend Charles Lindbergh was urging nonintervention in the late 1930s, Carrel denounced the Nazis as "rejecting classical culture, Christianity, the sacredness of human personality and liberty. . . ." He labeled Hitler a "prodigious phenomenon in the history of humanity—an uncanny and gigantic power more audacious than Tamerlane and Genghis Khan . . . a clairvoyant who senses the

future, who reaches his goal through cunning, crime and bloodshed with somnambulistic cruelty." W. S. Edwards, *Alexis Carrel: Visionary Surgeon* (Springfield, Ill.: Charles C. Thomas, 1974), p. 112.

132    INSURE HE HAD NOT FLED: Ibid., p. 121. For details on Carrel's last days, see also pp. 110–22.

133    A MEDICAL STOCKPILE HAD BEEN GROWING: Medical Department, United States Army, *Medical Supply in World War II* (Washington, D.C.: Office of the Surgeon General, Department of the Army, 1968), pp. 265–304; Medical Department, United States Army, *Surgery in World War II: Ophthalmology and Otolaryngology* (Washington, D.C.: Office of the Surgeon General, Department of the Army, 1957), p. 34.
"ORGANIZED CONFUSION": Medical Department, United States Army, *Medical Supply,* p. 274.
WHOLE BLOOD SERVICE COMMITTEE: For a detailed account of the blood-bank-related planning for D-Day, see Elliott Cutler, "Chapter II: The Chief Consultant in Surgery," in Medical Department, United States Army, *Surgery in World War II,* vol. 2, pp. 19–298; J. B. Mason, "Planning for the ETO Blood Bank," *Military Surgeon,* June 1948, pp. 460–68; Kendrick, *Blood Program in World War II,* pp. 485–568.
"A GREAT SHOUTER": Interview with Dr. William Crosby.

134    A STANDARD PARAMETER OF D+90: Mason, "Planning," p. 465.

134–35    "I AM WORRIED . . .": Cutler, "Chapter II," p. 151.

135    "CATCHING UP; THINGS MOVING . . .": Ibid., p. 181.
THEY SAT ON A DOCK IN NEW YORK: Ibid., pp. 154–55.

136    AN "ALARMING REDUCTION . . .": Mason, "Planning," p. 200.
THE OPENING OF THE BLOOD BANK WOULD SIGNIFY: Kendrick, *Blood Program in World War II,* pp. 548–49.

137    "I DON'T KNOW WHEN D-DAY IS . . .": Cutler, "Chapter II," p. 202.
250 BOTTLES OF WHOLE BLOOD THAT DAY: Ibid., p. 485.
"THE CONTINENTAL INVASION IS ON AT LAST . . .": Cutler, "Chapter II," p. 207.
"THE TREMENDOUS DEMAND FOR BLOOD . . .": Ibid., p. 234.

138    THE ARMY'S NEED FOR WHOLE BLOOD: Ibid.
"TO MAKE ONE MORE TRY . . .": Ibid., p. 239.
"THE SURGEON GENERAL IS DEFINITELY OPPOSED . . .": Ibid., p. 240; see also Crosby, "War Within a War," p. 36.
"BURDEN IS BEING IMPOSED . . .": Crosby, "War Within a War," p. 36.
ABOUT 250 PINTS A DAY: Cutler, "Chapter II," p. 244; Crosby, "War Within a War," p. 36.
". . . IF THE SURGEONS OF THE E.T.O. . . .": Cutler, "Chapter II," p. 244.

139    HE ESTABLISHED NATIONAL "DAYS OF BLOOD": J.-P. Cagnard, *La Transfusion sanguine française* (Paris: Ministre de la Santé, 1987).
AIR-DROPPED A THOUSAND BOTTLES OF BLOOD INTO WARSAW: W. N. B. Watson, *The Scottish National Blood Transfusion Association 1940–1965* (Edinburgh and London: E. & S. Livingstone, 1965), pp. 17, 31.
THE RED CROSS BEGAN COLLECTING TYPE O BLOOD: Robert H. Fletcher, *The History of the American National Red Cross,* vol. 32-A, *An Administrative History of the Blood Donor Service, American Red Cross*

*during the Second World War,* Red Cross internal monograph (Washington, D.C.: American National Red Cross, 1950), pp. 115–17, courtesy of American Red Cross.

140      "NO COMMAND . . .": Kendrick, *Blood Program in World War II,* pp. 602–5.

BLAKE ENCOUNTERED A YOUNG SOLDIER: "How Your Blood Saves Soldiers' Lives," San Francisco *Chronicle,* Dec. 15, 1944.

IT BECAME COMMON ON LUZON: Kendrick, *Blood Program in World War II,* pp. 634–35.

140–41    BATTLE OF IWO JIMA: *The History of the Medical Department of the United States Navy in World War II* (Washington, D.C.: U.S. Government Printing Office, 1950), vol. 1, pp. 89–104; B. D. Ross, *Iwo Jima: Legacy of Valor* (New York: Vintage Books, 1985).

141      THE MEDICS ROSE TO HEROIC LEVELS: *History of Medical Department,* pp. 92–93, 105.

142      "THE MOST PRECIOUS CARGO . . .": R. W. Myers, "Lifesaving Blood Flows on Iwo, Thanks to Last Month's Donors," *New York Times,* March 2, 1945.

PHARMACIST'S MATE JOHN H. WILLIS: Ross, *Iwo Jima,* pp. 241–42.

143      "COMBAT MEDICS IN BOUNCING JEEPS . . .": Kendrick, *Blood Program in World War II,* p. 59.

"HOLDING PLASMA BOTTLES . . .": "Plasma Paragraphs," *Red Cross Courier,* Apr. 1945, courtesy of American Red Cross.

"TO LEAVE HIM WITHOUT TREATMENT . . .": Kendrick, *Memoirs,* p. 637.

144      "THE CORPORAL HAS THE BLOOD OF A GENERAL": *New York Times,* Feb. 12, 1945.

HARRY STARNER WAS RECEIVING SOME PLASMA: "Gets His Own Blood," *Red Cross Courier,* Nov. 1943, courtesy of American Red Cross.

## CHAPTER 9: DR. NAITO

147      LIEUTENANT COLONEL MURRAY SANDERS: P. Williams and D. Wallace, *Unit 731: Japan's Secret Biological Warfare in World War II* (New York: Free Press, 1989). The definitive accounting of Japan's human experiments, and the attempts of Sanders and others to uncover them.

DR. RYOICHI NAITO: For biographical information on Dr. Naito, see ibid.; Midori Juji, *Midori Juji Sanjyu-nenshi* [*Thirty-Year History of Green Cross Company*] (Tokyo: Toppan Insatsu Kabushiki Gaisha, 1980); Fusao Ikeda, *Shiroi Ketsueki* [*White Blood*] (Tokyo: Ushio Shuppansha, 1995).

150      SHIRO ISHII HEADED A GERM-WARFARE RESEARCH FACILITY: Williams and Wallace, *Unit 731.*

151      "THE VERY STRONG RIGHT ARM OF DR. ISHII . . .": Ikeda, *Shiroi Ketsueki,* pp. 101–2.

THE SWISS RED CROSS DELIVERED PACKAGES OF DRIED AMERICAN PLASMA: Ibid., p. 107.

NAITO MEMOIRS: Ryoichi Naito, *Rou SL no Souon* [*Noises of an Old Locomotive*] (Tokyo: Daiwa Toppan Kougei Insatsu Kabushiki Gaisha, 1980).

151 "... THE ATMOSPHERE OF A FLOWER GARDEN": Ibid., p. 7.
"TYPHOON": A. Spaeth, "A Demanding Boss," *Forbes,* Oct. 29, 1979, p. 156.
"HIS MIND WORKED SO QUICKLY . . .": Interview with former Alpha Therapeutic Corporation executive.

152 ORDERED HIS SOLDIERS TO SANITIZE THE AREA: Williams and Wallace, *Unit 731,* pp. 84–85.
"JAPAN IS DEFEATED . . .": Ikeda, *Shiroi Ketsueki,* pp. 109–10. Naito kept his promise. He convinced a large Japanese corporation with a chemical division to open a Nigata branch, where most of the workers remained.

153 "... ONLY TO RESCUE OUR POOR, DEFEATED NATION . . .": R. Naito, "Private (Secret Information) to Colonel Sanders," reproduced in Williams and Wallace, *Unit 731,* app. A, pp. 257–61.
"... HE 'VOWS' THIS HAS NEVER BEEN THE CASE": Ibid., p. 133.
INGINBUREI: Ikeda, *Shiroi Ketsueki,* pp. 114–15.

154 THE AMERICANS RESPONDED: An account of the growth of Japan's post-war blood industry, and America's role, can be found in R. Naito, "Experiences in the Development of Plasma Derivatives in Japan," *Vox Sanguinis* 23 (1972): 35–37; interview with Dr. Chiaki Myoshi, Bureau of International Cooperation, International Medical Center of Japan, Ministry of Health and Welfare; Naito, *Rou SL no Souon;* Midori Juji, *Midori Juji;* C. F. Sams, "Medic," unpublished ms., Uniform Services, University of Health Science, n.d., pp. 498–558.
IN 1948 A WOMAN CONTRACTED SYPHILIS: Naito *Rou SL no Souon,* p. 10; interview with Dr. Myoshi.

155 "THE PRINCIPLE OF FREE BLOOD . . .": Z. S. Hantchef, "The Red Cross and Blood Transfusion," *Vox Sanguinis* 2 (1957): 138.
EACH NATION EMBARKED ON ITS OWN COURSE: For a detailed overview of the international postwar blood situation, see the informational booklet J. Julliard and Y. Menasché, *Organisation de la transfusion sanguine dans divers pays: Une Enquête internationale* (Bordeaux: Union Française d'Impression, n.d.).
IN GERMANY, BLOOD BANKS BECAME ESTABLISHED HAPHAZARDLY: O. Sifrin, "Die Entwicklung der Transfusionsmedizin in der Bundesrepublik Deutschland nach dem zweiten Weltkrieg am Beispiel von Marburg," unpublished Ph.D. thesis; Volkmar Sachs, "Bluttransfusionswesen heute und in Zukunft," *Soziale Medizin und Hygiene* 110 (1968): 218–24; interviews with Karen Buchner, Dr. Alfred Haessig, Ursula Lassen, Dr. Heinz Schmitt.
IN EAST GERMANY: W. Scheffler and K. Thomas, "Organization of the Blood Donor Service in the German Democratic Republic and Its Present Problems," *Bibliotheca Haematologica,* no. 38, pt. 2, 1971, pp. 13–14.

156 IN SWITZERLAND: Interview with Dr. Alfred Haessig.
THE DUTCH CONTINUED TO ASTOUND: Interviews with Dr. J. A. Loos and Dr. Johannes J. van Loghem.

157 THE CANADIANS DEVELOPED A BLOOD SYSTEM: Julliard and Menasché, *Organisation,* pp. 7–9; G. A. McVicar, "The Development of Plasma Derivatives in Canada," *Vox Sanguinis* 23 (1972): 33–34.
IN ITALY: Julliard and Menasché, *Organisation,* pp. 11–12; F. Peyretti, "In anno di esperienze alla 'Banca del Sangue e del Plasma della Città di

Torino," *Giornale Italiano di Anestesidegia* 15 (July–Sept. 1949): 181–88; "Rome Has Blood Black Market," *New York Times,* Oct. 27, 1951.

157 PEKING UNION MEDICAL COLLEGE: W. S. Lu, T. Fan, Y. L. Howe, "The Stored Blood Transfusion Service of the Peking Union Medical College Hospital," *Chinese Medical Journal* 66 (Oct. 1949): 555–67.

IN HONG KONG: R. N. Fraser, "Hong Kong," *How to Recruit Voluntary Blood Donors in the Third World?* (Geneva: League of Red Cross and Red Crescent Societies, 1984), p. 14.

IN THAILAND: "Thai Royalty to Give Blood," *New York Times,* Sept. 27, 1952.

158 "WHEN I USED TO GO TO ARMY UNITS . . .": G. W. Bird, "Observations . . . 1944–1964," in *How to Recruit Donors,* p. 4.

SOME DONORS FELL INTO A TRANCE: Ibid., p. 6.

"THEIR SOUL PUT INTO A BOTTLE": S. Baba, "Ivory Coast," in *IIIrd Red Cross International Seminar on Blood Transfusion* (Stockholm: 1964), p. 7.

"ONE NIGHT, AN AFRICAN PASSED BY A TENT . . .": G. Bolton, "Tanganyika," in *IIIrd Red Cross Seminar on Transfusion,* p. 9.

THE NEWLY ESTABLISHED NATIONAL BLOOD SERVICE: H. Gunson and H. Dodsworth, "Fifty Years of Blood Transfusion," *Transfusion Medicine* 6, suppl. 1 (1996): pp. 15–24.

159 NO COUNTRY APPROACHED FRANCE: J. P. Cagnard, *La Transfusion sanguine française* (Paris: Ministre de la Santé, Imprimerie Martinenq-Ivry, 1987), pp. 4–6; M. Aujaleu, and Laporte, "Le Développement de la transfusion sanguine en France au cours des dix dernières annees," *Bulletin de l'Académie Nationale de Médecine,* Nov. 8–15, 1955, pp. 495–97.

160–61 "WE PAID FOR HEALTH INSURANCE . . .": Naito, *Rou SL no Souon,* p. 15.

161 THE JAPANESE RED CROSS TOOK IN JUST OVER FIVE HUNDRED: Ikeda, *Shiroi Ketsueki,* p. 125.

A FOCAL POINT FOR THE DISENFRANCHISED: Ibid., p. 39ff.

162 "THE CONCEPT THAT . . .": Naito, *Rou SL no Souon,* pp. 13–14.

HE MADE AN AGREEMENT WITH CUTTER LABORATORIES: Midori Juji, *Midori Juji,* p. 28.

SALES EXCEEDED $500,000: Ibid., p. 40.

## CHAPTER 10: DR. COHN

164 "THE EVENT BEGAN FAULTLESSLY": J. L. Tullis, "Cellular Preservation and Interaction of Cells and Coagulation Proteins," in D. H. Bing, ed., *The Chemistry and Physiology of the Human Plasma Proteins: Proceedings of a Conference Held 19–21 November 1978 in Boston, Massachusetts, Sponsored by the Center for Blood Research* (New York: Pergamon Press, 1979), p. 11.

165 "ALMOST BEYOND ESTIMATE . . .": W. Laurence, " 'Life Elixirs' Seen in Blood Advances," *New York Times,* July 12, 1950.

"LOOK AT THE GOOD THINGS WE CAN DO . . .": Interview with Dr. Sam Gibson.

166 MEETING IN CLEVELAND IN JUNE 1947: G. F. McGinnes, "National Blood Program Inaugurated," *Red Cross Courier* 27, no. 1 (July 1947): 3, courtesy of American Red Cross.

166      THE NUMBER OF BLOOD GROUPS HAD GROWN DRAMATICALLY: Landsteiner believed that so many blood antigens would eventually be discovered that every person would be found to possess a unique blood type. That turned out not to be true, although more than two hundred blood types and subtypes have been found.

         RH PROTEIN TRIGGERS A MILD IMMUNE RESPONSE: Two disciples of Landsteiner's at the Rockefeller Institute, Drs. Philip Levine and Alexander Weiner, discovered Rh. They found it so difficult to share credit for their discovery that they remained bitter enemies for most of their lives.

167      "YOU'RE GOING TO GO DOWN . . .": L. K. Diamond, "The Reminiscences of Louis K. Diamond," unpublished ms., Columbia University Oral History Project, pp. 61–62. More on Diamond's work with erythroblastosis fetalis can be found in L. K. Diamond, "Erythroblastosis Fetalis, VII, Treatment with Exchange Transfusion," *New England Journal of Medicine* 244 (Jan. 11, 1951): 39–49; L. K. Diamond, "Historic Perspective of 'Exchange Transfusion,' " *Vox Sanguinis* 45 (1983): 333–35.

168      ONE AND A QUARTER MILLION PACKAGES: "Surplus Plasma," *Red Cross Courier,* April 1946, courtesy of American Red Cross.

         THE GRAND OPENING: G. Korson, "Rochester Makes History," *Red Cross Courier* 27, no. 9 (March 1948): 3–6; G. Korson, "Attica Does Its Part," ibid., pp. 8–10, all courtesy of American Red Cross; internal documents, courtesy of Rochester–Monroe County Chapter of the American Red Cross.

169      "WITH ALL THE TENDERNESS AT HER COMMAND . . .": Korson, "Rochester Makes History," p. 6.

         ". . . THERE IS NO DIFFERENCE IN THE BLOOD . . .": S. Gibson, "Racial Designation of Blood Donors in the Red Cross Blood Program," American Red Cross, internal memo, July 18, 1958; see also "Red Cross Plans Big Blood Supply," *New York Times,* June 10, 1947.

170      "COLLECT AND HOLD BLOOD . . .": P. Maas, "The Red Cross Answers Its Critics," *Look,* March 28, 1961, p. 86.

         "IN CONFORMITY WITH APPLICABLE STATE LAWS": Ibid.

171      ". . . THE SPIRIT OF THE PIONEER STILL LIVES . . .": "Other Regional Blood Centers Opened," *Red Cross Courier,* March 1948, p. 13, courtesy of American Red Cross.

         "AS YOUR ANCESTORS WERE PIONEERS . . .": Ibid.

         "THERE CAN BE NO RECESSION . . .": "ARC Prepares to Go Forward," *Red Cross Courier* 27, no. 1 (July 1947): 5, courtesy of American Red Cross.

         "UPON US [THE DOCTORS] RESTS THE RESPONSIBILITY . . .": J. Scudder, "The Blood Bank Controversy," letter, *New York Times,* Feb. 23, 1948.

172      "THEY'VE BOMBED PEARL HARBOR! . . .": Interview with Bernice Hemphill.

         "I DIDN'T KNOW WHERE MY HUSBAND WAS . . .": Ibid.

173      UPTON LAID THE GROUNDWORK: J. R. Upton, "An Integrated System of Community Blood Banks in California," *Western Journal of Surgery* 58 (1950): 380–85.

173–74    "WE HAVE A FEW PROBLEMS . . .": Ibid., p. 385.

174      "BIRTHRIGHT" . . . "OUTSIDE AGENCY": Ben Pearse, "What's Wrong with Our Blood Banks?" *Saturday Evening Post,* March 24, 1959, p. 96.

174    "NATIONALIZATION OF THE BLOOD BANKS": "National Blood Bank Is
        Voted by Doctors," *New York Times,* Nov. 18, 1947.
        NATIONAL BLOOD PROGRAM AS SOCIALISTIC: Pearse, "What's Wrong,"
        p. 96.
        "WHEREAS . . .": "Summary of Business Session, Wednesday, Nov. 19,
        1947," in Blood Bank Institute Sponsored by Wm. Buchanan Blood Cen-
        ter of Baylor Hospital: Baker Hotel, Dallas, November 17–19, 1947 (min-
        utes of meeting), p. 176. For a definitive account of the conflict between
        the American Red Cross and the AABB, see Louanne Kennedy, "Commu-
        nity Blood Banking in the United States from 1937–1975: Organizational
        Transformation and Reform in a Climate of Competing," unpublished
        Ph.D. thesis, New York University, Feb. 1978.

175    "IF THE TWO ORGANIZATIONS COULD ONLY . . .": B. Pearse, "What's
        Wrong," pp. 37ff.
        THE NATIONAL BLOOD CLEARINGHOUSE: Interview with Bernice
        Hemphill.

177    DRIED PLASMA HAD BEEN CONTAMINATED WITH HEPATITIS: Interview
        with Dr. John Ashworth; E. Cohn, "Facts About Blood and the Red Cross
        Program," summary of address given at the Massachusetts Regional Blood
        Center of the American National Red Cross, Aug. 5, 1950; letter from
        Norman T. Kirk, Surgeon General, to Basil O'Connor, Chairman, Ameri-
        can Red Cross, Feb. 13, 1946, courtesy of Francis A. Countway Library of
        Medicine; *Parker v. State,* Court of Claims of New York, 105 N.Y.S. 2d
        735 (1951) (court findings on the case of a patient infected with hepatitis
        from the surplus plasma).

178    CHARLES JANEWAY EXPLAINED HOW THE BLOOD ECONOMY COULD
        WORK: C. Janeway, "Clinical and Immunological Control of Biologic
        Products," paper presented at joint meeting of Protein Foundation, Inc.,
        Commission on Plasma Fractionation and Related Processes, and Univer-
        sity Laboratory of Physical Chemistry Related to Medicine and Public
        Health, Dec. 11, 1953, courtesy of Center for Blood Research.

179    "IF AN ATOMIC BOMB . . .": E. J. Cohn, "History of the Development of
        the Scientific Policies of the University Laboratory of Physical Chemistry
        and Public Health, Harvard University," 1951, p. 13, courtesy of Center
        for Blood Research.

179–80 NEARLY 140 SCIENTISTS: Interview with Dr. Douglas Surgenor; *The
        Preservation of the Formed Elements and the Proteins of the Blood,* "Con-
        ference called at the request of the Committee on Medical Sciences of the
        Research and Development Board of the National Military Establishment
        by the Committee on Blood and Blood Derivatives of the National
        Research Council," Jan. 6, 7, 8, 1949 (Boston: Harvard Medical School,
        1949); Cohn, "History of the Development," p. 13.

180    DR. CARL WALTER: Biographical information from interview with Dr.
        Surgenor; J. R. Brooks, "Carl W. Walter, MD: Surgeon, Inventor and
        Industrialist," *American Journal of Surgery* 148 (Nov. 1984): 555–58. For
        detailed information on the development of the blood bag, see C. W. Wal-
        ter, "Invention and Development of the Blood Bag," *Vox Sanguinis* 47
        (1984): 318–24.
        ". . . THERE MUST BE A BETTER WAY! . . .": "There Must Be a Better
        Way!," *Focus: News of the Harvard Medical Area,* June 18, 1981, p. 5.

181    HE MARCHED INTO COHN'S LAB: Interview with Dr. Surgenor; Lamar
       Souter and Douglas M. Surgenor, "Reflections on Blood Transfusion,"
       *American Journal of Surgery* 148 (Nov. 1984): 563. The throwing of the
       blood bags became a rich source of medical lore, with various people hav-
       ing assorted misadventures. On one occasion, Lloyd Newhouser, the
       navy's blood chief, demonstrated the bag to a visiting admiral by throwing
       a fully laden one at the official's feet. Unfortunately, that particular bag
       had a flaw, and blood splattered all over the admiral's white pants.
182    BAXTER LABORATORIES: Baxter had made its mark early in blood-
       transfusion equipment. Its Transfuso Vac container, the first vacuum-sealed
       blood-and-plasma bottle, had been standard equipment during the war.
       "VERITABLE WONDERLAND ON WHEELS": W. L. Laurence, "Mobile
       Blood Unit Held Defense Aid," *New York Times,* Oct. 12, 1950; see also
       "Vital Fractions," *Time,* Oct. 23, 1950, p. 95.
       HIS TEAM PRODUCED A DOZEN OF THESE PROTOTYPES: The machine,
       however, never saw wide use. Cohn's original technologies turned out to
       be more practical and less likely to transmit hepatitis.
183    "IT'S A GREAT LIFE . . .": John Lear, "You May Be Drafted to Give Blood,"
       *Collier's,* March 10, 1951, p. 59.
       ". . . 'GO AHEAD, SHOOT IT ALL IN' ": Interview with Dr. Sam Gibson.
       COLLEAGUES RECALL A COLLOQUIUM: Interviews with Dr. Henry
       Isliker, Dr. John Edsall.
184    "GEORGIE, I CAN'T HEAR YOU": G. Scatchard, "Edwin J. Cohn Lecture:
       Edwin J. Cohn and Protein Chemistry," *Vox Sanguinis* 17 (1969): p. 39.
       ". . . MY EMOTIONS WERE VERY MIXED": Interview with Dr. John Edsall.
185    NOT ONE MAN BUT TWO: Scatchard, "Cohn Lecture," p. 39.
       ". . . ONE OF THE GLORIES OF AMERICAN SCIENCE . . .": "What Blood
       Told Dr. Cohn," *New York Times,* Oct. 6, 1953.

## CHAPTER 11: THE BLOOD BOOM

186    MARGARET WAS A "REGISTERED NURSE": Even though Margaret Bass
       represented herself as a nurse and appeared in a nurse's uniform in her
       company's literature, she was not licensed to practice in Missouri. Several
       newspaper reporters were able to obtain nurse's degrees identical to hers
       merely by sending checks to a mail order address. Petitioner's Brief, *Com-
       munity Blood Bank of the Kansas City Area, Inc.* v. *Federal Trade Commis-
       sion,* 8th Cir. (1969), no. 18645, p. 17. For other details about the Basses'
       blood bank and the controversy it triggered, see also *In the Matter of
       Community Blood Bank of the Kansas City Area, Inc. et al.,* 70 FTC 728,
       744 (1966); *Petitioners' Brief, Community Blood Bank of the Kansas City
       Area, Inc.* v. *Federal Trade Commission,* 8th Cir. (1969), no. 18645; inter-
       views with Dick H. Woods and Ross D. Eckert.
187    "ISN'T THAT" . . . "YES!": *In the Matter of . . . ,* p. 919.
188    "GREED, WASTE, CHAOS AND DANGER": A. H. Groeschel, "Statement,"
       U.S. Congress, Senate Committee on the Judiciary, Subcommittee on
       Antitrust and Monopoly, *A Bill to Amend the Antitrust Laws to Provide
       That the Refusal of Nonprofit Blood Banks and Hospitals and Physicians to
       Obtain Blood and Blood Plasma from Other Blood Banks Shall Not Be*

*Deemed to Be Acts in Restraint of Trade and For Other Purposes,* 88th Congress, 2nd session (Washington, D.C.: U.S. Government Printing Office, 1964), pp. 98–108.

189 "THAT'S ONE OF OUR DONORS!": "City Blood Program Is Termed Chaotic," New York *Journal-American,* May 3, 1963, cited in ibid., p. 183.

194 TRANSFUSION PROCESS WAS A MEDICAL *SERVICE:* There is a rich literature on liability and other legal questions in blood banking. Some useful sources include G. Clark, ed., *Legal Issues in Transfusion Medicine* (Arlington, Va.: American Association of Blood Banks, 1986); "Medicine and the Law," *JAMA* 163 (1957): 283–89; interviews with Dick H. Woods and Ross D. Eckert.

"IS HUMAN BLOOD A COMMODITY . . .": M. Goldman, "Letter to Dr. Dreskin," *Transfusion* 5 (1964): 207–8.

195 "MONSTROUS! . . .": "Threat to Community Blood Bank Posed by FTC Ruling Is Termed 'Monstrous,' " Newport News *Press,* June 14, 1964, cited in U.S. Congress, *Bill to Amend* (1964), pp. 208–9.

NOTHING SHORT OF "SACRILEGIOUS": B. A. Myhre, "Letter to Hon. Edward V. Long," Aug. 3, 1964, cited in ibid., p. 227.

"BLOOD, EYES, BONE FOR GRAFTS, ETC. . . .": C. M. Poser, "Letter to Hon. Edward V. Long," Aug. 7, 1964, cited in ibid., p. 228.

". . . WILL CEASE TO EXIST": M. A. Meservey, "Letter to Hon. Edward V. Long," July 21, 1964, cited in ibid.

IN GENEVA, ILLINOIS: Ibid., p. 233.

196 "THIS DESTRUCTIVE AND EXPENSIVE LITIGATION . . .": Thomas O'Donnell, "Keynote Address, 16th Annual Meeting of the American Association of Blood Banks," Nov. 5, 1963, cited in ibid., pp. 198–201.

"MUST BE ROOTED IN CHARITY . . .": Cited in ibid.

". . . CAN IT BE CONSIDERED PROPER . . .": T. Greenwalt, "Letter to Hon. Edward V. Long," July 24, 1964, cited in ibid., pp. 231–32.

"TO TAMPER WITH A LIFESAVING MEDICAL PRACTICE": Cited in ibid., p. 4.

197 THE ISBT'S EIGHTH ANNUAL MEETING: Fusao Ikeda, *Shiroi Ketsueki* (Tokyo: Ushio Shuppanska, 1995), pp. 127–30.

198 "IF KIDNEY TRANSPLANTATION BECOMES POPULAR . . .": Ibid., p. 128.

"MY HEART SANK TO THE BOTTOM . . .": Ibid., p. 129.

MANY DONORS DRANK HOME-BREWED CONCOCTIONS: Ibid., 135–36. On tako, see "Tokyo Is Talking About 'The Blood Sellers," San Francisco *Chronicle,* Dec. 5, 1962.

VOLUNTARY DONATIONS CLIMBED: Ikeda, *Shiroi Ketsueki,* p. 160.

199 ". . . YOU CAN LEAVE": Ibid., p. 133.

"WHERE IS THIS MAN GOING? . . .": E. O. Reischauer, *My Life Between Japan and America* (New York: Harper & Row, 1986), p. 262. A full account of the stabbing and the aftermath can be found on pp. 262–75; also in *Japan Times,* in several months of coverage starting March 24, 1964.

THE YOUNG MAN PLUNGED A BUTCHER KNIFE: The assailant was a deranged nineteen-year-old named Norikazu Shioya.

199–200 THE FUJI ORGAN PHARMACEUTICAL COMPANY: Ikeda, *Shiroi Ketsueki,* p. 134.

200 THE BLOOD HE HAD RECEIVED MUST HAVE CONTAINED: Even though he returned to Japan and enjoyed a long and distinguished career, the

ambassador never fully recovered from hepatitis. He eventually died of complications, at the age of seventy-nine.

200 SETTING ASIDE 85 MILLION YEN: Ikeda, *Shiroi Ketsueki,* p. 164.

"THIS IS THE COMPANY THAT I FOUNDED . . .": Ibid., p. 186.

201 "IS HUMAN BLOOD PROPERLY . . .": U.S. Congress, *Bill to Amend* (1964), p. 4.

202 "HELL, WE WERE THE *ANTI*TRUST SUBCOMMITTEE . . .": Interview with S. Jerry Cohen.

"IS IT YOUR FEELING . . .": U.S. Congress, *Bill to Amend* (1964), pp. 5–25, 33–49.

205 "ATYPICAL TO THE POINT OF FREAKISHNESS. . . .": *In the Matter of Community Blood Bank,* p. 958.

FALSIFYING EXPIRATION DATES: "Jury Convicts Pair of Blood Mislabeling," Dallas *Morning News,* April 19, 1966, cited in U.S. Congress, Senate, Committee on the Judiciary, Subcommittee on Antitrust and Monopoly, *A Bill to Amend the Antitrust Laws . . . ,* 90th Congress, 1st session (Washington, D.C.: U.S. Government Printing Office, 1968), p. 89.

BUYING BLOOD FROM KNOWN HEPATITIS CARRIERS: "Private Blood Bank Surrenders License," Patterson *News,* Jan. 15, 1963, cited in ibid., p. 186.

A FEDERAL APPEALS COURT RULED: *Community Blood Bank of the Kansas City Area, Inc.* v. FTC, 405 F. 2d 1011 (1969).

206 BLOOD "SHIELD" LAWS: Interviews with James McPherson, Jane Starkey, and Ross D. Eckert.

## CHAPTER 12: BAD BLOOD

207 WELL ABOVE SIX MILLION PINTS A YEAR: National Academy of Sciences–National Research Council, *An Evaluation of the Utilization of Human Blood Resources in the United States* (Washington, D.C.: 1971), Component Therapy Institute, p. 13.

208 104 TIMES A YEAR: J. Palmer, "Large Scale Application of Plasmapheresis," *Vox Sanguinis* 8 (1963): 96.

"EXPLOITING FOR ITS PROTEINS . . .": Tibor J. Greenwalt, "Plasmapheresis as a Source of Human Protein," in *Conference on Plasmapheresis,* "XXth Scientific Meeting of Protein Foundation, Inc." (Boston, April 7, 1966), p. 2.

HYLAND DIVISION: Hyland Laboratories, one of the original fractionators in Edwin Cohn's project, was purchased by Baxter Laboratories in 1952 and became the company's Hyland division. In subsequent years Baxter itself underwent a couple of name changes, becoming Baxter Travenol Laboratories in 1976 and Baxter International in 1987. These name changes are reflected throughout the text.

"ABSOLUTE DEAD CENTER, SKID ROW . . .": Interview with Tom Asher.

209 "THE PAIN OF INSERTION . . .": Stuart Bauer, "Blood Farming," *New York,* May 19, 1975, p. 62.

210 DOROTHY GARBER OF MIAMI, FLORIDA: *U.S.* v. *Garber,* 589 F. 2d 843,845 (5th Cir. 1979).

". . . EITHER ILLITERATE OR FUNCTIONALLY ILLITERATE . . .": Letter from G. Laub, Medical Director of the Columbia, S.C., Plasmacenter, to Ms. Mary Rae, Mr. William Tarleton, Cutter Biological Labs, Berkeley, Calif.

211 "DROPPING LIKE FLIES": W. Rugaber, "Prison Drug and Plasma Projects Leave Fatal Trail," *New York Times,* July 29, 1969; see also J. Farmer, "Inmate's Death Linked to Cummins Prison Injections?," Pine Bluff *Commercial,* Jan. 15, 1969.

212 HEPATITIS HAS A HISTORY: An enormous literature exists on post-transfusion hepatitis and its history. Three useful general references are: R. H. Purcell, "Hepatitis B: A Scientific Success Story (Almost)," in R. Y. Dodd and L. F. Barker, eds., *Infection, Immunity, and Blood Transfusion,* Proceedings of the XVIth Annual Scientific Symposium of the American Red Cross, Washington, D.C., May 9–11, 1984 (New York: Alan R. Liss, 1985), pp. 11–43; L. B. Seeff, "Transfusion-Associated Hepatitis B: Past and Present," *Transfusion Medicine Reviews* 2 (Dcc. 1988): 204–14; A. J. Zuckerman, "Twenty-five Centuries of Viral Hepatitis," *Rush-Presbyterian Medical Bulletin* 15 (1976): 57–82; interview with Dr. Alan Kliman.
"THE ETIOLOGY OF THESE EPIDEMICS . . .": A. Lürman, "Eine Icterusepidemie," *Berliner klinische Wochenschrift* 22 (Jan. 12, 1885): 20ff., trans. in J. G. Allen, *The Epidemiology of Posttransfusion Hepatitis* (Stanford, Calif.: Commonwealth Foundation, 1972), pp. 4–7.

214 FROM TWO UNIMPEACHABLE SOURCES: Documentation for this incident includes W. A. Sawyer, K. F. Meyer, M. D. Eaton, "Jaundice in Army Personnel in the Western Region of the United States and Its Relation to Vaccination Against Yellow Fever," *American Journal of Hygiene* 39 (Jan. 1944): 337–432; 40 (July 1944): 35–105; "Jaundice Following Yellow Fever Vaccination" (editorial), *JAMA* 119 (Aug. 1, 1942): 1110; "Yellow Fever Vaccination," Ibid., p. 1114.
"PROBABLY TRANSFUSED . . .": E. Standlee, Headquarters, Mediterranean Theater of Operations, United States Army, "Hepatitis and Transfusions," memo, n.d., courtesy of Francis A. Countway Library of Medicine.
"THE SERGEANT WHO MADE THE DONATION . . .": Douglas B. Kendrick, *Blood Program in World War II* (Washington, D.C.: Office of the Surgeon General, Department of the Army, 1989), pp. 678–79.

215 THIS ONE-DAY SNAPSHOT OF THE DISEASE: Ibid., pp. 675–77.
THE RED CROSS ATTEMPTED TO LIMIT THE PROBLEM: Ibid., p. 674.
THE BRITISH TOOK ANOTHER APPROACH: L. Hogben, "Risk of Jaundice Following Transfusion with Pooled Plasma or Serum," *British Journal of Social Medicine* 1 (1947): 209.
"SAVING THE PATIENT'S LIFE . . .": Kendrick, *Blood Program in World War II,* p. 677.
"I AM WRITING YOU . . .": Norman T. Kirk, Letter to Hon. Basil O'Connor, Chairman, American Red Cross, Feb. 13, 1946, courtesy of Countway Library.

216 RED CROSS RECALLED THE THOUSANDS OF CANS: Interview with Dr. John Ashworth; E. J. Cohn, "Facts About Blood and the Red Cross Blood Program," summary of address given at the Massachusetts Blood Center of the American Red Cross, Aug. 9, 1950, courtesy of Countway Library.
NEARLY 22 PERCENT: V. M. Sborov et al., "Incidence of Hepatitis Follow-

ing Use of Pooled Plasma," *A.M.A. Archives of Internal Medicine* 92 (Nov. 1953): 678–83.

216       PLASMA POOLS HAD NOW GROWN TO FOUR HUNDRED UNITS: Kendrick, *Blood Program in World War II,* p. 781.
THE NATIONAL INSTITUTES OF HEALTH RECOMMENDED: Ibid.
"TO SUPPORT BLOOD VOLUME": Ibid., p. 782.

217       DR. J. GARROTT ALLEN: Biographical information from interviews with Barry Allen and Dr. Edward Stemmer.

218       LONG-TERM GENTLE HEATING ELIMINATED THE VIRUSES: J. G. Allen et al., "Homologous Serum Jaundice and Its Relation to Plasma Storage," *JAMA* 144 (Nov. 25, 1950): 1069; J. G. Allen et al., "Homologous Serum Jaundice and Pooled Plasma: Attenuating Effect of Room Temperature Storage on Its Virus Agent," *Annals of Surgery* 138 (Sept. 1953): 476–86.
A 50 PERCENT HEPATITIS REDUCTION: R. Murray, "Effect of Storage at Room Temperature on Infectivity of Icterogenic Plasma," *JAMA* 155 (May 1, 1954): 13–15.
DR. ALLAN G. REDEKER FOUND ALMOST NO REDUCTION: A. G. Redeker et al., "A Controlled Study of the Safety of Pooled Plasma Stored in the Liquid State at 30–32 C for Six Months," *Transfusion* 8 (March–April 1968): 60–64.
"A FULL 10°C. LOWER . . .": J. G. Allen, "Importance of Requirements to Produce Minimal Risk Plasma," *Archives of Surgery* 98 (May 1969): 558–65 (full critique).
"SERIOUS DOUBT . . .": Committee on Plasma and Plasma Substitutes, Division of Medical Sciences, National Research Council, "Statement on Normal (Whole, Pooled) Human Plasma," *Transfusion* 8 (March–April 1968): 57–59.

219       ". . . THE TRUTH WOULD FINALLY COME OUT": Interview with Dr. Edward Stemmer.
RATE OF HEPATITIS HAD GROWN ALMOST IN LOCK-STEP: Allen, *Epidemiology of Posttransfusion Hepatitis.*

220       250 PAGES OF A BOOK: Ibid.
"IMPRACTICAL, UNWORKABLE, AND CAUSE FOR CONCERN": E. R. Jennings and J. J. Palmer, "Control of Post-Transfusion Hepatitis," letter, *California Medical Association,* Aug. 1966, p. 3.
". . . YOU MAY AS WELL POUR IT DOWN THE DRAIN": C. Holden, "Blood Banking: Tangled System Resists Swift Change," *Science* 175 (March 31, 1972): 1444.
AT LEAST THREE TIMES THE RISK: Letter from R. Murray, Director, Division of Biologics Standards, National Institutes of Health, to J. Garrott Allen, M.D., July 23, 1971.

221       RESIDENTS OF SKID ROW: Letter from J. G. Allen to James W. Mosley, M.D., Sept. 9, 1975.
"TRANSFUSION ROULETTE": L. K. Altman, "Use of Commercial Blood Increases with Shortages in U.S.," *New York Times,* Sept. 5, 1970.

221–22    PHILIP CAPUTO PEDDLED HIS BLOOD: P. Caputo, "Blood Banks: Pay Stations," *Chicago Tribune,* Sept. 14, 1971.

222       "PROCUREMENT OF BLOOD PLASMA . . .": "Chronolog," NBC-TV, transcript, p. 20.

222    HEMOPHILIACS: In recent years the hemophilia community has come to prefer the term "person with hemophilia," since it connotes an individual rather than a syndrome. I apologize for using the outmoded designation, but found it necessary to do so for economy of language.

223    "NO MORE PAIN! NO MORE PAIN! . . .": R. and S. Massie, *Journey* (New York: Ballantine Books, 1975), p. 69.

A WHITE RESIDUE: J. G. Pool and A. E. Shannon, "Production of High-Potency Concentrates of Antihemophilic Globulin in a Closed-Bag System," *New England Journal of Medicine* 273 (Dec. 30, 1965): 1443–47.

224    "ONE COULD NOT STRAY FAR FROM THE DEEP FREEZE . . .": Massie, *Journey*, p. 227.

A NEW AND MORE CONCENTRATED FORM OF FACTOR VIII: K. M. Brinkhous, E. Shanbrom et al., "A New High-Potency Glycine-Precipitated Antihemophilic Factor (Ahf) Concentrate," *JAMA* 205 (Aug. 26, 1968): 613–17.

"TODAY BOBBY HANDLES . . .": Massie, *Journey*, p. 255.

225    "IS THERE ANY GOOD REASON . . .": Letter from J. Garrott Allen, Professor of Surgery, Stanford University School of Medicine, to Elliott L. Richardson, Secretary of Department of Health, Education and Welfare, June 21, 1971. Courtesy of Barry Allen.

"ALL WE NEED TO DO . . .": Letter from J. Garrott Allen, Professor of Surgery, Stanford University School of Medicine, to Elliott L. Richardson, Secretary of Department of Health, Education and Welfare, July 26, 1971. Courtesy of Barry Allen.

"THE FACTS . . .": Letter from J. Garrott Allen, Professor of Surgery, Stanford University School of Medicine, to Elliott L. Richardson, Secretary of Department of Health, Education and Welfare, July 27, 1971. Courtesy of Barry Allen.

"CHICAGO IS BY NO MEANS . . .": Letter from J. Garrott Allen, Professor of Surgery, Stanford University School of Medicine, to Elliott L. Richardson, Secretary of Department of Health, Education and Welfare, Sept. 23, 1971. Courtesy of Barry Allen.

226    "A NUMBER OF LEGAL PROBLEMS": Letter from I. Mitchell, Department of Health, Education, and Welfare, to J. Garrott Allen, Nov. 1, 1971.

STARFISH: Interview with Barry Allen.

227    "BROADLY RESEMBLE THE POPULATION . . .": R. M. Titmuss, *The Gift Relationship: From Human Blood to Social Policy* (New York: Pantheon, 1971).

"COMMERCIALIZATION OF BLOOD . . .": Titmuss, *The Gift Relationship*, p. 245.

228    A CARICATURE: By classifying donors according to the various incentives they had received, Titmuss ruled out many we would think of as voluntary. He asserted that most donors were "tied" to giving blood by replacement policies or family credit deposit arrangements. He categorized such people as "Responsibility Fee" or "Family Credit" donors. Using such categories, he claimed that of the six million units collected annually in America from 1965 to 1967, about 62 percent effectively had been coerced, 29 percent had been paid for with money, and only 9 percent were truly volunteer. In contrast, the National Academy of Sciences estimated that only 15 percent of the American blood supply came from pro-

fessional donors, while the rest could broadly be assumed to come from volunteers. Titmuss, *The Gift Relationship,* p. 94; National Academy of Sciences, *An Evaluation of the Utilization of Human Blood Resources in the United States,* p. 37.

228      "NO-SURF MURPH": F. Kent, "The Blood Business," Miami *Herald,* Nov. 11, 1973.

ROBERT IRBY: P. Caputo, "Find Blood of Paid Donors Polluted with Hepatitis," Chicago *Tribune,* Sept. 12, 1971.

"A UNIQUE NATIONAL RESOURCE": Information about this episode and the runup to the National Blood Policy is drawn from P. J. Schmidt, "National Blood Policy, 1977: A Study in the Politics of Health," *Progress in Hematology* 10 (1977): 151–72; D. Surgenor, "Progress Toward a National Blood System," *New England Journal of Medicine* 291 (July 4, 1974): 17–22; I. Mitchell, "Developments Leading to a National Blood Policy," March 15, 1978, memorandum, Department of Health, Education and Welfare (National Blood Policy Papers, MS C 393, in the History of Medicine Division, National Library of Medicine).

229      RICHARDSON . . . MADE A SEAT-OF-THE-PANTS DECLARATION: U.S. Congress, Senate Subcommittee on Executive Reorganization and Government Research of the Committee on Government Operations, Testimony of Elliot Richardson. *Consumer Safety Act of 1972,* 92d Congress, 2nd sess., 1972, pp. 172–201.

BLOOD BANKING HAD BECOME UNDISCIPLINED AND WASTEFUL: *NHLI's Blood Resource Studies,* vols. 1 and 2 (Bethesda, Md.: U.S. Department of Health, Education and Welfare, 1972).

BLOOD WAS SO POORLY DISTRIBUTED: I. Mitchell et al., "Blood Banking: Major Findings by HEW Task Force," July 12, 1973, memorandum, Department of Health, Education and Welfare (National Blood Policy Papers, MS C 393, in the History of Medicine Division, National Library of Medicine).

THIRTY-FIVE HUNDRED A YEAR: Comptroller General of the United States, Feb. 13, 1976, *Hepatitis From Blood Transfusions: Evaluation of Methods to Reduce the Problem,* report to Congress.

## CHAPTER 13: WILDCAT DAYS

231      SOMOZA FAMILY'S LEGACY HAD BLOSSOMED: L. and P. Pezullo, *At the Fall of Somoza* (Pittsburgh: University of Pittsburgh Press, 1993); "Nicaragua: How the Local Boys Made Good," *Latin America Economic Report,* Jan. 27, 1978, p. 27.

232      COMPAÑÍA CENTROAMERICANA DE PLASMAFÉRESIS: Peter Davis, "Mirror of Our Midlife Crisis: United States and Nicaragua," *Nation* 238 (Jan. 28, 1984): 76.

*CASA DE VAMPIROS:* Ibid.

DEMAND CLIMBED BY DOUBLE-DIGIT PERCENTAGES: Division of Blood Diseases and Resources, National Heart, Lung and Blood Institute, *Study to Evaluate the Supply-Demand Relationships for AHF and PTC Through 1980,* DHEW publication no. (NIH) 77-1274, p. 51.

232      "GOOD PICKINGS IN THE UNITED STATES . . .": Interview with Fred Marquart.

233      A LIVELY TRADE WAS ALREADY UNDER WAY: "Foreign Establishments That Provide Materials Under the Short Supply Provisions of Sec. 73.240 of the PHS Regulations," attachment to letter from Sam T. Gibson, Acting Director, Division of Biologics Standards, National Institutes of Health, to Hon. Victor V. Veysey, U.S. House of Representatives, Nov. 10, 1971 (this letter, in response to a congressional inquiry, lists nine Third World sources of blood fractions); interviews with Tom Asher, John Ashworth, Dr. Jean C. Emmanuel, Dr. Ben G. Grobbelaar, Dr. Alfred Haessig, Tom Hecht, Fred Marquart.

      HEMO CARIBBEAN: R. Severo, "Impoverished Haitians Sell Plasma for Use in the U.S.," *New York Times,* Jan. 28, 1972.

      MEXICO, BELIZE: S. Gibson, "Foreign Establishments," letters to the Hon. Christopher Shays, re Subcommittee Inquiry of Dec. 18, 1995, from Bayer Healthcare Corporation, Armour Pharmaceutical Company, Alpha Therapeutic Corporation, in U.S. Congress, House of Representatives, Committee on Government Reform and Oversight, Subcommittee on Human Resources and Intergovernmental Affairs, *Protecting the Nation's Blood Supply from Infectious Agents: New Standards to Meet New Threats,* 104th Cong., 1st sess., Oct. 12 and Nov. 2, 1995 (Washington, D.C.: U.S. Government Printing Office, 1996), pp. 202–15.

233–34  "WE WERE PEDDLING . . .": Interview with Fred Marquart.

234      CENTERS ALONG THE UNITED STATES' SOUTHERN BORDER: P. J. Hagen, *Blood: Gift or Merchandise* (New York: Alan R. Liss, 1982), p. 171; G. M. Gaul, "America: OPEC of Global Plasma Industry," Philadelphia *Inquirer,* pt. 5 of series running Sept. 24–28, 1989.

      ". . . HE WAS SO SMOOTH": Interview with Tom Asher.

      SOME CAME FROM LESOTHO: Interview with Dr. Grobbelaar.

235      THERE HE BOUGHT PLASMA FOR $5 A DONATION: Grobbelaar had to overcome a problem: He could not ship the plasma directly to his customers, because an antiapartheid boycott was in place. Instead, he sent the frozen plasma to importing companies in England and Switzerland, where the boycott was less rigidly enforced. These companies would change the labels to disguise the port of origin and forward the plasma to its final destination. "There's a thousand and one ways to bypass a boycott," he said in an interview with the author.

      "I FIND FAULT . . .": Letter from B. G. Grobbelaar, to the author, April 15, 1996.

236      ". . . FROM THIS POCKET OF BLOOD . . .": C. Mérieux, *Le Virus de la découverte* (Paris: Robert Laffont, 1988), p. 124.

      AT THE HEIGHT OF PRODUCTION: Institut Mérieux stopped producing gamma globulin in 1992.

      5 PERCENT OF THE PLACENTAS: Mérieux, *Virus,* p. 183.

      WAY STATIONS NEAR AIRPORTS IN MEXICO AND MIAMI: E. Harriman, "Blood Money," *New Scientist,* March 13, 1980, pp. 858–59.

      "WE SHIPPED HUGE QUANTITIES . . .": Interview with Tom Hecht.

236–37  MONTREAL AND ZURICH: A. Picard, "Canada Still Lacks Control on Plasma Trade, Inquiry Told," *Globe and Mail,* Dec. 14, 1995.

237      A WOMAN FROM ONE OF THE BARRIOS COMPLAINED: "Desapaeció: Iba

a Plasmaféresis," *Prensa,* Sept. 26, 1977; "El 'donante' no aparece . . ." *Prensa,* Sept. 27, 1977.

237        "NINETY PERCENT OF THE DONORS . . .": "Quiénes están detrás de este 'negocio'?," *Prensa,* Sept. 28, 1977.
"WHY ARE WE INVOLVED IN THIS BLOODY BUSINESS . . .": Headlines from *Prensa,* Nov. 17, Sept. 13, and Nov. 19, 1977.

238        "BUT WHEN THE NEEDLE . . .": "Incidente en Plasmaféresis por la 'ayuda,' " *Prensa,* Oct. 7, 1977.
"PROFOUND ANEMIA AND MALNUTRITION": " 'Donantes' cuestan miles a hospitales," *Prensa,* Oct. 11, 1977.
A DISCOURAGING COLLECTION OF DERELICTS AND ALCOHOLICS: "World in Action: Blood Money," Granada Television Limited, 1975, transcript, pp. 7–16.
"AN OFFENSE TO HUMAN DIGNITY": Letter from A. Z. Zuckerman to Prof. J. Garrott Allen, Sept. 6, 1976. Courtesy of Barry Allen.
"REJECTED STRAIGHTAWAY": "World in Action," transcript, p. 15.

239        "THE PHYSICAL PLANT IMPRESSED ME . . .": Interview with Tom Hecht.
"SHAME ON THE NATION . . .": P. J. Chamorro, "Detràs de la sangre," *Prensa,* Nov. 18, 1977.
". . . WEEKENDS AND SUNDAYS TOO": "Más repudio a tráfico de plasma," *Prensa,* Oct. 6, 1977.

240        A QUARTER OF THE ALLOWABLE AMERICAN LEVELS: L. Aledort and S. H. Goodnight, "Hemophilia Treatment: Its Relationship to Blood Products," *Progress in Hematology* 12 (1981): 134.
"DISORGANIZED SHAMBLES": J. Cash, "The Blood Transfusion Service and the National Health Service," *British Medical Journal,* Sept. 12, 1987, p. 617. General information about the progress of the British transfusion service from interviews with J. H. Cash, H. H. Gunson.
FAILED TO MEET EVERY DEADLINE: The National Health Service's delays in expanding fractionation capacity became an important issue in medical circles and the media. To understand the dimensions and passions of the controversy, see A. King, "Lives Too Expensive for Britain to Save," Yorkshire *Post,* Jan. 17, 1975; A. King, "Now MP's Take Up the Blood-Disease Scandal," Yorkshire *Post,* Feb. 19, 1975; P. Jones, "Factor VIII Supply and Demand," letter, *British Medical Journal,* June 21, 1980, pp. 1531–32.
THE COUNTRY IMPORTED MORE THAN HALF ITS SUPPLY: Commercial plasma concentrates, which were first imported to England and Wales in 1972 (Scotland has an autonomous system), occupied about a third of the market in 1974 and an estimated two-thirds by 1985 (Cash, "Blood Transfusion Service"; R. Biggs, "Haemophilia Treatment in the United Kingdom from 1969 to 1974," *British Journal of Haematology* 35 (1977): 487–504; Jones, "Supply and Demand."
FRANCE IMPORTED AS MUCH AS 26 PERCENT: J. P. Allain, "Production of Antihemophilic Factor in France," *Scandinavian Journal of Haematology* 33, suppl. 40 (1984): 502.
JAPAN IMPORTED AN ASTONISHING 98 PERCENT: "America the Blood Bank," *Economist,* Oct. 17, 1981, p. 87.
". . . SO DOES THE U.S. BLEED FOR THE WORLD . . .": T. Drees, "Examination of International Plasma Resources," address at American Blood Resources Association Plasma Forum III, Feb. 25, 1980.

241 "HOW THEY [THE EUROPEANS] . . .": Letter from Grobbelaar to author.
SOLD FOR AT LEAST TRIPLE THAT AMOUNT: Bundeskartellamt
"Beschluss: In dem Kartellverwaltungsverfahren," B3-43 21 90-T-42/81.
This document is an order issued in 1981 by the Bundeskartellamt, the
equivalent of the American Federal Trade Commission, against German
clotting-factor suppliers for alleged abuse of German pharmaceutical trade
laws. The document comprises two sections. The first orders the suppliers
to supply detailed information about their business practices; the second is
an "opinion" describing the results of the government's investigation to
date. It details many of the questionable practices cited in the text.
IN WEST GERMANY: The estimates for German use vary, from twice that
of their American counterparts to four times (Jones, "Supply and
Demand"; E. R. Koch, *Böses Blut: Die Geschichte eines Medizin-Skandals*
[Hamburg: Hoffmann und Campe, 1990], p. 183).
DR. HANS EGLI: For detailed information on Egli, the activities of the
Bonn hemophilia center, and the related financial scandal, see Koch, *Böses
Blut.*

242 BRACKMANN REPORTED THAT, OF TWENTY "HIGH RESPONDERS":
Other physicians agreed with Brackmann in principle but did not neces-
sarily go along with his practice of using massive quantities of Factor VIII,
because of uncertainty over long-term effects on the patients' immune systems.
"AT THE AGE OF SIXTEEN . . .": Interview with Dr. Werner Kalnins.

242–43 "BROUGHT HEMOPHILIA INTO THE AGE OF THE COMPUTER . . .": Frank
Schnabel, "Report on Schnabel's visit to the Bonn Hemophilia Centre,"
Nov. 8, 1977, memorandum, World Federation of Hemophilia.

243 "COMPETENT DOCTOR IN THE LOCALITY": Koch, *Böses Blut,* p. 184.
ANOTHER DOCTOR QUESTIONED: The question of the prophylactic use
of Factor VIII concentrates remains open for discussion, although many
hemophiliacs accept it. Even the most ardent advocates, however, recommend barely a third of the dosages that Egli's group administered.
"HUNDREDS OF THOUSANDS OF UNITS . . .": Ibid., p. 193.
"WE ALL REGARD PRECISE DOCUMENTATION . . .": G. I. C. Ingram et al.,
"Different Plans of Treatment," unpublished memorandum, World Federation of Hemophilia, October 1977.
THE COUNCIL OF EUROPE DENOUNCED: Koch, *Böses Blut,* p. 205; see
also *Preparation and Use of Coagulation Factors VIII and IX for Transfusion* (Strasbourg: Council of Europe, 1980).
MOST LIBERAL SYSTEMS FOR THE HANDICAPPED: H. H. Brackmann et
al., "Home Care of Hemophilia in West Germany," *Thrombosis and
Haemostasis* 35 (1976): 544–51; H. Egli, "The Situation of the Haemophiliac: Yesterday and Today," *Haemostasis* 10, suppl. 1 (1981): 1–10.

244 "HAD NO RECOGNIZABLE PURPOSE . . .": Koch, *Böses Blut,* p. 236.
". . . AN OUTRIGHT INTEREST IN KEEPING . . . PRICES UP": Bundeskartell-
amt, "Beschluss."
MOST OF THE WORLD FEDERATION OF HEMOPHILIA'S BUDGET: Inter-
view with Sheila Brading, WFH.
15 TO 25 PERCENT OF ITS OPERATING BUDGET: Deposition, Alan P.
Brownstein, *Wadleigh v. Rhone-Poulenc Rover, Inc., et al.,* July 28, 1994.
RECEIVED TENS OF THOUSANDS OF DOLLARS: The issue of the medical

directors' financial ties to industry is disputed and controversial. One former NHF medical director, Dr. Louis M. Aledort, has testified that his industry subsidies amounted to $10,000–12,000 a year; but Thomas C. Drees, former president of Alpha Therapeutics, says his company alone funded Aledort in excess of $25,000 annually. (Deposition of Louis M. Aledort, M.D., July 9, 1986. Superior Court, State of California, County of Santa Clara, *Michele Gallagher et al.* v. *Cutter Laboratories et al.,* no. 548947; interview with T. Drees.)

245 $15 MILLION IN INDUSTRY REBATES SINCE 1975: Koch, *Böses Blut,* p. 236.

AOK WAS PAYING $133 MILLION ANNUALLY: Bundeskartellamt, "Beschluss."

"THE MOST EXPENSIVE PATIENT IN TOWN": Interview with Dr. Werner Kalnins.

"BEGAN WITH FOG . . .": H. Egli, "The Situation of the Haemophiliac: Yesterday and Today," *Haemostasis* 10, suppl. 1 (1981): 9.

"FRESH HOPE IS EXPANDING . . .": F. Schnabel, "Dear Professor Egli," *World Federation of Hemophilia,* special issue, 25th anniversary, 1980.

"A MESS OF ENORMOUS DIMENSIONS . . .": Koch, *Böses Blut,* p. 246.

246 ". . . FURTHER DEPLETION OF AN ALREADY DEPLETED POPULATION": Letter from J. G. Allen to Dr. René J. Dubos, Jan. 31, 1972. Courtesy of Barry Allen.

"A NEW CLINICAL PHARMALOGICAL [*SIC*] INDUSTRY . . .": Serocenter of America, Inc., "Proposal," n.d. (given to Dr. Maurice Shapiro of the South African Blood Transfusion Service).

"THIS NEW MODALITY . . .": Hagen, *Blood,* p. 165.

OF THE TWELVE NATIONS, ELEVEN SAID: World Health Organization, "Utilization and Supply of Human Blood and Blood Products: Information Provided by the Director-General," internal circular, May 1, 1975.

THE DELEGATES UNANIMOUSLY VOTED TO APPROVE IT: Z. S. Hantchef, "Red Gold—or the Blood Trade," *Transfusion Today,* March 1989, p. 10.

247 ". . . THE FAT LITTLE DICTATOR . . .": Chamorro, "Detrás."

"CONTRIBUTING TO THE DISGRACE OF THE COUNTRY": C. W. Flynn and R. E. Wilson, "An Interview with Somoza's Foe, Now Dead," *New York Times,* Jan. 13, 1978.

"SPECIAL-HELP PLAN": "Incidente en Plasmaféresis."

248 IT PAID NO TAXES: Flynn and Wilson, "Interview"; "Salud decide investigar a Plasmaféresis," *Prensa,* Nov. 1, 1977.

CHAMORRO'S DEATH: Accounts of Chamorro's assassination and the ensuing riots from *Prensa,* Jan. 10, 13, 18, 1978; "Protesting Nicaraguans Riot," Washington *Post,* Jan. 13, 1978; "New Rioting Erupts in Nicaraguan Capital," *New York Times,* Jan. 13, 1978.

HE CALLED THE CHARGES "STUPID": *Prensa,* Jan. 18, 1978.

"PEDRO JOAQUÍN CHAMORRO . . .": A. Somoza, *Nicaragua Betrayed* (Boston: Western Islands, 1980), p. 115.

249 A JURY TRIED RAMOS *IN ABSENTIA*: "Jury Convicts 9 in Murder of Nicaraguan Publisher," Miami *Herald,* June 11, 1981.

MANAGING A PLASMA-FOR-EXPORT CENTER IN BELIZE: "Belize: The Enemy May Have a Foot in the Door," *Latin America Regional Reports,* Jan. 9, 1981; Harriman, "Blood Money."

249 "YOU GOT TIRED . . .": Interview with Tom Hecht.
"A MAJOR POINT OF REFERENCE . . .": J. Reasor, "Reasor on Demand," *Plasma Quarterly,* March 1981, p. 9.

## CHAPTER 14: THE BLOOD-SERVICES COMPLEX

251 THE POLICY INCLUDED A LIST OF TEN GOALS: P. Schmidt, "National Blood Policy, 1977: A Study in the Politics of Health," *Progress in Hematology* 10: 165.
"THE MOST SIGNIFICANT . . .": Ad Hoc Committee to Establish the American Blood Commission, news release, March 28, 1975.
"THINK FOR A MOMENT . . .": Transcript of Proceedings American Blood Commission (unpublished), April 4, 1975, pp. 10–12 (National Blood Policy Papers, MS C 393, in the History of Medicine Division, National Library of Medicine).
"IN A WORD, IT WAS HOSTILE . . .": Interview with Dr. Byron Myhre.

252 MORE THAN FIVE MILLION UNITS A YEAR: J. Cook, "Blood and Money," *Forbes,* Dec. 11, 1978, p. 37.
"EXCESSES OF REVENUE . . .": Andrea Rock, "Inside the Billion Dollar Business of Blood," *Money,* Mar. 1986, p. 158.
MORE THAN $9 MILLION . . . $27 MILLION: G. Gaul, "Red Cross: From Disaster Relief to Blood," Philadelphia *Inquirer,* Sept. 27, 1989. Over the years journalists have made much of the fact that the American Red Cross, a nonprofit agency, realized enormous revenues from its blood program. The truth is that the agency's fortunes varied wildly, depending on regulatory and market conditions.
FRIGHTENED BERNICE HEMPHILL: Interview with Bernice Hemphill.

253 THE IMBROGLIO OVER THE NONREPLACEMENT FEE: Interviews with Suzanne Gaynor, James McPherson, Jane Starkey; L. Kennedy, A. W. Drake, S. N. Finkelstein, and H. M. Sapolsky, "Community Blood Banking in the United States from 1937–1975: Organizational Transformation and Reform in a Climate of Competing," Ph.D. thesis, New York University, Feb. 1978; *The American Blood Supply* (Cambridge, Mass.: M.I.T. Press, 1982).
"TANTAMOUNT TO SELLING BLOOD": American Blood Commission, "Recommendation for Unified Donor Recruitment," July 26, 1977.
"I HAD HIGH HOPES . . .": Minutes, Board of Directors Meeting, American Blood Commission, Sept. 28, 1977, p. 25 (National Blood Policy Papers, MS C 393, in the History of Medicine Division, National Library of Medicine).

254 THE CASH EQUIVALENT OF ABOUT $300,000: R. Eckert and E. L. Wallace, *Securing a Safer Blood Supply: Two Views* (Washington, D.C.: American Enterprise Institute, 1985), p. 41.
"WE BELIEVED . . . WE WOULD BE ENTITLED . . .": "Statement of Col. Melvin W. Ormes . . . ," U.S. Congress, Senate, Committee on Labor and Human Resources, Subcommittee on Health and Scientific Research, *Oversight on Implementation of National Blood Policy, 1979,* 96th Cong., 1st sess. (Washington, D.C.: U.S. Government Printing Office, 1979), p. 11.

254 "FRAGMENTATION . . .": Letter from B. Hemphill, American Association of Blood Banks, to Mr. George M. Elsey, President, American National Red Cross, Oct. 4, 1976.
TO SEEK OUT AABB-AFFILIATED BLOOD BANKS: B. Hemphill, AABB, "Memo re: Important Guidelines and Announcement of Special Meeting on Termination of AABB-ANRC Interorganizational Meeting," Oct. 4, 1976 (National Blood Policy Papers, MS C 393, in the History of Medicine Division, National Library of Medicine).

255 "THE CONFUSION AND ANXIETY . . .": Letter from R. G. Wick, Vice-President, the American National Red Cross, to John J. Corson, President, the American Blood Commission, Oct. 28, 1976.
"IF WE CAN PROVE IT TO THE COUNTRY . . .": D. Zimmerman, "Happy Birthday Red Cross . . . ," *Investigator,* Sept. 1981, p. 62.
PROFITEERING FROM DONATED BLOOD: Information on this episode from interview with Bernice Hemphill; "State Sues Irwin Blood Bank," press release, California Department of Consumer Affairs, June 1, 1977; Irwin Memorial Blood Bank of San Francisco Medical Society, statement, June 1, 1977; J. Lynch, "Blood Bank Wins Fight over Fees," San Francisco *Chronicle,* Jan. 5, 1979; R. Coglan, editorial, KGO-TV, Feb. 2, 1979.
"FRACTURED NATURE . . .": "Statement of . . . Dr. Alvin Drake, Massachusetts Institute of Technology," in U.S. Congress, *Oversight,* p. 24.

256 "DISAGREEMENT BETWEEN . . .": General Accounting Office, *Problems in Carrying Out the National Blood Policy,* HRD-77-150 (Washington, D.C.: U.S. Government Printing Office, 1978).
SENATOR RICHARD SCHWEIKER MADE ONE MORE ATTEMPT: U.S. Congress, *Oversight.*
DR. BARUCH BLUMBERG: Blumberg et al., "A 'New' Antigen in Leukemia Sera," *JAMA* 191 (Feb. 15, 1965): 541–46.
ABOUT 15 PERCENT EFFECTIVE: Comptroller General of the United States, Feb. 13, 1976, *Hepatitis from Blood Transfusions: Evaluation of Methods to Reduce the Problem,* Report to Congress, pp. 40–42.

257 THE HIGHEST LEVELS OF PROFESSIONALISM: Interview with Dr. Howard Taswell.
LABEL THEIR BLOOD BAGS AS "PAID" OR "VOLUNTEER": "Transfusion Blood Soon Must Indicate Volunteer Donors," *New York Times,* Jan. 14, 1978.
AN ESTIMATED 180,000 TRANSFUSION RECIPIENTS: "Disease Burden from Viral Hepatitis A, B, and C in the United States," Centers for Disease Control and Prevention Web site (http://www.cdc.gov/ncidod/diseases/hepatitis/heptab3.htm).
DANNY KAYE: "Danny Kaye, 74, Dies . . . ," Los Angeles *Times,* March 3, 1987.

258 BY THE LATE 1970S: Interviews with Tom Asher, John Ashworth, Tom Hecht, John McCray, Robert Reilly; IFPMA Working Group, *A Study of Commercial and Non-Commercial Plasma Procurement and Plasma Fractionation* (Zurich: IFPMA, 1980). "Reasor on Demand," *Plasma Quarterly,* March 1981.
GREEN CROSS BOUGHT ALPHA THERAPEUTIC: "The Red Cross: Drawing Blood from Its Rivals," *Business Week,* Sept. 11, 1978, p. 113.

258    TOM HECHT PURCHASED NABI: Interview with Tom Hecht.

259    "... THIS WAS NOT UNIQUE TO CUTTER ...": Interview with John Ash-
       worth.

260    COUNTLESS *AD HOC* ARRANGEMENTS: This and other background infor-
       mation about blood banking in the 1970s and '80s from interviews with
       James McPherson, Dr. Robert Westphal, Dr. Tom Zuck.
       "BUNDLING" OR "TYING": U.S. Congress, House of Representatives,
       Committee on Energy and Commerce, Subcommittee on Oversight and
       Investigations, *Blood Supply Safety,* 102nd Cong., 1st sess. (Washington,
       D.C.: Government Printing Office, 1990), pp. 168–74.
       A PULITZER PRIZE–WINNING SERIES: G. M. Gaul, "The Blood Brokers,"
       Philadelphia *Inquirer,* series of articles running Sept. 24–28, 1989.
       CONGRESSIONAL HEARINGS: U.S. Congress, *Blood Supply Safety.*

261    "WHAT'S GOING DOWN THERE? ...": A. Kellner, "Self-Sufficiency:
       Lessons from the Euroblood Experience," in J. L. McPherson, ed., *Ade-
       quacy of the Nation's Blood Supply* (Washington, D.C.: Council of Com-
       munity Blood Centers, 1990), pp. 60.

262    "IT'S A STINKING SHAME ...": P. J. Hagen, *Blood: Gift or Merchandise*
       (New York: Alan R. Liss, 1982), p. 143.
       MORE THAN A THIRD OF NEW YORK CITY'S SUPPLY: Kellner, "Self-
       Sufficiency."
       "YOU DO WHAT YOU HAVE TO DO ...": Ibid.

263    THE ARRANGEMENT PROVED LUCRATIVE: "Red Cross," *Business Week.*
       "WE'RE NOT AFTER ANYONE'S SHARE ...": Ibid.

263–64  ERODING THE OLD PARTNERSHIP: A. Picard, *The Gift of Death: Con-
       fronting Canada's Tainted Blood Tragedy* (Toronto: HarperCollins, 1995),
       pp. 34–35, 84–91.

264    AT LEAST HALF THE PLASMA PRODUCTS CIRCULATING IN CANADA:
       Ibid., p. 98.

265    SELECTIVELY INCLUDE PLASMA FROM HIGH-RISK POPULATIONS: The
       drug companies could have been more careful about the risks associated
       with taking plasma from gays. As early as 1975, Wolf Szmuness, one of
       America's most respected hepatitis researchers, wrote in the *Annals of
       Internal Medicine* that because of the risk of spreading the disease gays
       "should be advised to refrain from blood donation."

## CHAPTER 15: OUTBREAK

266    PENTAMIDINE: Starting in 1967, the CDC was the nation's only distribu-
       tor of pentamidine. After the drug became commonly used for AIDS-
       related pneumonia, the government licensed it for broad distribution.
       From that point on, in 1985, any drugstore could fill pentamidine pre-
       scriptions, and the CDC stopped distributing the drug. Now more effec-
       tive drugs are used—in particular, Bactrim, the brand name for the
       combination sulfa drug sulfamethoxazol-trimotheprim. Pentamidine is
       used for patients—representing a minority of PCP cases—who are allergic
       to sulfa drugs.
       DR. BRUCE EVATT GOT A CALL: Interview with Dr. Bruce Evatt.

267    "IF IT'S REAL THERE WILL BE MORE OF THEM": E. W. Etheridge, *Sentinel*

*for Health: A History of the Centers for Disease Control* (Berkeley: University of California Press, 1992), p. 331.

267    THE DISEASE STRUCK TWO MORE HEMOPHILIACS: "*Pneumocystis Carinii* Pneumonia Among Persons with Hemophilia A," *Morbidity and Mortality Weekly Reports* (hereafter *MMWR*) 31 (July 16, 1982): 365–67.

HE LAID OUT HIS FINDINGS: Ibid.

267–68    ON JULY 16, 1982: Interview with Dr. Bruce Evatt; Institute of Medicine, *HIV and the Blood Supply: An Analysis of Critical Decisionmaking* (Washington, D.C.: National Academy Press, 1995) p. III-9.

268    DISEASE NOW AFFLICTED MORE THAN 440: Letter from W. H. Foege, Centers for Disease Control, to L. G. Hershberger, Cutter Laboratories. July 9, 1982 (courtesy of the Institute of Medicine).

ABOUT A FIFTH OF THE IRWIN BLOOD BANK'S SUPPLY: There is considerable disagreement about how much blood gays actually provided. In 1983 and 1984, the directors at Irwin operated under the working assumption that gays provided 20 percent; years later, they re-examined their donor records and revised the estimate to a mere 4 percent. In a February 1983 memo, Red Cross economist P. Cumming estimated that 25 percent of the agency's blood donors were gay or bisexual.

269    ALEDORT FELT THAT HE NEEDED MORE DATA: Institute of Medicine "Fact Finding Interview Summary: Louis M. Aledort," Jan. 9, 1995 (courtesy of the Institute of Medicine).

PASSED A RESOLUTION: Medical and Scientific Advisory Council: resolution to National Hemophilia Foundation Board of Directors, Oct. 2, 1982 (courtesy of the Institute of Medicine).

GAYS, IV DRUG USERS, AND HAITIANS: These groups, along with hemophiliacs, were the first to show elevated AIDS rates in America and were considered at risk for the disease. In 1984, for example, among the nearly 7,000 AIDS cases in America, 5,038 were homosexuals, 1,190 were IV drug users, 145 were hemophiliacs or transfusion recipients, and 249 were Haitians (the rest were heterosexual women or "unknown"). The high rate among Haitians proved a mystery, because no one understood why they should be a high-risk group. Some scientists theorized that Haiti, as a center for gay tourist cruises, might have been an early point of incubation. Others thought that HIV might have traveled to Haiti with workers who had been living in the Congo, another early center for the disease. Still others thought that the classification itself was erroneous because those Haitians who had had occasional homosexual encounters did not report themselves as gay. Whatever the causes, the focus on Haiti proved to be politically and scientifically unpopular, and the CDC dropped the "Haitian" category from its AIDS surveys.

RISK OF AIDS IN FACTOR VIII WAS MINIMAL: National Hemophilia Foundation, "Medical Bulletin #5," Jan. 17, 1983, "Medical Bulletin #7," May 11, 1983 (courtesy of the Institute of Medicine).

"KNOCKED OFF [HIS] CHAIR": Interview with Tom Drees.

270    "HE IS NOT BASING THIS REQUEST . . .": J. Hink, Cutter Biological, "Immunodeficient Syndrome," memo, Aug. 30, 1982.

A BABY IN SAN FRANCISCO HAD DIED: "Possible Transfusion-Associated Acquired Immune Deficiency Syndrome (AIDS)—California," *MMWR* 31

(Dec. 10, 1982): 652ff; A. J. Amman et al., "Acquired Immunodeficiency in an Infant: Possible Transmission by Means of Blood Product," *Lancet,* April 30, 1983, pp. 956–58.

270 DAY-LONG SESSION AT CDC HEADQUARTERS: This meeting was so controversial that it was important to obtain and confirm details and quotations from multiple sources, including "Summary Report on Workgroup to Identify Opportunities for Prevention of Acquired Immune Deficiency Syndrome," CDC internal document, Jan. 4, 1983 (courtesy of the Institute of Medicine); B. D. Colon (*Newsday* reporter), unpublished reporter's notes provided to author; W. E. Check, "Preventing AIDS Transmission: Should Blood Donors Be Screened?" *JAMA* 249 (Feb. 4, 1983): 567–70; J. L. Marx, "Health Officials Seek Ways to Halt AIDS," *Science* 219 (Jan. 21, 1983): 271–72; statements given to the Institute of Medicine by meeting participants (courtesy of the Institute of Medicine).

271 "A SNAP": Interview with Dr. Bruce Evatt.
"STIGMATIZE . . .": Marx, "Health Officials," p. 569.
"SCREENING BLOOD, NOT PEOPLE": Colon, "reporter's notes."
"I DON'T THINK ANYONE": Marx, "Health Officials," p. 569.
"WE DON'T HAVE ANYTHING ELSE TO OFFER": Ibid.

272 "I DISAGREE VEHEMENTLY": Ibid.
"STONEWALLED" IT AS "IMMATERIAL": J. Hink, Cutter Biological, memorandum, "AIDS Meeting Jan. 4th at CDC Atlanta," Jan. 6, 1983.
"TO EXCLUDE SUCH PLASMA": Ibid.
"SURE IT'LL COST MORE": Colon, "reporter's notes."
"A VERY SERIOUS PROBLEM": Ibid.

273 "A MAJOR WORRY": Deposition of Joseph R. Bove, M.D., Feb. 17, 1994, *Jane Doe and John Doe v. Belle Bonfils Memorial Blood Center,* District Court, City and County of Denver, Col., case no. 93CV393, pp. 155–56.
"WHAT DO WE HAVE . . .": Marx, "Health Officials," p. 568.
"WE ARE CONTEMPLATING": Ibid.
"CONCERNED ABOUT THE CONCEPT": Colon, "reporter's notes."
"HOW MANY PEOPLE HAVE TO DIE?": Interview with Dr. Donald Francis.
"I JUST COULDN'T BELIEVE": Ibid.
"A BEND IN THE TRAIN TRACK": Video deposition of Dr. Donald R. Francis, July 3, 1992, District Court, Denver, Col., *Chris and Susie Quintana* v. *United Blood Services,* case no. 86CV11750, p. 45.
"I THINK THEY WERE LISTENING": Interview with Bruce Evatt.

274 "IT HAS LONG BEEN NOTED . . .": P. Cumming, American Red Cross, "Dr. Katz's 1/26/83 AIDS memo," Feb. 5, 1983 (courtesy of the Institute of Medicine).
". . . FOR HEMOPHILIACS I FEAR IT MIGHT BE TOO LATE": D. Francis, Department of Health and Human Services, "Opportunities for Eliminating Blood Donors at Risk for AIDS," memo, Jan. 6, 1983 (courtesy of the Institute of Medicine).

275 JOINT STATEMENT: American Red Cross, American Association of Blood Banks, Council of Community Blood Banks, "Joint Statement on Immune Deficiency Syndrome (AIDS) Related to Transfusion," Jan. 13, 1983 (courtesy of the Institute of Medicine).
". . . THERE IS LITTLE DOUBT IN MY MIND . . .": J. Bove, "Report to the Board, Committee on Transfusion Transmitted Diseases," Jan. 24, 1983.

275    "LAPEL-GRABBING, FINGER-POINTING QUESTIONS": Interview with Tom Asher.

276    "THINK ABOUT THE UNTHINKABLE . . .": Quoted in S. M. I. Gaynor, "Decisions Without Data: An Analysis of Decision Making Concerning the U.S. Blood Supply During the AIDS Crisis," unpublished Ph.D. thesis, University of Michigan, 1991, p. 68.
THE BLOOD SISTERS PROJECT: R. Shilts, *And the Band Played On: Politics, People and the AIDS Epidemic* (New York: Viking Penguin, 1988), p. 456.
DONATIONS HAD DROPPED BY 25 PERCENT: Gaynor, "Decisions Without Data," p. 78.
KELLNER OPENLY SCOFFED: Ibid., p. 187.
PERKINS REBUFFED A PLEA: D. Perlman, "Blood Bank Rebuffs UC on Test for AIDS," San Francisco *Chronicle*, Feb. 4, 1983.
U.S. PUBLIC HEALTH SERVICE AIDS RECOMMENDATIONS: U.S. Public Health Service, "Prevention of Acquired Immune Deficiency Syndrome (AIDS): Report of Inter-Agency Recommendations," *MMWR* 32 (March 4, 1983): 101–3.

277    SUSIE QUINTANA: Details on Quintana case from interviews with Bruce Jones and Maureen Witt; *Chris and Susie Quintana* v. *United Blood Services,* Denver *Post,* Jan. 5, 1988; July 16–Aug. 6, 1992.

278    "THERE ARE NO GAYS . . .": Testimony of Ron Quintana, May 6, 1988, *Chris and Susie Quintana* v. *United Blood Services,* p. 19.
ONE DONOR READ THE SHEETS AND ANSWERED THE QUESTIONS: Written interrogatory, "To the Donor of Blood Unit No. 12-308721," in ibid., p. 25.
DANA KUHN: Biographical information from D. Kuhn, address to annual meeting of National Hemophilia Foundation, Indianapolis, Ind., Nov. 1992, author's notes.
COREY DUBIN: Biographical information from interviews with Corey Dubin.
"THE INITIAL REACTION . . .": Interview with David Watters.

279    PUTTING THE FACILITY ON ROUND-THE-CLOCK SHIFTS: V. Berridge, *AIDS in the UK: The Making of Policy, 1981–1994* (Oxford: Oxford University Press, 1996), p. 42.
ABOUT HALF THEIR FACTOR VIII FROM AMERICAN FIRMS: Department of Health and Social Security, "AIDS—and Blood Donation," press release, Alexander Fleming House, Sept. 1, 1983.
DR. JOHN CRASKE PUBLISHED TWO STUDIES: J. Craske et al., "An Outbreak of Hepatitis Associated with Intravenous Injection of Factor-VIII Concentrate," *Lancet,* Aug. 2, 1975, pp. 221–23; J. Craske et al., "Commercial Factor VIII associated hepatitis, 1974–75, in the United Kingdom: A Retrospective Survey," *Journal of Hygiene* 80 (1978): 327–36.
"IN THE LAP OF THE GODS": "AIDS Poses Threat to Haemophiliacs," *Hospital Doctor* C3 (May 12, 1983), p. 19.
A PATIENT AT THE BONN CLINIC DIED: E. R. Koch, *Böses Blut: Die Geschichte eines Medizin-Skandals* (Hamburg: Hoffmann und Campe, 1990), p. 21.

280    THE NUMBER IN GERMANY HAD RISEN TO SIX: Horace Krever, *Commis-*

*sion of Inquiry on the Blood System in Canada* (Ottowa: Canadian Government Publishing, 1997), vol. 3, p. 848.

280 THE COUNCIL OF EUROPE URGED ITS MEMBER NATIONS: Koch, *Böses Blut,* p. 65.

"... A LIMITATION OF IMPORTS ...": Ibid., p. 70. The import limitation was not the only action rejected at the meeting. Dr. Joanna L'Age Stehr, an epidemiologist at the Robert Koch Institute, had visited the United States to learn more about the AIDS epidemic. Convinced of a pending disaster in the blood system, she and others made several suggestions, including surrogate testing, using cryoprecipitate, or reduction in the size of plasma pools. After lengthy discussion, none of these were supported.

"EXCLUSIVELY IN AMERICA ...": Ibid., p. 53.

280–81 "AS SOON AS I HEARD ABOUT AIDS ...": Interview with Dr. Werner Kalnins.

281 "... ALARMINGLY LARGE": Koch, *Böses Blut,* p. 125.

EVATT FELT HIMSELF SET UP IN A WAY: Interview with Dr. Bruce Evatt.

"PRUDENT ...": J. M. Jason, B. L. Evatt et al., "Acquired Immunodeficiency Syndrome (AIDS) in Hemophiliacs," *Scandinavian Journal of Haematology* 33, suppl. 40 (1984): 355.

DIETRICH HAD A COMPLICATED EXPERIENCE WITH AIDS: Institute of Medicine, "Fact Finding Interview Summary: Dr. Shelby Dietrich (January 11, 1995)" (courtesy of the Institute of Medicine).

282 "THERE IS INSUFFICIENT EVIDENCE ...": "WFH General Assembly: Karolinska Institute—Stockholm, Sweden" *World Federation of Hemophilia,* June 29, 1983, p. 11.

THE WORDING OUTRAGED THE DUTCH REPRESENTATIVES: Interview with Cees Smit.

DR. TAKESHI ABE: Some useful English-language overviews of this episode include E. Feldman, "Blood and Bureaucracy in Japan: Law, Conflict and Compromise," unpublished ms.; D. P. Hamilton, "Japan AIDS Scandal Raises Fear That Safety Came Second to Trade," *Wall Street Journal,* Oct. 9, 1996; E. H. Updike, "Anatomy of a Tragedy," *Business Week,* March 11, 1996, pp. 44–45. Interview sources for background on the Japanese situation include Eric Feldman, Shinichi Tokunaga, and staff members from the *Asahi Shimbun, Yomiuri Shimbun,* and *Mainichi Daily News.*

283 "... IT SEEMED VERY SIMILAR ...": Interview with Tom Spira.

THE PATIENT HAD DIED OF "QUASI-AIDS": Hamilton, "Japan AIDS Scandal."

284 ABE SENT FORTY-EIGHT BLOOD SAMPLES: "Green Cross Knew Danger of Unheated Products," *Yomiuri Shimbun,* Aug. 25, 1996.

"... THE NEED TO WORRY ABOUT AIDS IS SLIGHT ...": Feldman, "Blood and Bureaucracy," p. 29.

"[HE] WAS A HOMOSEXUAL ...": Ibid., p. 2.

"... WE PREFER TO HIDE THE REAL DATA ...": Ibid., p. 41.

285 JEAN PÉRON-GARVANOFF: Biographical information from interview with Jean Péron-Garvanoff.

ALLAIN HAD IMPECCABLE CREDENTIALS: J.-P. Allain, *Le SIDA des hémophiles: Mon témoignage* (Paris: Editions Frison-Roche, 1993).

"... HE'S A SAINT!": Ibid., p. 126.

286     "I REMEMBER THAT DATE . . .": Interview with Jean Péron-Garvanoff.
        SURVEY OF TWENTY-THREE HUNDRED HEMOPHILIACS: Commission
        Consultative de la Transfusion Sanguine, "Séance du 9 Juin 1983: Procès-
        verbal," in J. Géronimi and M. Lucas, *Rapport d'Enquête Sur Les Col-
        lectes de Sang en Milieu Pénitentiare* (Paris: Inspection Générale des
        Services Judiciaires, Inspection Générale des Affaires Sociales, 1992),
        annex 56.
        *"JE NE SUIS PAS RASSURÉ . . .":* Interview with Jean Péron-Garvanoff.
287     JEAN PIERRE SOULIER: Interview with Dr. Jean Pierre Soulier; J. P.
        Soulier, *Transfusion et SIDA: Le Droit à la vérité* (Paris: Editions Frison-
        Roche, 1992).
        "PROTEST TO THE POINT OF THREATENING": A. M. Casteret, *L'Affaire
        du sang* (Paris: Editions la Découverte, 1992), p. 68.
287–88  "TEMPER THEIR ENTHUSIASM": Soulier, *Transfusion et SIDA,* p. 40.
288     "DONORS WERE LIKE GODS": Interview with Claudine Hossenlopp,
        CNTS.
        "FAGGOTS—AN UNDESIRABLE BLOOD GROUP?": Quoted in Soulier,
        *Transfusion,* pp. 173–74.
        ATTEMPTED A "SMOOTHER" APPROACH: Ministère des Affaires Sociales
        et de la Solidarité Nationale, "Circulaire DGS/3B No. 569 du 20 Juin
        1983," author's correspondence from CNTS official, Dec. 24, 1996.
289     "FEUDAL SYSTEM CONSISTING OF MULTIPLE BARONIES": J. Géronimi
        and M. Lucas, *Collectes,* p. 94.
        LIFESTYLE SCREENING WAS "SYSTEMATICALLY FORGOTTEN": M. Set-
        bon, "Politique de santé et information: L'Affaire du sang contaminé,"
        *Récherche* 24 (May 1993): 624–27.
        "IMMEDIATE AND STRICT" APPLICATION OF SCREENING: Ministère
        des Affaires Sociales et de la Solidarité Nationale, in Géronimi and
        Lucas, *Collectes,* amend 97; "Lettre-Circulaire DGS/3B/80 du 16 Janvier
        1985."
        PRISONERS WHO GAVE BLOOD ENJOYED IT: Géronimi and Lucas, *Col-
        lectes,* pp. 11–20.
        PRISON BLOOD NEVER EXCEEDED .5 PERCENT: Ibid., p. 24.
        DR. LUC NOËL EXAMINED 212: Interview with Dr. Luc Noël; Noël et al.,
        "Marqueurs du VHB, bêta 2 microglobuline at anti HTLV dans une popu-
        ation de donneurs de sang en milieu carcéral," *Revue française de transfu-
        sion et immuno-hématologie* 27 (1984): 537–41.
290     DOCTORS IN STRASBOURG AND TOULOUSE FOUND HIGH VIRAL MARK-
        ERS: A. Falkenrodt, "Explorations biologiques et recherches de déficits
        immunitaires chez les donneurs de sang en milieu carcéral," *Revue
        française de transfusion et immuno-hématologie* 27 (1984): 525–29; J.
        Ducos, "Etudes des marqueurs de l'HBV et des populations lymphocy-
        taires chez le polytoxicomane asymptomatique," *Revue française de trans-
        fusion et immuno-hématologie* 27 (1984): 549–53.
        A BUREAUCRATIC SIDESHOW IN THE DEPARTMENT OF JUSTICE: Géro-
        nimi et al., *Collectes,* pp. 133–37.
        "A LITTLE PUBLIC-HEALTH RESEARCH": Interview with Dr. Pierre
        Espinoza.
291     CONVEYED THEIR SUSPICIONS TO JUSTICE MINISTER EZRATTY:

Espinoza's repeated attempts to get action from Roux and Ezratty is documented in Géronimi and Lucas, *Collectes,* pp. 150–62, with exhibits.

291 ESPINOZA AND DUEDARI COMPLETED THEIR SURVEY: P. Espinoza, "Don du sang au grand quartier du Centre Pénitentiaire de Fresnes," letter, June 20, 1985, in Géronimi and Lucas, *Collectes,* annex 112.

292 DR. EDGAR ENGLEMAN: Engleman's experiences and impressions recounted in testimony of Dr. Edgar Engleman, *Chris and Susie Quintana v. United Blood Services;* testimony of Edgar G. Engleman, July 13, 1990, U.S. Congress. *Blood Supply Safety,* pp. 33–37.

"DECISIONS WITHOUT DATA": Gaynor, "Decisions Without Data."

293 *"A RECALL ACTION SHOULD NOT CAUSE ANXIETY . . .":* "Medical Bulletin #7," National Hemophilia Foundation, May 11, 1983.

"MY POSITION IS BUSINESS AS USUAL . . .": "Highlights of the June 1983 Plasma Forum Program," *Plasma Quarterly,* Winter 1983, p. 106.

CHRISTOPHER WHITFIELD: Details of the Whitfield episode from J. F. Huxsoll, "Recall on Koate and Konyne with AIDS Donor—Status Report No. 1," Cutter Biological memo, Nov. 4, 1983, J. F. Huxsoll, "Recommendation for Corrective Action—Koate and Konyne With AIDS Donor," Cutter Biological memo, Nov. 11, 1983; R. J. Modersbach, "Inquiry from Austin American Statesman," Cutter memo, Jan. 13, 1984; Austin *American-Statesman,* Oct. 25, 30, 1983; Jan. 4, 1984; Nov. 2, 1988; courtesy of Austin History center.

"THERE IS NO EVIDENCE . . .": R. Sullivan, "Blood Plasma Is Withdrawn as AIDS Link," *New York Times,* Nov. 2, 1983.

294 "NO CUTTER CENTERS IN NEW YORK . . .": Cutter Biological, "Cutter Laboratories Announces Plasma Donor Screening Program," news release, Feb. 23, 1983.

THEY SHIPPED AT LEAST SOME: "Plasma Shipping-Receiving Report," Alpha Therapeutic Corp. no. 11648, Dec. 6, 1982.

"ONE IN A MILLION": American Red Cross, American Association of Blood Banks, Council of Community Blood Banks, "Joint Statement on Directed Donations and AIDS," June 22, 1983, courtesy of the Institute of Medicine.

BASED ON A BACK-OF-THE-ENVELOPE CALCULATION: Interview with James McPherson.

AS HE LATER EXPLAINED TO AUTHOR RANDY SHILTS: Shilts, *Band Played On,* p. 398.

295 DR. HERBERT PERKINS CONDUCTED A TRIAL: Deposition of Herbert Perkins, M.D., Jan. 11, 1994, *Marietta Advincula v. United Blood Services, Inc.,* Circuit Court of Cook County, Ill., case no. 89L7199.

CDC SCIENTISTS SCORNED PERKINS'S EXPERIMENT: Interviews with Drs. Donald Francis and Bruce Evatt.

THEY RAISED THE ISSUE AT A MEETING: Office of Biologics Research and Review, Food and Drug Administration, "Summary Minutes—Meeting 10, Blood Products Advisory Committee," Dec. 15–16, 1983.

"THE GENERAL THRUST . . .": S. J. Ojala, Cutter Biological, "Trip Report, FDA-NIH Non-Specific Testing Meeting Dec. 15–16, 1983," memo, Dec. 19, 1983.

HEPATITIS B CORE TESTING WAS "NOT APPROPRIATE": Hepatitis B Core

Antibody Testing Study Group, "Interim Study Statement," internal memo to the FDA Blood Products Advisory Council, March 6, 1984.

295 "WE NOW HAD PATIENTS . . .": Deposition of Herbert Perkins, p. 294.

296 CUTTER AND ALPHA EXPERIMENTED WITH CORE TESTS: Institute of Medicine, "Meeting Summary: Alpha Therapeutics," Sept. 8, 1994; letter from M. Carr, Alpha Therapeutic Corporation, to John C. Petricciani, Food and Drug Administration, March 15, 1983; letter from J. C. Petricciani, Food and Drug Administration, to Marietta Carr, Alpha Therapeutic Corporation, May 3, 1983, all courtesy of the Institute of Medicine.

HE HAD PREVENTED THIRTY-THREE RECIPIENTS: Institute of Medicine, *HIV and the Blood Supply: An Analysis of Critical Decisionmaking* (Washington, D.C.: National Academy Press, 1995), p. V-15.

"IT TURNS TO GLUE": Information on early attempts at heat treatment from interviews with John Ashworth, Charles Hildebrant, Edward Shanbrom; see also next note.

297 "THAT *IF* AIDS WERE CAUSED BY A VIRUS . . .": Institute of Medicine, "Expert Report—Milton M. Mozen," Cutter Biological, Oct. 5, 1994, p. 17 courtesy of the Institute of Medicine.

THE FOUNDATION URGED THAT DOCTORS "STRONGLY CONSIDER": National Hemophilia Foundation, "Medical Bulletin #15," Oct. 13, 1984.

298 ". . . ANOTHER MIRACLE . . .": Shilts, *Band Played On,* pp. 450–51.

FORTY-NINE AMERICANS HAD BEEN INFECTED: Institute of Medicine, *HIV and Blood Supply,* pp. 3–6.

CONTRARY TO HECKLER'S PREDICTION: Under normal circumstances the difference between six months and a year might not seem great. However, in this situation people were actively pushing for the use of a surrogate test, which could have been put in place right away. The press conference convinced people that the surrogate test was unnecessary, because a "real" AIDS test would soon be ready. As it turned out, the test wasn't ready for another year. During that time, 121 additional hemophiliacs and transfusion recipients were diagnosed with AIDS.

## CHAPTER 16: "ALL OUR LOTS ARE CONTAMINATED"

300 THE SECOND LIMITATION OF ELISA: The ELISA test had a 3 percent false-positive rate when it was commercially introduced in the spring of 1985. Since then the test has been improved: The false-positive rate is now only 0.5 percent. Still, that is high enough so that all positives must be confirmed with the more specific Western Blot test. Interview with Charles Schable, chief of the HIV Serology Section, CDC.

"JUST FANTASTIC.": L. K. Altman, "Blood Supply Called Free of AIDS," *New York Times,* Aug. 1, 1985.

301 "THERE WAS A TERRIBLE PERIOD . . .": V. Berridge, *AIDS in the UK: The Making of Policy, 1981–1994* (Oxford: Oxford University Press, 1996), p. 47.

"WE WROTE A LETTER TO CLINICIANS . . .": Interview with David Watters.

301 NOT UNTIL 1987: "Blood: A Commodity, After All," *Economist,* Sept. 19, 1987.

302 "THE CHANCE TO STEP INTO THE BUSINESS . . .": P. Marsh, "Testing Time for AIDS Screening," *Financial Times,* July 31, 1985; see also "Ministers Delayed Launch of AIDS Test," *New Scientist,* Aug. 8, 1985, p. 16; J. A. F. Napier, "Delayed AIDS Testing," rebuttal letter, *New Scientist,* Aug. 22, 1985, p. 55.

RED CROSS DECIDED TO REDIRECT ITS PLASMA TO CUTTER: A. Picard, *The Gift of Death: Confronting Canada's Tainted Blood Tragedy* (Toronto: HarperCollins, 1995), p. 175.

MORE THAN ELEVEN MILLION UNITS: Ibid., p. 120.

302–3 FIFTY-FIVE TRANSFUSION RECIPIENTS: Ibid., p. 135.

303 ". . . 'HERE'S A NEW PRODUCT, A BETTER ONE' . . .": Interview with Werner Kalnins.

"PEOPLE KEPT TAKING THE OLD STUFF": Years later, in 1994, an Investigative Committee of the German Parliament confirmed these allegations. According to the committee's report, the federal government never recalled the unheated Factor VIII that was still in the marketplace, even as the heated product came into use. Indeed, the shelf life of Factor VIII was such that contaminated factor could have remained for two years in a patient's refrigerator before he unthinkingly injected it. Yet the government never collected this material, or notified patients to throw it away. This failure to inform, according to the committee, represented "negligent behavior on the part of the attending physicians." Horace Krever, *Commission of Inquiry on the Blood System in Canada* (Ottowa: Canadian Government Publishing, 1997), vol. 3, p. 860.

A SMALL PROPORTION OF THE SWISS BLOOD TESTED POSITIVE: M. Simons, "Swiss Red Cross Faces AIDS Probe," *New York Times,* May 22, 1994.

"THIS WAS DONE BY OUR BEAUTIFUL RED CROSS . . .": Ibid.

304 THE JAPANESE CONSUMED: "Executive Pipeline Links Ministry, Plasma Firms," Mainichi *Daily News,* March 19, 1988; "The Market for Blood Products in Japan," *COMLINE Daily News: Biotechnology and Medical Technology,* Aug. 30, 1988.

"AIDS IS A MATTER OF SERIOUS CONCERN.": D. P. Hamilton, "Japan AIDS Scandal Raises Fear That Safety Came Second to Trade," *Wall Street Journal,* Oct. 9, 1996.

"ONE MUST ALSO BE READY . . .": A. Noguchi, "The Fall of Japan's Top Blood Company," *Tokyo Business Today,* Sept. 1988, p. 59.

*AMAKUDARI*: E. Feldman, "Blood and Bureaucracy in Japan: Law, Conflict and Compromise," prepared for p. 94.

305 "THE CORPORATE CULTURE DISTINCTLY CHANGED . . .": Interview with former Alpha Therapeutic executive.

"THERE IS NO GUARANTEE . . .": "1983 Document Shows Green Cross Knew Risk," *Yomiuri Shimbun,* Sept. 22, 1996.

"THE RISK IS NEARLY ZERO . . .": Ibid.

"UNAVOIDABLE . . . SUSPICIONS . . . UNFORTUNATE . . .": Hamilton, "Japan AIDS Scandal."

305–6 "WITHIN THE SUBCOMMITTEE . . .": Feldman, "Blood and Bureaucracy," p. 29.

306 MORE THAN EIGHTEEN HUNDRED JAPANESE HEMOPHILIACS: "Opinion Regarding the Settlement Recommendation," Oct. 6, 1995, Tokyo District Court, 15th Department of Civil Cases.
GOVERNMENT INVESTIGATIONS EVENTUALLY REVEALED: "Green Cross Falsified Report on Recalling Unheated Blood Products," *Yomiuri Shimbun,* Feb. 28, 1996; Feldman, "Blood and Bureaucracy," pp. 37–39; "Opinion Regarding Settlement Recommendation."

307 ANYWHERE FROM 10 TO 30 PERCENT: A. M. Casteret, *L'Affaire du sang* (Paris: Editions la Découverte, 1992), p. 31.
TRAVENOL APPROACHED CNTS: Letter from C. Cibault, Directeur des Affaires Scientifiques, Travenol, to Dr. M. Garretta, May 10, 1983, in Michel Lucas, *Transfusion Sanguine et SIDA en 1985* (Paris: Inspection Générale des Affaires Sociales, 1991), annex 4.

308 "FOR ME IT WAS SIMPLE . . .": Casteret, *L'Affaire,* pp. 107–9; J.-P. Allain, *Le SIDA des hémophiles: Mon témoignage* (Paris: Editions Frison-Roche, 1993), pp. 40–42.
"A GENUINE ALARM CRY": Sénat, 2d session ordinaire de 1991–1992, *Rapport de la commission d'enquête sur le système transfusionel français en vue de son éventuelle réforme,* no. 406 (Paris: Imprimerie du Sénat, 1992), p. 105.
DR. JEAN BRUNET REPORTED: Commission Consultative de la Transfusion Sanguine, "Séance du 22 Novembre 1984: Procès-verbal," p. 7, in J. Géronimi and M. Lucas, *Rapport d'Enquête Sur Les Collectes de Sang en Milieu Penitentiaire* (Paris: Inspection Générale des Services Judiciaires, Inspection Générale des Affaires Sociales, 1992), annex 95.
DR. JACQUES LEIBOWITCH: J. Leibowitch et al., "Expérience d'un dépistage systématique anticorps anti-HTLV III/LAV chez des donneurs de sang," in Géromini and Lucas, *Les Collectes de sang en milieu pénitentiaire* (Paris: Inspection Générale des Services Judiciaires, Inspection Générale des Affaires Sociales, 1992), app. 106.

309 ALL THE PLASMA PRODUCTS WERE "CURRENTLY INFECTED": Letter from Dr. J. Brunet "à l'attention de Monsieur Roux," March 12, 1985.
COME "TO SWITCH TO HEAT-TREATED FACTOR VIII CONCENTRATES . . .": "Blood Transfusion, Haemophilia, and AIDS," editorial, *Lancet,* Dec. 22, 1984, pp. 1433–35.
IN EARLY 1985: Letter from J. P. Allain to Prof. J. Ruffié and Dr. M. Garretta, Jan. 16, 1985, J.-P. Allain, *Le SIDA des hémophiles: Mon témoinage* (Paris: Éditions Frison-Roche, 1993), pp. 157–58.
STUDY OF EIGHTEEN "VIRGIN" HEMOPHILIACS: C. Rouzioux et al., "Absence of Antibodies to AIDS Virus in Haemophiliacs Treated with Heat-Treated Factor VIII Concentrate," letter, *Lancet,* Feb. 2, 1985, p. 271.
"IMPORTS PROVE VERY COSTLY . . .": "Judgement Hearing on the 23rd of October 1992, 16th Division," p. 77.
PÉRON-GARVANOFF FOUND HIMSELF THWARTED: Interview with Jean Péron-Garvanoff.

310 A MEMORABLY AWKWARD SCENE: Allain, *SIDA,* p. 47.
"WORKING FROM THE INSIDE": "Judgment Hearing on the 23rd of October, 1992, 16th Division," p. 76.

311      NATIONAL PUBLIC HEALTH LABORATORY DELAYED ABBOTT'S APPLICATION: The sequence of events in the delay, along with exhibits, is detailed in M. Lucas, *Transfusion Sanguine et SIDA en 1985* (Paris: Inspection Générale des Affaires Sociales, 1991), pp. 45–51.

"THE MOMENT THE TESTS ARE AUTHORIZED . . .": Secrétariat Général du Gouvernement, "Compte-rendu de la réunion interministérielle tenue de 9 mai 1985," in Lucas, *Transfusion,* annex 17.

". . . IT DOES NOT SEEM POSSIBLE TO DELAY . . .": Note from Dr. Robert Netter to Dr. Weiselberg and E. Hervé, "Enregistrement de réactifs pour le diagnostic du SIDA," April 25, 1985, in ibid., annex 14.

312      "ABSOLUTE URGENCY . . .": Letter from M. Garretta to Madame M. T. Pierre, Ministère des Affaires Sociales et de la Solidarité Nationale, May 9, 1985, ibid., annex 16.

"THE ENTIRE STOCK OF 'CONTAMINATED PRODUCTS' . . .": "Judgement Hearing on the 23rd of October 1992, 16th Division," trans. of trial transcript, p. 30.

"ALL OUR LOTS ARE CONTAMINATED . . .": B. Girault, "Compte rendu de la réunion du 29 mai 1985," in Lucas, *Transfusion,* annex 18.

313      "MASSIVE AND TRANSITORY": "Judgement Hearing on the 23rd of October 1992, 16th Division," p. 66.

". . . FOR HOW MUCH AIDS WILL WE THEREFORE BE RESPONSIBLE?": Ibid.

"THE DISTRIBUTION OF NONHEATED PRODUCTS . . .": Lucas, *Transfusion* (summary report), p. 41.

"INSISTENT REQUEST . . .": Dr. B. Habibi, CNTS, "Distribution de fractions coagulantes durant les mois de juillet et août," memo, July 3, 1985, in Lucas, *Transfusion,* annex 28.

"THERE'S ONLY A LITTLE . . .": Interview with Jean Péron-Garvanoff.

314      FIFTEEN STILL TESTED POSITIVE FOR AIDS: P. Espinoza, "Don du sang au grand quartier du Centre Pénitentiaire de Fresnes," letter, July 26, 1985.

"BASICALLY HE'S RIGHT": Géronimi et al., *Collectes,* p. 169.

HEALTH MINISTRY BUREAUCRATS DELETED THE OFFENDING PARAGRAPHS: Ibid., p. 170.

DUEDARI SENT THE DATA: Letter from N. Duedari to M. Edmond Hervé, Aug. 2, 1985; N. Duedari, "Réflexion sur les collectes de sang en milieu carcéral," Aug. 2, 1985.

315      TO USE THE WORDS "PRISON" AND "AIDS": Géronimi et al., *Collectes,* p. 176.

BAHMAN HABIBI SENT A QUESTIONNAIRE: B. Habibi, "Rapport à la Commission Consultative Nationale de la Transfusion Sanguine," Nov. 7, 1985.

PRISON PLASMA: Although fractionators stopped using prison plasma for clotting factors, some continued using it for other products, such as immune globulins. Alpha Therapeutic continued the practice until 1985, and Cutter did so until 1989. Under current FDA guidelines prison plasma can no longer be used. Letter to Hon. Christopher Shays from Dan McIntyre, Bayer Corporation, Jan. 19, 1996, in U.S. Congress, House of Representatives, Subcommittee on Human Resources and Intergovernmental Relations of the Committee on Government Reform and Oversight, *Pro-*

*tecting the Nation's Blood Supply from Infectious Agents: New Standards to Meet New Threats,* 104th Cong., 1st sess., October 12 and November 2, 1995 (Washington, D.C.: Government Printing Office, 1996), pp. 202–3; "Baxter Healthcare Corporation's Response to the Human Resources and Intergovernmental Relations Subcommittee," ibid., p. 208; letter to Hon. Christopher Shays from Edward A. Colton, Alpha Therapeutic Corporation, Dec. 26, 1995, ibid., p. 211; interview with Anne-Marie Finley.

315    JOINT LETTER WARNING AGAINST PRISON BLOOD: "6me Lettre commune A.D.T.S.-S.N.T.S."
       ALMOST ALL LOCAL BLOOD BANKERS ABANDONED THE PRISONS: Géronimi et al., *Collectes,* p. 177.

316    "... 'GEE, THE WHOLE WORLD IS GETTING BETTER' ...": Interview with Dr. Robert Westphal.
       "WHEN A MOVIE STAR WAS DIAGNOSED ...": R. Shilts, *And the Band Played On: Politics, People and the AIDS Epidemic* (New York: Viking Penguin, 1988), p. xxi.
       "OUR SOCIETY IS GENERALLY THREATENED": W. F. Buckley, "Crucial Steps in Combating the AIDS Epidemic: Identify All the Carriers," *New York Times,* March 18, 1986.

317    27 PERCENT: "The Blood Frights Put in Perspective," editorial, Kansas City *Star,* July 26, 1987.
       THE CENTER SENT OUT TWO UNTESTED PINTS: Interview with Peter Smith.
       OFFICIALS FOUND HUNDREDS OF VIOLATIONS: Testimony of Gerald V. Quinnan, Jr., Acting Director, Center for Biologics Evaluation and Research, Food and Drug Administration, and Mary Carden, National Expert Investigators for Biologics, Buffalo District Office, U.S. Congress, House of Representatives, Committee on Energy and Commerce, Subcommittee on Oversight and Investigations, *Blood Supply Safety,* 102nd Cong., 1st sess. (Washington, D.C.: U.S. Government Printing Office, 1990), pp. 40–81, 98–135.

317–18 INSPECTORS FOUND A "SYSTEMATIC FAILURE ...": Food and Drug Administration, "Inspectional Observations ... Los Angeles–Orange County Region," inspection report, Sept. 12–Nov. 18, 1988.

318    "... CLEAR THE PIPELINE THROUGH NORMAL SALES ...": T. Johnson, Cutter Biological, "Unscreened Inventory for HTLV-III," memo, March 19, 1986.
       UGLY SITUATION IN COSTA RICA: Institute of Medicine, "Meeting with Miles Legal Counsel," Aug. 3, 1994, courtesy of the Institute of Medicine; S. Benesch, "U.S. Firm Accused in Latins' AIDS," St. Petersburg *Times,* Jan. 11, 1993.

318–19 ARMOUR PHARMACEUTICAL COMPANY: Krever, *Commission of Inquiry,* vol. 1, pp. 488–502.

319    "THE COMPANY WAS A LITTLE SLOW ...": J.-Y. Nau and F. Nouchi, "Un Entretien avec M. Alain Mérieux," *Monde,* Nov. 2, 1992.
       PLASMA PHARM SERA SENT CONTAMINATED BLOOD PRODUCTS TO PORTUGAL: "O escândolo do sangue em Portugal," *Semanário,* July 4, 1992; "Infected Blood Problems in Portugal," SCRIP no. 1743, Aug. 12, 1992, W. Boryli, p. 7; "Une Firme autrichienne aurait exporté des produits sanguins contaminés par le SIDA," *Monde,* July 7, 1992.

320     "STRIKING AND ALARMING" RESULT: J. Jason et al., "Human T-Lymphotropic Retrovirus Type III/Lymphadenopathy-Associated Virus Antibody," *JAMA* 253 (June 21, 1985): 3409–15.

IN ENGLAND: "Blood Transfusion, Haemophilia, and AIDS," *Lancet*, Dec. 22/29, 1984, p. 143.

IN CANADA: Picard, *The Gift of Death*, p. 106.

FRENCH SCIENTISTS: Sénat, 2d session ordinaire de 1991–1992, *Rapport de la commission d'enquête sur le système transfusionel français en vue de son éventuelle réforme*, no. 406 (Paris: Imprimerie du Sénat, 1992), p. 100.

IN DENMARK: B. Evatt et al., "Coincidental Appearance of LAV/HTLV-III Antibodies in Hemophiliacs and the Onset of the AIDS Epidemic," *New England Journal of Medicine* 312 (Feb. 21, 1985): 483.

YASUNORI AKASE LEARNED: Interview with Shinichi Tokonaga.

321     COREY DUBIN LEARNED: Interview with Corey Dubin.

DANA KUHN FOUND IT DIFFICULT TO GET TESTED: D. Kuhn, Address at National Hemophilia Foundation annual meeting, Oct. 4, 1993, Indianapolis.

". . . SHORT, THEY'VE GIVEN ME AIDS": Testimony of Ron Quintana, May 6, 1988, *Chris and Susie Quintana v. United Blood Services*, District Court, Denver, Col., case no. 86CV11750, p. 31.

## CHAPTER 17: JUDGMENT

322     HE FOUND HER SITTING ALONE: Video Deposition of Marshall Donnelly Burke, June 27, 1992, *Chris and Susie Quintana v. United Blood Systems*, District Court, Denver, Col., case no. 86CV11750.

323     "THIS WAS A TRAGEDY . . .": D. Barr, address at 45th annual meeting of the American Association of Blood Banks, San Francisco, Nov. 9, 1992.

"I DID NOT KNOW . . .": Written interrogatory, "To the Donor of Blood Unit No. 12-308721," *Quintana v. United Blood Services*.

324     "COMMUNICATION PRINCIPLES": H. Pankratz, "Gay Thought Only Promiscuous Got AIDS," *Denver Post*, July 22, 1992.

"OFFENSIVE AND UNTHINKABLE": H. Pankratz, "Witness: Sex Query 'Unthinkable' in '83" *Denver Post*, July 28, 1992.

"WE FINALLY CAME TO AN ACCORD . . .": S. Garnaas, "AIDS Victim Loses Blood-Bank Lawsuit," *Denver Post*, June 5, 1988.

"A TOUGH DECISION . . .": "Blood Bank Is Cleared in Colorado Case," *Associated Press*, June 5, 1988.

"NO MEDICAL, ECONOMIC OR SOCIAL REASON . . .": Deposition of J. Garrott Allen, March 23, 1987, *Stella Mae McKee et al. v. Cutter Laboratories et al.*, U.S. District Court, Eastern District of Kentucky, case no. 62-248.

325     THE ONLY CLEAR VICTORY: *Steven and Jason Christopher, Brenda Walls v. Cutter Laboratories, Armour Pharmaceutical Company*, U.S. Court of Appeals, 11th Circuit, case no. 93-3212, June 2, 1995.

MICHAEL ROSENBERG FORMED A SPIN-OFF GROUP: Interview with Michael Rosenberg.

326     "IT IS LIKE A MOB SCENE . . .": M. McLeod, "Bad Blood," *Florida Magazine* (Sunday supplement to Orlando *Sentinel*), Dec. 19, 1993, p. 22.

327 "I SIMPLY CANNOT UNDERSTAND . . .": Noryasu Akase, "Statement," July 25, 1989, courtesy of M. Kobayashi.

ALLOWED THEM TO TESTIFY FROM BEHIND A SCREEN: "Opinion Regarding the Settlement Recommendation," Oct. 6, 1995, Tokyo District Court, 15th Dept. of Civil Cases.

328 "I DON'T WANT TO DIE . . .": E. Feldman, "Blood and Bureaucracy in Japan: Law, Conflict and Compromise," p. 17.

"MY CD4 COUNT . . .": Testimony of Yoshiati Ishida, Feb. 13, 1992, 14 oral proceeding Tokyo litigation, notes courtesy of M. Kobayashi.

REYUHEI KAWADA ROSE TO TAKE ISHIDA'S PLACE: A. Pollack, "Dying of AIDS, Japanese Youth Wants Apology," *New York Times,* Oct. 29, 1995.

329 "TAKING DOWN THE WALL AROUND THE BUREAUCRATS . . .": S. WuDunn, "Japanese Aide Gains Favor by Fighting the Bureaucrats," *New York Times,* Nov. 11, 1996.

". . . I MAKE A HEARTFELT APOLOGY . . .": Feldman, "Blood and Bureaucracy," p. 31.

EXECUTIVES OF THE DRUG COMPANIES BOWED IN SHAME: "Japanese HIV Suit Settled," Associated Press, March 14, 1996; Feldman, "Blood and Bureaucracy," pp. 35–36.

330 HE HAD FOUND IT NECESSARY TO HAVE A BODYGUARD: Interview with Anne-Marie Garretta.

PÉRON-GARVANOFF HAD BEEN SUFFERING TERRIBLY: Interview with Jean Péron-Garnavoff.

331 *"ASSASSINS!":* J. Kramer, "Bad Blood," *New Yorker,* Oct. 11, 1993, p. 78.

ROUND-ROBIN OF FINGER-POINTING: "Judgement Hearing on the 23rd of October 1992, 16th Division."

332 *LANCET* RAN AN EDITORIAL: "Palais d'Injustice," *Lancet,* July 24, 1993, p. 188.

". . . HIGH LEVEL OF PROFESSIONAL COMPETENCE . . .": P. J. Lachmann et al., "Statement of the Royal College of Pathologists on the Matter of Professor Jean-Pierre Allain," Nov. 19, 1992, courtesy of Dr. Robin.

ACTIVISTS RUSHED DR. BAHMAN HABIBI: Interview with Dr. Ronald Gilcher.

333 "GUILTY BUT NOT RESPONSIBLE": Kramer, "Bad Blood," p. 80.

JUDGES FOUND THREE OF THE FOUR DEFENDANTS GUILTY: C. Tastemain, "Three Physicians Convicted in French 'Blood-Supply Trial,' " *Science* 258 (Oct. 30, 1992): 735.

"HE WHO ABSTAINS . . .": A. Dorozynski, "French Tainted Blood Affair Continues Beyond Court Sentence," *British Medical Journal,* Oct. 31, 1992, p. 1047.

334 UNNECESSARY CONTAMINATION OF 70 TO 350 HEMOPHILIACS: Interviews with Drs. Jean Pierre Soulier and Helen Lee.

PENITENTIARIES ACCOUNTED FOR 25 PERCENT: Géronimi et al., *Les Collectes de sang en milieu pénitentiaire* (Paris: Inspection Générale des Services Judiciaires, Inspection Générale des Affaires Sociales, 1992), p. 64.

WOULD NOT HAVE BEEN "CONVENIENT": Interview with Dr. Pierre Espinoza.

334    ELIMINATED 86 PERCENT OF THE HIGH-RISK DONORS: H. Perkins, "The Safety of the Blood Supply: Making Decisions in Transfusion Medicine," in S. J. Nance, ed., *Blood Safety: Current Challenges* (Bethesda, Md.: American Association of Blood Banks, 1992), p. 141.

335    A COMPARATIVE STUDY OF AIDS TRANSFUSION RATES: Interview with Michel Setbon; M. Setbon, "Politique de santé et information: L'Affaire du sang Contaminé," *Recherche* 24 (May 1993): 624–27.
       "THE ENTIRE BLOOD INDUSTRY WAS NEGLIGENT . . .": H. Pankratz, "AIDS Expert: Blood Bank Negligent in '83," Denver *Post,* July 16, 1992.
       "INABILITY TO ACCEPT REALITY . . .": Video deposition of Dr. Donald R. Francis, July 3, 1992, *Quintana* v. *United Blood Services.*

336    "ALL I CAN SAY IS THAT IT'S TERRIBLE . . .": H. Pankratz "Plaintiff in AIDS Case Dies," Denver *Post,* Aug. 1, 1992.
       "WORSE THAN I'D EVER SEEN HER": Ibid.

336–37 "ONE SIMPLE QUESTION . . .": Ibid.

337    THE BLOOD BANK'S ATTORNEYS APPEALED: H. Pankratz, "$8 Million Awarded in AIDS Trial," Denver *Post,* Aug. 2, 1992.
       INSTITUTE OF MEDICINE RELEASED AN ANALYSIS: Institute of Medicine, *HIV and the Blood Supply: An Analysis of Critical Decisionmaking* (Washington, D.C.: National Academy Press, 1995).

338    ". . . OUR ENTIRE PUBLIC HEALTH SYSTEM MISSED OPPORTUNITIES . . .": Statement of Donna Shalala, Secretary, Health and Human Services. U.S. Congress, House of Representatives, Committee on Government Reform and Oversight, Subcommittee on Human Resources and Intergovernmental Relations, *Protecting the Nation's Blood Supply from Infectious Agents: New Standards to Meet New Threats,* 104th Cong., 1st sess., Oct. 12 and Nov. 2, 1995 (Washington, D.C.: Government Printing Office, 1996), p. 9.
       "MANY OF THE IOM'S FINDINGS . . .": Testimony of James Reilly, American Blood Resource Association, *Protecting the Nation's Blood Supply,* p. 181.
       "IN REALITY . . .": T. F. Zuck and M. E. Eyster, "Blood Safety Decisions, 1982 to 1986: Perceptions and Misconceptions," *Transfusion* 36 (1996): 928.

339    ONE IN FOUR HUNDRED DONORS TESTED POSITIVE: "Results of Human T-Lymphotropic Virus Type III Test Kits Reported from Blood Collection Centers—United States, April 22–May 19, 1985," *MMWR* 34 (June 28, 1985). This survey did not specifically identify AIDS carriers, because no confirmatory test was available at the time. Jane Starkey of America's Blood Centers asserts that as many as 90 percent of these tests may have been false positives.
       ". . . WE HADN'T A CLUE . . .": Interview with James McPherson.

340    "IT APPEARS TO ME TO BE ADVISABLE . . .": G. Gaul, "Judge Allows Use of AIDS Memo in Hemophiliacs' Suit," Philadelphia *Inquirer,* May 16, 1990.
       "I WOULD LIKE TO OFFICIALLY AND OPENLY APOLOGIZE . . .": E. Shanbrom, address at International Conference on AIDS in the Blood Supply, Kobe, Japan, Nov. 1996, author's notes.

341    "YOU ARE SURROUNDED . . .": Author's notes.
343    ". . . *KILLED MY SON!*": Ibid.

343          "TODAY, I GOT UP TO SHOWER . . .": Author correspondence.
344          "THEY KILLED OUR SONS!": Author's notes.

## EPILOGUE: BLOOD IN A POST-AIDS SOCIETY

345          "IN RECENT YEARS . . .": R. Westphal, "Donors and the U.S. Blood Sup-
             ply," *Transfusion* 37 (1997): 237–41.
             MOST ESTIMATES: U.S. General Accounting Office, *Blood Supply: Trans-
             fusion-Associated Risks* (Washington, D.C.: Government Accounting
             Office, Feb. 1997), p. 10.
346          HALF OF AMERICA'S HEMOPHILIA POPULATION: Institute of Medicine,
             *HIV and the Blood Supply: An Analysis of Critical Decisionmaking* (Wash-
             ington, D.C.: National Academy Press, 1995), p. 1.
             WORLDWIDE MORE THAN FORTY THOUSAND HEMOPHILIACS: Cour-
             tesy World Federation of Hemophilia.
             COUNTRIES WITH THE HIGHEST RATES: S. Franceschi et al., "Trends in
             Incidence of AIDS Associated with Transfusion of Blood and Blood Prod-
             ucts in Europe and the United States, 1985–93," *British Medical Journal*
             311 (Dec. 1995): 1534–36.
346–47       "MOST . . . UTILIZE PROFESSIONAL PAID DONORS . . .": R. Westphal,
             *Blood Programme Department: Report of a Mission to Central America,*
             internal report, League of Red Cross and Red Crescent Societies, Dec. 8,
             1992.
347          IN PAKISTAN "LITTLE IF ANY TESTING . . .": Interview with Robert West-
             phal.
             "LYING IN THE SUN": J. Leikola, "Blood Transfusion in Developing Coun-
             tries: Problems and Progress," *Vox Sanguinis* 46 (1984), p. 53.
             40 PERCENT OF AIDS CASES IN PAKISTAN: S. A. Mujeeb and A. Hafeez,
             "Blood Transfusion Services: A Potential Source of AIDS Spread in Pak-
             istan," *Transfusion Today* 14 (1992): p. 10.
             10 PERCENT OF CASES THROUGHOUT AFRICA: S. J. Heymann et al.,
             "The Problem of Transfusion-Associated Acquired Immunodeficiency Syn-
             drome in Africa: A Quantitative Approach," *American Journal of Infection
             Control* 20 (1992): 256–62.
             IN INDIA 95 PERCENT OF THEIR BLOOD IS NOT SAFE: P. Kandela, "India:
             HIV Banks," *Lancet,* Aug. 17, 1991, pp. 436–37. The figure refers to haz-
             ards from all causes; researchers put the transfusion-related HIV rate at
             about 16 percent.
             THE FRENCH: J. Ruffie, *Rapport: Enseignement—formation—recrutement
             en transfusion sanguine* (Paris: Ministre de l'Education Nationale et de la
             Culture, Ministre de la Santé et de l'Action Humanitaire, Feb. 18, 1993).
             THE BRITISH ESTABLISHED: "UK National Blood Authority Goes Ahead,"
             *SCRIP* 1777 (Dec. 8, 1992): 9.
347–48       THE CANADIAN GOVERNMENT TOOK THE BLOOD PROGRAM AWAY:
             "Canadian Government Revokes Red Cross Authority Over Blood," *AABB
             Weekly Report* 2, no. 33 (Sept. 13, 1996).
348          THE FALL OF THE ONCE POWERFUL GREEN CROSS: A. Noguchi, "The
             Fall of Japan's Top Blood Company," *Tokyo Business Today,* Sept. 1988,
             pp. 58–59; "Japan's Yoshitomi, Scandal-Ridden Green Cross to Merge,"

*Agence France-Presse,* Feb. 24, 1997; "Japan's Scandal-Tainted Drug Maker . . ." Reuter, Feb. 24, 1997.

348     TO LOWER THEIR POOL SIZES: Interview with Anne-Marie Finley. "HURL THE INDUSTRY INTO BANKRUPTCY . . .": "Supreme Court Declines to Hear Appeal of Decision Decertifying Hemophilia AIDS Class Action," Oversight, Subcommittee on Human Resources and Intergovernmental Relations, U.S. Congress, House of Representatives, Committee on Government Reform and *CCBC Newsletter,* Oct. 6, 1995, p. 3.

349     "BLOOD SAFETY MUST NEVER AGAIN BE A SECONDARY ISSUE": Statement of Donna Shalala, Secretary, Health and Human Services, *Protecting the Nation's Blood Supply from Infectious Agents: New Standards to Meet New Threats,* 104th Cong., 1st sess., Oct. 12 and Nov. 2, 1995 (Washington, D.C.: Government Printing Office, 1996), p. 10.
P24 ANTIGEN TEST: Since the p24 antigen test became required in 1995, only three cases of HIV-tainted blood have been found that otherwise would not have been.
"I WAS ABSOLUTELY STUNNED . . .": Statement of Corey Dubin, ibid., p. 45.

350     "TIMIDITY IN CONFRONTING THE AIDS THREAT . . .": Letter from Christopher Shays to David A. Kessler, M.D., July 12, 1995, in ibid., p. 3.
$287 MILLION: "Statement of Brian McDonough, Chief Operating Officer . . . American Red Cross Blood Services," U.S. Congress, House of Representatives, Committee on Government Oversight and Reform, Subcommittee on Human Resources, *Public Health 2000: Immune Globulin Shortages: Causes and Cures,* May 7, 1998, 105th Cong., 2nd sess., p. 1.
CREUTZFELDT-JACOB DISEASE: Dr. Fred Darr, address to American Blood Resources Association, Washington, D.C., June 19, 1997; industry interviews; P. Brown, "Can Creutzfeldt-Jacob Disease Be Transmitted by Transfusion?," *Current Opinions in Hematology* 2 (1995): 472–77; "Blood Products Recalled," *FDA Consumer* 29, no. 2 (March 1995): 3.
$130 MILLION: "Statement of Brian McDonough," p. 4.
LOST $113 MILLION IN 1995 ALONE: D. Frantz, "Elizabeth Dole: Her Power as Leader of Red Cross," *New York Times,* May 30, 1996.

351     ONLY TWO HAVE CONTINUED TO OPERATE: Author correspondence with Institut National de la Transfusion Sanguine.
UB PLASMA SET UP IN BUCHAREST: Interview with Romanian Health Ministry official; D. Gow et al., "Bad Blood on Their Hands," *Guardian,* Nov. 6, 1993; S. Kinzer, "German AIDS Blood Scandal Spills Across Europe," *New York Times,* May 5, 1993; "Blood Case Spurs Murder Charges," Associated Press, Aug. 10, 1995.

352     DR. RYOICHI NAITO CONDUCTED HIS MOST DARING EXPERIMENT: R. Naito, *Rou SL No Souon* (Memoirs of an Old Locomotive) (Tokyo: Daiwa Toppan Kougei Insatsu Kabushiki Gaisha, 1980), pp. 423–24.
A SUBSTITUTE FOR RED BLOOD CELLS: There is a rich literature about artificial hemoglobin. Useful sources include *A New Generation of Oxygen Therapeutics* (Stamford, Conn.: Stover & Associates, Sept. 1996); R. Lewis, "Companies Investigate a Range of Options in Manufacture of Red Cell Substitutes," *Genetic Engineering News* 10, no. 5 (May 1990).

353     BIOPURE LOBBIED THE MILITARY: R. Pool, "Blood, Money and the Pentagon," *Science* 250 (Dec. 21, 1990): 1656.

353    THE MILITARY USED A TOTAL OF TWO THOUSAND PINTS: Interview with Dr. Anthony Polk.

354    ALTERING THE GENETIC CODING OF BACTERIA: Interview with Dr. Charles Scoggin.

355    LEISHMANIA: Interview with Dr. Anthony Polk; R. G. Westphal, "Parasitic Disease and Blood Transfusion," in S. J. Nance, ed., *Blood Safety: Current Challenges* (Bethesda, Md.: American Association of Blood Banks, 1992), pp. 106–9.
CHAGAS' DISEASE: R. G. Westphal, "Parasitic Disease and Blood Transfusion," in ibid., pp. 100–104.
HEPATITIS C CASES TRANSMITTED BY IMMUNE GLOBULINS: "FDA Begins Testing Immunoglobulin Products for HCV RNA," *CCBC Newsletter,* Jan. 13, 1995; T. M. Burton, "A Drug from Baxter Is Said to Have Posed a Risk of Hepatitis," *Wall Street Journal,* July 20, 1995.

# ACKNOWLEDGMENTS

I'd like to thank Bernie Daina, who long ago gave me the germ of an idea, and Ellen Ruppel Shell, who helped me see that the idea could become a book and relentlessly prodded me to write it. My agent, Kris Dahl, along with Gordon Kato, deserves credit for helping shape a vague idea into a focused proposal. My editor, Jonathan Segal, guided this project with a wise and patient hand.

In the course of my travels and interviews over the years, several people went beyond the traditional role of interview subject and almost became tutors, as they patiently answered my endless and repetitive questions. I found such people in every aspect of this history, from the profit and nonprofit sectors, from science and the military; from those whose lives depend on using blood products and from those to whom the products transmitted HIV. I would like to thank these voluntary mentors, including John "Newt" Ashworth, Dr. Sam Gibson, Tom Asher, Dr. Robert Westphal, Jane Starkey, Suzanne Gaynor, David Bing, Pat Gilbo, John McCray, Corey Dubin, Patrick Robert, William Schneider, Dr. Byron Myhre, Robert and James Reilly, Dr. J. Worth Estes, Stephen C. Redhead, and Dr. Steven Tahan. I appreciate the help of several others who, because of their sensitive positions, cannot be thanked publicly. I thank the many members of the hemophilia community who shared their insights and stories with me. I would like to posthumously thank certain individuals who passed away during the course of this project—Bernice Hemphill and General Douglas Kendrick, who lived long and full lives; and Michael Rosenberg and Loras Goedkin, who died tragically of AIDS. You could not imagine a more disparate group of individuals, yet they all shared an honesty and strength of character that I admired.

This book took me to several countries, where local researchers and journalists selflessly shared their own work and views. I could not have completed this project without the help of Egmont Koch in Germany; Claudine Hossenlopp, Franc Nouchi, and Marie-Angèle Hermitte in France; André Picard and Elizabeth Carlton in Canada; reporters too numerous to mention from the *Asahi Shimbun, Yomiuri Shimbun,* and *Mainichi Daily* newspapers in Japan; and Donna Shaw in the United States, among others. I received translation help from Peggy Conant, Lisbeth Fog, Kim Fujimoto, Ronnel Nel, Azlin Perdomo, and Mary Tunnel. Eric Feldman and Masami Kobayashi provided generous help with translations and insights into the events in Japan. Dr. Vitaly Korotich's translations of Russian medical texts and informal tutorials on Soviet medical history were generous, fascinating, and invaluable.

I would like to thank the librarians and archivists at several facilities, including Richard Wolfe and the staff of the rare books department of the Countway Medical Library at Harvard Medical School; Douglas Surgenor, who kindly allowed me access to the Center for Blood Research in Boston and shared his recollections, even as he was writing a book of his own; Richard Steele of the Mount Sinai Medical Center Archive in New York; the helpful staffs of the Rockefeller Archive Center in New York and the Wellcome Library in London; and the tireless research librarians in the Boston University library system.

My graduate-student researchers over the years have included Kimberly Ridley, Barbara Moran, Neil Savage, Jeff Baliff, Neil Andrews, and Ellen Bailey Pippinger. Not only did they ease the drudgery of research, but they, like their classmates, were a continual source of cheer and inspiration.

Several people took the time to read all or parts of my unwieldy manuscript, including the keen-eyed yet tactful Marcia Bartusiak, Robert Westphal, John Ashworth, George Seage, and Bernie Daina. Financial support at critical moments came from the Fund for Investigative Journalism and the Freedom Forum. I would like to thank my colleagues at Boston University, who tolerated my absences and obsession. Ellen Ruppel Shell, Larry Kahaner, and David Danforth deserve special praise for years of faithful advice and support.

Whenever I have read serious books in the past, I've been struck by the part in the acknowledgments in which the author thanks his or her family for tolerating long periods of preoccupation and stress. Surely they must dramatize, I used to imagine: What can be so difficult for a family when one of its members sits down to write a book? Now I understand. The cumulative stress can be corrosive to personal relationships, and I'm deeply grateful to my family and friends for putting up with me when I was not at my best. In particular, I would like to thank my parents, Arnold and Ruth Starr, for their unwavering enthusiasm; and my sons, Gordon and Gregory, for filling our lives with affection and humor, especially when this project was approaching its grimmest. Most of all, I'd like to thank my wife, Monica Sidor, a person of great substance if ever there was one. A busy and respected professional in her own right, she contributed far more than her share toward maintaining our family's stability, harmony, and sense of fun. She never forgets what's important in life. To her I owe the deepest debt of gratitude and love.

# INDEX

## ILLUSTRATION CREDITS

*Following page 78:* Man receiving blood from a lamb: Courtesy of the National Library of Medicine. / Illustrations from a 1679 treatise: Courtesy of the National Library of Medicine. / Transfusion techniques (Elsholtz): Courtesy of the National Library of Medicine. / Bossé's *The Bloodletting:* Courtesy of the Francis A. Countway Library of Medicine. / Bloodletting instruments: Courtesy of the Division of Science, Medicine and Society, the Smithsonian Institution. / Transfusion at La Pitié hospital: Courtesy of the Francis A. Countway Library of Medicine. / Dr. Karl Landsteiner: Courtesy of the Rockefeller Archive Center. / Dr. Alexis Carrel: Courtesy of the Rockefeller Archive Center. / Dr. Richard Lewisohn: Courtesy of Archives of the Mount Sinai Medical Center. / Dutch doctors transfusing blood arm to-arm: Courtesy of the Central Laboratory of the Netherlands Red Cross Blood Transfusion Service, Amsterdam. / Dr. Norman Bethune in Málaga: Courtesy of the Communist Party of Canada. / American medic in Sicily: Courtesy of the National Library of Medicine. / Dr. Charles Drew: Courtesy of the U.S. National Archives. / Dr. Edwin J. Cohn lecturing: Courtesy of the Francis A. Countway Library of Medicine. / Russian poster: Courtesy of the National Library of Medicine. / "Your Blood Can Save Him": Courtesy of the American Red Cross. All rights reserved in all countries. / British poster: Courtesy of Blackwell Science, Ltd.

*Following page 174:* Mobile blood collection center in Bristol: Courtesy of Blackwell Science, Ltd. / Nuns in Milwaukee: Courtesy of the American Red Cross. All rights reserved in all countries. / Shipping container for whole blood: Courtesy of the National Museum of Health and Medicine, Armed Forces Institute of Pathology. / Dried plasma and albumin transported in Italy and New Guinea: Courtesy of the American Red Cross. All rights reserved in all countries. / "Journée du Sang" poster: Courtesy of the Institut National de la Transfusion Sanguine, Paris. / Automated plasmapheresis: Courtesy of Haemonetics Corporation. / Dr. J. Garrott Allen: Courtesy of Barry W. Allen, Ph.D. / Modern fractionation facility: Courtesy of the CLB/Sanguin Blood Supply Foundation, the Netherlands. / Susie Quintana: Courtesy of the Quintana family. / Michael Rosenberg: Photograph by Masami Kobayashi. / Corey Dubin and friend: Courtesy of Corey Dubin. / Michel Garretta and Jacques Roux: AP/Wide World Photos. / Demonstration in Osaka: Kyodo News International. / Takehiko Kawano bowing in apology: Kyodo News International.

A NOTE ABOUT THE AUTHOR

Douglas Starr is an associate professor of journalism and codirector of the Graduate Program in Science Journalism at Boston University. A former newspaper reporter and field biologist, he has written on the environment, medicine, and science for a variety of publications, including *Smithsonian, Audubon, Sports Illustrated, National Wildlife,* and *American Health.* He was science editor of "Bodywatch," a health series that ran for three years on PBS. He lives near Boston with his wife and two sons.

## A NOTE ON THE TYPE

The text of this book was set in Sabon, a typeface designed by Jan Tschichold (1902–1974), the well-known German typographer. Based loosely on the original designs by Claude Garamond (c. 1480–1561), Sabon is unique in that it was explicitly designed for hot-metal composition on both the Monotype and Linotype machines as well as for filmsetting. Designed in 1966 in Frankfurt, Sabon was named for the famous Lyons punch cutter Jacques Sabon, who is thought to have brought some of Garamond's matrices to Frankfurt.

*Composed by North Market Street Graphics, Lancaster, Pennsylvania*

*Printed and bound by Quebecor Printing, Martinsburg, West Virginia*

*Designed by Robert C. Olsson*